Editors

Lecture Notes in Chemistry

Edited by G. Berthier M. J. S. Dewar H. Fischer
K. Fukui G. G. Hall H. Hartmann H. H. Jaffé J. Jortner
W. Kutzelnigg K. Ruedenberg E. Scrocco

26

Salvatore Califano
Vincenzo Schettino
Natale Neto

Lattice Dynamics of Molecular Crystals

Springer-Verlag
Berlin Heidelberg New York 1981

Authors

Salvatore Califano
Vincenzo Schettino
Istituto di Chimica Fisica, Università di Firenze
Via G. Capponi 9, 50121 Firenze, Italy

Natale Neto
Istituto di Chimica Organica, Università di Firenze
Via G. Capponi 9, 50121 Firenze, Italy

ISBN-13: 978-3-540-10868-9 e-ISBN-13: 978-3-642-93186-4
DOI: 10.1007/978-3-642-93186-4

2152/3140-543210

PREFACE

The lattice dynamics of molecular crystals has undergone an enormous progress in these last twenty years or so. The experimental and theoretical advances have been realized by two different approaches. From one side molecular spectroscopists have been primarily interested in the vibrational properties of the molecules themselves subjected to the perturbing influence of the crystal environment. From the other side the lattice dynamical theory familiar in solid state physics for atomic lattices has been extended to molecular arrays. Although the overlap between the two approaches has been considerable the reference material is rather scattered in specialized papers. The purpose of this book is to partly fill this gap and to discuss the lattice dynamical theory of molecular crystals in a compact and specialized form. As such,the book is not intended exclusively for researchers and specialists in the field but also for graduate students entering an activity in solid state molecular spectroscopy.

The size imposed to the volume has prevented the treatment of all the aspects of the subject and the authors have been forced to the selection of some of the fundamental topics. The emphasis is on the intermolecular potentials and their use for the calculation of intermolecular coupling constants in terms of molecular coordinates and for the calculation of harmonic and anharmonic vibrational properties,including crystal frequencies,infrared and Raman intensities,anharmonic shifts, band widths and two-phonon band shapes.The discussion of specific molecular crystals has been considered as an illustration and clarification of the theory but no attempt has been made to review all the growing experimental and theoretical literature. However,the references included should allow the reader an easy access to the most relevant published works on the topics discussed.

CONTENTS

CHAPTER 1

LATTICE DYNAMICS

1.1 INTRODUCTION

The usual empirical classification of crystalline solids considers four types of materials:a) ionic;b) covalent;c) metallic and d) molecular crystals. This classification is based on differences in chemical and physical properties. As such the distinction among the various types of solids is not sharply defined. In fact the physical properties cover a continuous range of values and there are materials which are exactly in between the various categories of crystals. Nevertheless, the above classification is very convenient and in the present volume we shall be concerned primarily with the vibrational properties of the last type of crystals, the molecular crystals. It is, however, well understood that several aspects of both the experimental and the theoretical approach may be common to the various types of crystals.

The distinguishing properties of a molecular crystal are the low cohesion energy, low melting point, softness and deformability, poor electrical and thermal conductivity and poor mechanical properties. From our point of view it is, however, convenient to characterize a molecular crystal on the basis of the structural arrangement of the atoms and of the nature and strength of the interaction forces.

Molecular crystals are characterized by the occurrence of well defined entities, the molecules themselves, repeated through space in the pattern of a Bravais lattice. The persistence in the solid of the molecular identity is due to the fact that the binding forces between atoms of the same molecule are much stronger then those holding together different molecules in the crystal. A direct consequence of this situation is that in the crystal the molecules maintain structures that are normally very close to the ones they possess in the gas phase. Minor variations in the bond lengths and angles are generally observed between the gas and the crystal phase. These are due to the perturbation that the intermolecular field produces on the intramolecular bonds. In some particular cases, however, especially for molecules containing single bonds with low rotational barriers, it may happen that the intermolecular interactions constrain the molecular structure into a conformation that is not stable in the gas phase. Typical examples are those of biphenyl[1] which is planar in the crystal and twisted in the gas phase and of ethylenediamine [2] that has a gauche conformation in the gas phase but assumes the trans configuration in the crystal.

The molecular nature of these crystals is clearly shown by their vibrational spectra that can be illustrated, in their gross features, by means of the schematic drawing of Fig. 1.1.

The idealized spectrum of a gas, taken as representative of an assembly of non-interacting molecules, is reported in the upper part of the figure. For simplicity the rotational structure is not shown.

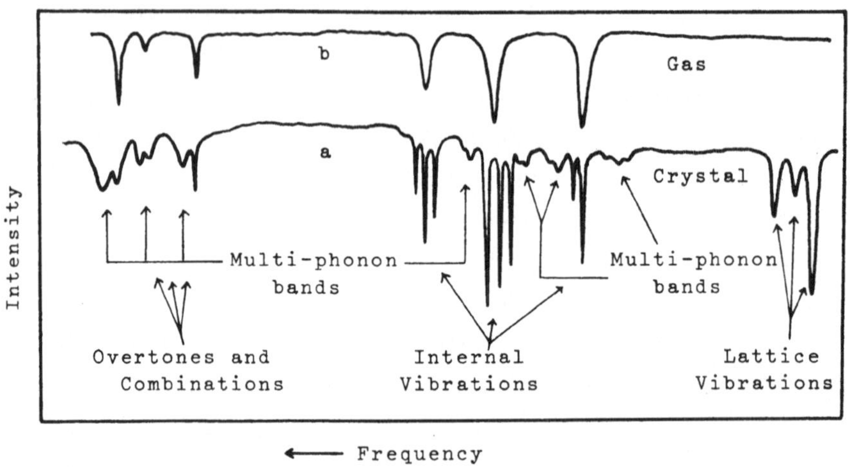

Fig. 1.1. Schematic drawing of the vibrational spectrum of a molecular crystal a), compared with the spectrum of the gas phase b).

In the lower part of the figure the corresponding spectrum of the crystal is schematically represented. In the spectrum of the gas we observe the fundamental modes of vibration of the molecules as well as their overtones and combinations. In the spectrum of the crystal the same bands are observed but they are shifted in frequency and split into a number of components which depends on the number of molecules per unit cell and on the unit cell symmetry.

These shifts and splittings are normally of the order of few wavenumbers but in some specific cases, when strong directional interactions occur between the molecules, they can be of the order of 100 cm^{-1} and even larger.

An important feature of the crystal spectrum is that in the low-frequency region new bands appear that are absent in the gas spectrum. These bands are due to vibrations of the lattice in which the molecules perform small translational and rotational motions about their equilibrium positions. In the gas phase the molecules are thus free to translate and to rotate whereas in the crystal these motions are subjected to a restoring force due to the presence of the intermolecular interactions. It is customary to classify them as "external" or "lattice" motions in contrast to the molecular vibrations that are called "internal". The same classification is applied to the corresponding bands in the vibrational spectrum.

Normally the internal and the external vibrations are well separated in frequency and practically uncoupled among each other.In this case the classification in internal and external vibrations has a true physical meaning and permits to separate the dynamical problem into two well distinct parts, one relative to the external and one to the internal vibrations.

The lattice dynamical treatment of the external crystal vibrations in which the internal modes are considered separately is called the rigid body approximation since it corresponds to assume that the molecules behave as rigid bodies with no degree of internal freedom. This approximation is largely utilized in actual calculations owing to its simplicity. It offers the advantage of reducing to six the number of degrees of freedom of each non-linear molecule. For crystals of very large molecules it represents the only possible approach to the dynamical problem. Obviously for linear molecules only two rotations can be defined and thus only five degrees of freedom per molecule must be considered.

The classification into internal and external vibrations is so convenient that it is used even for crystals for which the rigig body approximation breaks down and all degrees of freedom must be considered at the same time. This is the case for crystals in which normal modes of internal nature possess low frequencies and thus overlap with the lattice modes. The reason for which it is still convenient to maintain the separation between internal and external vibrations for crystals with low frequency internal modes, is connected with the use of appropriate models for the crystal potential and will be discussed later.

The rigid body approximation emphasizes the study of external vibrations. The main reason for concentrating the attention on them is that these crystal vibrations are completely controlled by the intermolecular potential and can be thus utilized to understand the physical nature of the interactions among molecules and to test the validity of model potentials. For the internal vibrations instead, the intermolecular interactions act only as a small perturbation on the internal force field. For this reason the shifts and the splittings of the internal bands are less utilized for model calculations.

In some cases it is convenient to consider a single internal vibration and to study in detail the perturbation caused by the crystal field on this specific mode. This is possible when the vibration is weakly coupled to the other internal modes and can be done using the so called vibrational "exciton model" to be described in Chapters 5 and 6.

In addition to the external and internal bands, the crystal spectra show a series of broad absorptions as side bands of the internal vibrations (see Fig.1.1). These bands are due to combinations of internal and lattice modes and are called "multiphonon" bands. They will be treated in details in Chapter 5. Multiphonon absorptions occur also in the region of the overtones and combinations of internal modes and correspond mainly to the contemporary excitation of different internal modes on the different molecules in the unit cell. These multiphonon bands are also discussed in Chapter 5.

1.2 THE DYNAMICAL EQUATIONS IN CARTESIAN COORDINATES

The theory of lattice dynamics can be formulated in any convenient set of displacement coordinates.It is clear from the previous Section that in the case of a molecular crystal one would naturally be inclined to select a set of "molecular" coordinates as best suited to describe the vibrations of the crystal lattice.This requires an extension of the treatment first introduced by Born in terms of Cartesian displacements of the atoms for atomic crystals.

Before introducing the molecular coordinates it is appropriate to present the basic concepts of the lattice dynamics of molecular crystals in terms of Cartesian displacement coordinates,by a simple extension of Born theory since it is simpler in this way to introduce the general features of the dynamical problem.

The classical treatment is discussed in details in several books and noticeably in those by Born and Huang [3] and by Maradudin et al.[4]. It is thus not necessary to discuss it extensively here,reserving rather our attention to the problems encountered in the extension of the theory to molecular crystals.

In this Section the equations of motion of a vibrating molecular crystal will be set up in terms of independent displacements referred to Cartesian axes fixed to the crystal.The Cartesian displacements of each atom in the crystal,referred to this frame,can be written as

$$U_{\alpha}^{m\mu i} = X_{\alpha}^{m\mu i} - \bar{X}_{\alpha}^{m\mu i} \qquad 1.2.1$$

where $X_{\alpha}^{m\mu i}$ is the αth component (X_{α}= X,Y,Z) of a vector from the crystal-fixed origin to the instantaneous position of atom i belonging to the molecule μ in the unit cell m, $\bar{X}_{\alpha}^{m\mu i}$ being the corresponding equilibrium value.

In the following discussion we consider a crystal containing Z molecules per unit cell and N atoms per molecule,such that there are 3ZN independent vibrational displacements per unit cell.In order to exploit the periodicity of the crystal lattice,we assume that the crystal extends indefinitely in space.The transition to a finite size crystal will be afforded later by means of appropriate boundary conditions.

In terms of the Cartesian displacements 1.2.1 the total crystal potential V can be expanded in a power series about the equilibrium configuration as

$$V = V_0 + \sum_\alpha \sum_m \sum_\mu \sum_i \Phi_\alpha (m\mu i) U_\alpha^{m\mu i} + \frac{1}{2} \sum_{\alpha\beta} \sum_{mn} \sum_{\mu\nu} \sum_{ij} \Phi_{\alpha\beta} \binom{m\mu i}{n\nu j} U_\alpha^{m\mu i} U_\beta^{n\nu j}$$

$$+ \frac{1}{6} \sum_{\alpha\beta\gamma} \sum_{mnp} \sum_{\mu\nu\pi} \sum_{ijk} \Phi_{\alpha\beta\gamma} \begin{pmatrix} m\mu i \\ n\nu j \\ p\pi k \end{pmatrix} U_\alpha^{m\mu i} U_\beta^{n\nu j} U_\gamma^{p\pi k} + \dots \qquad 1.2.2$$

where expansion coefficients are derivatives of V evaluated at the equilibrium configuration of the crystal

$$\Phi_\alpha (m\mu i) = \left(\frac{\partial V}{\partial U_\alpha^{m\mu i}}\right)_0$$

$$\Phi_{\alpha\beta} \binom{m\mu i}{n\nu j} = \left(\frac{\partial^2 V}{\partial U_\alpha^{m\mu i} \partial U_\beta^{n\nu j}}\right)_0 \qquad 1.2.3a$$

$$\Phi_{\alpha\beta\gamma} \begin{pmatrix} m\mu i \\ n\nu j \\ p\pi k \end{pmatrix} = \left(\frac{\partial^3 V}{\partial U_\alpha^{m\mu i} \partial U_\beta^{n\nu j} \partial U_\gamma^{p\pi k}}\right)_0$$

etc.

The general expression 1.2.2 will be used in Chapter 4 when the problem of anharmonic vibrations will be considered. Here we shall confine ourselves to the harmonic approximation and the expansion will be truncated after the third term in 1.2.2. Furthermore, since V_0 is the potential energy of the static crystal, we shall eliminate it from our expressions, owing to the fact that this simply amounts to a shift of the origin of the energy scale. Finally, since the coordinates in 1.2.1 were assumed to be independent, in order that the crystal is at equilibrium in the static configuration V must have a minimum so that

$$\Phi_\alpha (m\mu i) = 0 \qquad \text{for all values of } \alpha, m, \mu \text{ and } i \qquad 1.2.3b$$

Thus in the harmonic approximation 1.2.2 reduces to

$$V = \frac{1}{2} \sum_{\alpha\beta} \sum_{mn} \sum_{\mu\nu} \sum_{ij} \Phi_{\alpha\beta} \binom{m\mu i}{n\nu j} U_\alpha^{m\mu i} U_\beta^{n\nu j} \qquad 1.2.4$$

Force constants in 1.2.4 have several important properties which greatly reduce the number of independent parameters to be used in the expansion of V. We first note that, due to the periodic nature of the crystal lattice, force constants cannot depend on the absolute position of unit cells but only upon their relative positions so that

$$\Phi_{\alpha\beta}\binom{m\mu i}{n\nu j} = \Phi_{\alpha\beta}\binom{m+l\ \ \mu i}{n+l\ \ \nu j} \qquad 1.2.5a$$

where m+l and n+l identify unit cells obtained from cells m and n through the same lattice translation l. Hence, if we set l = -m and l = -n we obtain

$$\Phi_{\alpha\beta}\binom{m\mu i}{n\nu j} = \Phi_{\alpha\beta}\binom{0\ \ \ \mu i}{n-m\ \ \nu j} = \Phi_{\alpha\beta}\binom{m-n\ \ \mu i}{0\ \ \ \nu j} \qquad 1.2.5b$$

The force constants as second derivatives of the scalar V must also be symmetric in the indices $\alpha m\mu i$ and $\beta n\nu j$

$$\Phi_{\alpha\beta}\binom{m\mu i}{n\nu j} = \Phi_{\beta\alpha}\binom{n\nu j}{m\mu i} \qquad 1.2.6$$

The conditions 1.2.5 and 1.2.6 are fundamental properties of the Cartesian force constants for a crystal. They limit the number of independent quantities and will be used extensively in the following treatment of the equations of motion.

In terms of the Cartesian components $\dot{U}_\alpha^{m\mu i}$ of the velocities the kinetic energy assumes the simple form

$$T = \frac{1}{2}\sum_\alpha \sum_{m\mu i} m_i (\dot{U}_\alpha^{m\mu i})^2 \qquad 1.2.7$$

Using 1.2.4 and 1.2.7 we can now derive the equations of motion of the crystal as

$$\sum_\beta \sum_{n\nu j} \Phi_{\alpha\beta}\binom{m\mu i}{n\nu j} U_\beta^{n\nu j} = -\, m_i \ddot{U}_\alpha^{m\mu i} \qquad 1.2.8$$

where m_i is the mass of atom i and $\ddot{U}_\alpha^{m\mu i}$ are time derivatives of the velocity components. It may be convenient to introduce mass-weighted displacements defined as

$$W_{\alpha}^{m\mu i} = m_i^{1/2} U_{\alpha}^{m\mu i} \qquad 1.2.9$$

and to rewrite the set of differential equations 1.2.8 in terms of mass-dependent force constants

$$\Phi_{\alpha\beta}\binom{m\mu i}{n\nu j} = (1/m_i m_j)^{1/2} \phi_{\alpha\beta}\binom{m\mu i}{n\nu j} \qquad 1.2.10$$

as

$$\sum_{\beta}\sum_{n\nu j}\Phi_{\alpha\beta}\binom{m\mu i}{n\nu j} W_{\beta}^{n\nu j} = -\ddot{W}_{\alpha}^{m\mu i} \qquad 1.2.11$$

The equations of motion 1.2.11 and 1.2.8 constitute an infinite set of linear differential equations.We look,as customary in such cases, for periodic solutions.Since the kinetic and the potential energy are both quadratic forms,it is possible to find a linear transformation to a new set of coordinates (normal coordinates) that completely decouples the dynamical equations.For this it is first convenient to exploit the periodicity of the lattice in order to reduce the infinite set of equations to a finite number.We choose therefore solutions periodic in time and space,of the type

$$W_{\alpha}^{m\mu i} = A_{\alpha}^{\mu i} e^{i(\mathbf{k}\cdot R_m - \omega t)} \qquad 1.2.12$$

This solution describes a general motion of the lattice in which all equivalent atoms in different unit cells experience periodic displacements of the same frequency $\omega/2\pi$ and amplitude $A_{\alpha}^{\mu i}$ with a phase difference determined by the vector **k**.In other words,1.2.12 represents a travelling wave of wavelength $2\pi/|\mathbf{k}|$,where **k** is the wavevector which gives the direction of propagation.

By substitution of 1.2.12 in 1.2.11 we obtain

$$\sum_{\beta}\sum_{\nu j} D_{\alpha\beta}\left(\begin{smallmatrix}\mu i\\ \nu j\end{smallmatrix}\middle|\mathbf{k}\right) A_{\beta}^{\nu j} = \omega^2 A_{\alpha}^{\mu i} \qquad 1.2.13$$

where

$$D_{\alpha\beta}\left(\begin{smallmatrix}\mu i\\ \nu j\end{smallmatrix}\middle|\mathbf{k}\right) = \sum_n \Phi_{\alpha\beta}\left(\begin{smallmatrix}0\mu i\\ n\nu j\end{smallmatrix}\right) e^{i\mathbf{k}\cdot\mathbf{R}_n} \qquad 1.2.14$$

The quantities $D_{\alpha\beta}\left(\begin{smallmatrix}\mu i\\ \nu j\end{smallmatrix}\middle|\mathbf{k}\right)$ defined by 1.2.14 are independent on m because of the lattice periodicity which leads to the conditions 1.2.5 which state that the Cartesian force constants depend only on the relative position of the unit cells.

The infinite set of equations 1.2.11 is thus reduced,by means of 1.2.12,to a set of 3ZN linear equation for each value of **k**.The system of equations 1.2.13 has non-trivial solutions only if the determinant of the coefficients vanishes

$$\left|D_{\alpha\beta}\left(\begin{smallmatrix}\mu i\\ \nu j\end{smallmatrix}\middle|\mathbf{k}\right) - \omega^2\delta_{\alpha\beta}\delta_{\mu\nu}\delta_{ij}\right| = 0 \qquad 1.2.15$$

We notice that the system of linear equations 1.2.13 does not determine the coefficients $A_\alpha^{\mu i}$ uniquely but only their ratios.It is therefore convenient to utilize "normalized" amplitudes and to write for each of the 3ZN solutions ω_{pk} of 1.2.13 the more useful expression

$$\sum_\beta\sum_{\nu j} D_{\alpha\beta}\left(\begin{smallmatrix}\mu i\\ \nu j\end{smallmatrix}\middle|\mathbf{k}\right) e_\beta(\nu j|p\mathbf{k}) = \omega_{pk}^2 e_\alpha(\mu i|p\mathbf{k}) \qquad 1.2.16$$

where the normalized coefficients $e_\alpha(\mu i|p\mathbf{k})$ bear now the label **pk** in addition to the label μi,to show that they refer to the solution ω_{pk}. They obey the conditions

$$\sum_\alpha\sum_\mu\sum_i e_\alpha^*(\mu i|p\mathbf{k}) e_\alpha(\mu i|p'\mathbf{k}) = \delta_{pp'} \qquad 1.2.17a$$

$$\sum_p e_\beta^*(\nu j|p\mathbf{k}) e_\alpha(\mu i|p\mathbf{k}) = \delta_{\alpha\beta}\delta_{\mu\nu}\delta_{ij} \qquad 1.2.17b$$

where the star indicates the complex conjugate.

Equation 1.2.16 is a typical eigenvalue equation and is thus more compactly written in matrix form

$$\mathbb{D}(\mathbf{k})\mathbb{E}(\mathbf{k}) = \mathbb{E}(\mathbf{k})\Omega(\mathbf{k})^2 \qquad 1.2.18$$

where $\Omega(\mathbf{k})^2$ is the diagonal matrix of the eigenvalues ω_{pk}^2 and $\mathbb{E}(\mathbf{k})$ is the matrix of the eigenvectors $e_{\alpha}(\mu i|p\mathbf{k})$.The Fourier transformed matrix $\mathbf{D}(\mathbf{k})$,with elements $D_{\alpha\beta}(\begin{smallmatrix}\mu i\\ \nu j\end{smallmatrix}|\mathbf{k})$, is of dimension 3ZN×3ZN and is called the "dynamical matrix".

From the definition 1.2.14 it is clear that the dynamical matrix has the property

$$\mathbf{D}(\mathbf{k})^{*}=\mathbf{D}(-\mathbf{k}) \qquad 1.2.18a$$

Furthermore,using the symmetry properties of the force constants expressed by 1.2.6 it can be seen that the matrix $\mathbf{D}(\mathbf{k})$ is also Hermitian and thus

$$\mathbf{D}(\mathbf{k})^{\dagger}=\mathbf{D}(\mathbf{k}) \qquad 1.2.18b$$

As pointed out before,the use of the lattice periodicity implies that we are dealing with an infinite crystal.We can,however, adapt the treatment to a finite crystal containing L unit cells,by imposing the so called "cyclic boundary conditions" first introduced in lattice dynamics by Born and Von Karman [5].If $\mathbf{t}_1$,$\mathbf{t}_2$ and $\mathbf{t}_3$ are basic translations of the crystal and L_1,L_2 and L_3 are integers representing the number of unit cells in the three directions of the crystal axes ($L_1 \times L_2 \times L_3 = L$), the cyclic boundary conditions can be imposed by requiring that movements of equivalent atoms separated by $L_i t_i$ must be identical.When introduced in 1.2.12 these conditions imply that

$$e^{i\mathbf{k}\cdot\mathbf{t}_1 L_1} = e^{i\mathbf{k}\cdot\mathbf{t}_2 L_2} = e^{i\mathbf{k}\cdot\mathbf{t}_3 L_3} = 1 \qquad 1.2.19a$$

The cyclic boundary conditions set therefore a limit to the allowed values of $\mathbf{k}$.As will be discussed in the next Chapter,the values of $\mathbf{k}$ can be represented as vectors of the reciprocal space.If we define a reciprocal lattice with basic vectors given by

$$\mathbf{t}_i\cdot\mathbf{b}_j = 2\pi\delta_{ij} \qquad 1.2.19b$$

it can be seen that the periodic boundary conditions are satisfied by wavevectors of the form

$$\mathbf{k} = (l_1/L_1)\mathbf{b}_1 + (l_2/L_2)\mathbf{b}_2 + (l_3/L_3)\mathbf{b}_3 \qquad 1.2.19c$$

where l_1 ,l_2 and l_3 are integers that can assume L_1,L_2 and L_3 consecutive values,respectively.It will be shown in the next Chapter that it is convenient to choose l_1, l_2 and l_3 in the range

$$\begin{aligned} -L_1/2 &< l_1 \leqslant L_1/2 \\ -L_2/2 &< l_2 \leqslant L_2/2 \\ -L_3/2 &< l_3 \leqslant L_3/2 \end{aligned} \qquad 1.2.19d$$

Anticipating the results of the space group theory of Chapter 2, we can state that the **k** vectors label the irreducible representations of the translational group. It is then possible to construct Cartesian symmetry coordinates belonging to the kth irreducible representation of the translational group

$$W_\alpha(\mu i|\mathbf{k}) = \left(\frac{1}{L}\right)^{1/2} \sum_m e^{-i\mathbf{k}\cdot\mathbf{R}_m} W_\alpha^{m\mu i} \qquad 1.2.20$$

The inverse relation is easily obtained by multiplying both sides of 1.2.20 by $\exp(i\mathbf{k}\cdot\mathbf{R}_n)$,summing over **k** and using the symmetry property of the lattice sums (see Chapter 2, Eq.2.2.23b)

$$\sum_{\mathbf{k}} e^{i\mathbf{k}\cdot(\mathbf{R}_n - \mathbf{R}_m)} = L\delta_{mn} \qquad 1.2.21$$

The inverse relation is then

$$W_\alpha^{m\mu i} = \left(\frac{1}{L}\right)^{1/2} \sum_{\mathbf{k}} e^{i\mathbf{k}\cdot\mathbf{R}_m} W_\alpha(\mu i|\mathbf{k}) \qquad 1.2.22$$

It is easy to show that the elements of the Fourier transformed dynamical matrix represent the force constants in the symmetrized coordinates

$$D_{\alpha\beta}\left(\begin{smallmatrix}\mu i\\ \nu j\end{smallmatrix}\middle|\mathbf{k}\right) = \left(\frac{\partial^2 V}{\partial W_\alpha(\mu i|-\mathbf{k})\,\partial W_\beta(\nu j|\mathbf{k})}\right) \qquad 1.2.23$$

The reduction of the infinite set of equations 1.2.11 to sets of 3ZN equations, one for each value of **k**, is thus easily interpreted as a direct consequence of the translational symmetry, which leads to a factorization of the dynamical matrix into blocks of dimension 3ZN×3ZN, one for each value of **k**. The block form corresponds to the fact that there cannot be coupling in the potential energy between coordinates belonging to different irreducible representations labelled by **k** and **k'**.

Using the normalized eigenvectors defined in 1.2.16 and 1.2.17, it is then possible to write the transformation to crystal normal coordinates in the form

$$W_\alpha(\mu i|\mathbf{k}) = \sum_p e_\alpha(\mu i|p\mathbf{k})Q_{p\mathbf{k}} \qquad 1.2.24$$

Using then 1.2.17a it is easily seen that the inverse transformation is

$$Q_{p\mathbf{k}} = \sum_{\alpha\mu i} e^*_\alpha(\mu i|p\mathbf{k})W_\alpha(\mu i|\mathbf{k}) \qquad 1.2.25$$

By substitution of 1.2.24 in 1.2.22, we obtain then the transformation from normal to Cartesian displacement coordinates

$$W_\alpha^{m\mu i} = \left(\frac{1}{L}\right)^{1/2} \sum_{p\mathbf{k}} e_\alpha(\mu i|p\mathbf{k})e^{i\mathbf{k}\cdot\mathbf{R}_m}\, Q_{p\mathbf{k}} \qquad 1.2.26$$

The kinetic and the potential energy of the crystal can be now expressed in terms of the normal coordinates. Using 1.2.17 and the lattice sum property (see Chapter 2, Eq. 2.2.23a)

$$\sum_m e^{i(\mathbf{k}-\mathbf{k}')\cdot\mathbf{R}_m} = L\delta(\mathbf{k}-\mathbf{k}') \qquad 1.2.27$$

we obtain

$$2T = \sum_{p\mathbf{k}} \dot{Q}^*_{p\mathbf{k}}\dot{Q}_{p\mathbf{k}} \qquad 1.2.28$$

and

$$2V = \sum_{p\mathbf{k}} \omega^2_{p\mathbf{k}} Q^*_{p\mathbf{k}} Q_{p\mathbf{k}} \qquad 1.2.29$$

The classical Hamiltonian for this system of 3ZNL independent oscillators is then

$$H = \frac{1}{2}\sum_{p\mathbf{k}}[\dot{Q}^*_{p\mathbf{k}}\dot{Q}_{p\mathbf{k}} + \omega^2_{p\mathbf{k}}Q^*_{p\mathbf{k}}Q_{p\mathbf{k}}] \qquad 1.2.30$$

or in terms of the conjugate momenta $P_{p\mathbf{k}} = \dot{Q}_{p-\mathbf{k}} = \dot{Q}^*_{p\mathbf{k}}$

$$H = \frac{1}{2}\sum_{p\mathbf{k}}[P_{p\mathbf{k}}P^*_{p\mathbf{k}} + \omega^2_{p\mathbf{k}}Q^*_{p\mathbf{k}}Q_{p\mathbf{k}}] \qquad 1.2.31$$

The corresponding quantum-mechanical operator is easily obtained from 1.2.31 by substitution of the classical quantities with the corresponding operators.These obey the usual commutation relations

$$[Q_{p\mathbf{k}}, P_{p'\mathbf{k}'}] = i\hbar\delta_{pp'}\delta(\mathbf{k} - \mathbf{k}') \qquad 1.2.32a$$

$$[Q_{p\mathbf{k}}, Q_{p'\mathbf{k}'}] = [P_{p\mathbf{k}}, P_{p'\mathbf{k}'}] = 0 \qquad 1.2.32b$$

The Schrödinger equation for the vibrating crystal is thus

$$\frac{1}{2}\sum_{p\mathbf{k}}[P_{p\mathbf{k}}P^*_{p\mathbf{k}} + \omega^2_{p\mathbf{k}}Q^*_{p\mathbf{k}}Q_{p\mathbf{k}}]\Psi = E\Psi \qquad 1.2.33$$

and can be separated into independent harmonic oscillator wave equations

$$\frac{1}{2}(P_{p\mathbf{k}}P^*_{p\mathbf{k}} + \omega^2_{p\mathbf{k}}Q^*_{p\mathbf{k}}Q_{p\mathbf{k}})\phi(n_{p\mathbf{k}}) = \varepsilon_{p\mathbf{k}}\phi(n_{p\mathbf{k}}) \qquad 1.2.34$$

The vibrating crystal is thus represented by a collection of 3ZNL independent harmonic oscillators each with energy

$$\varepsilon_{p\mathbf{k}} = (n_{p\mathbf{k}} + \frac{1}{2})\hbar\omega_{p\mathbf{k}} \qquad 1.2.35$$

The total crystal energy is then

$$E = \sum_{p\mathbf{k}}(n_{p\mathbf{k}} + \frac{1}{2})\hbar\omega_{p\mathbf{k}} \qquad 1.2.36$$

and the crystal wave function

$$\Psi = \Pi_{p\mathbf{k}}\phi(n_{p\mathbf{k}}) \qquad 1.2.37$$

The vibrational state of the crystal is specified in this representation by a state vector

$$|n_{1\mathbf{k}_1'}, n_{2\mathbf{k}_2'}, \ldots, n_{p\mathbf{k}}, \ldots\rangle \qquad 1.2.38$$

which depends on 3ZNL quantum numbers giving the degree of excitation of each oscillator.

For many purposes, as discussed in Chapter 4, it is convenient to define creation and annihilation operators through the relations

$$a^{\dagger}_{p\mathbf{k}} = (1/2\hbar\omega_{p\mathbf{k}})^{\frac{1}{2}}[\omega_{p\mathbf{k}}Q_{p-\mathbf{k}} - iP_{p\mathbf{k}}] \qquad 1.2.39$$

$$a_{p\mathbf{k}} = (1/2\hbar\omega_{p\mathbf{k}})^{\frac{1}{2}}[\omega_{p\mathbf{k}}Q_{p\mathbf{k}} + iP_{p-\mathbf{k}}] \qquad 1.2.40$$

In terms of these operators the Hamiltonian assumes the form

$$H = \sum_{p\mathbf{k}}\hbar\omega_{p\mathbf{k}}[a^{\dagger}_{p\mathbf{k}}a_{p\mathbf{k}} + \frac{1}{2}] \qquad 1.2.41$$

The basic properties of the operators $a_{p\mathbf{k}}$ and $a^{\dagger}_{p\mathbf{k}}$ are seen from their effect on the state vector 1.2.38

$$a_{p\mathbf{k}}|\ldots, n_{p\mathbf{k}}, \ldots\rangle = (n_{p\mathbf{k}})^{\frac{1}{2}}|\ldots, n_{p\mathbf{k}} - 1, \ldots\rangle \qquad 1.2.42$$

$$a^{\dagger}_{p\mathbf{k}}|\ldots, n_{p\mathbf{k}}, \ldots\rangle = (n_{p\mathbf{k}} + 1)^{\frac{1}{2}}|\ldots, n_{p\mathbf{k}} + 1, \ldots\rangle \qquad 1.2.43$$

The quantum number $\mathbf{n}_{p\mathbf{k}}$ was used in the previous discussion to define the state of excitation of the oscillator $p\mathbf{k}$. Rather then thinking of oscillators excited to different quantum states, we can consider $n_{p\mathbf{k}}$ as counting a number of "particles" with energy $\hbar\omega_{p\mathbf{k}}$ in the vibrational state pk. These quasi-particles are called "phonons" and obey the Bose-Einstein statistics. In the harmonic approximation used so far they are independent. In this way the vibrational state of the crystal is described by giving the number of phonons of each type present in the crystal. This formalism is very useful when dealing with anharmonic properties of crystals and will be fully exploited in Chapters 4 and 5.

1.3 DISPERSION CURVES.ACOUSTIC AND OPTICAL MODES

In the previous Section we have seen that for a piece of crystal containing L unit cells and fulfilling the cyclic boundary conditions the dynamical matrix factorizes into blocks of dimensionality 3ZN,one for each of the L distinct values of the wave vector $\mathbf{k}$.

We can follow the dispersion in frequency of each of the 3ZN eigenvalues $\omega^2_{p\mathbf{k}}$ of the dynamical matrix $\mathbb{D}(\mathbf{k})$ as $\mathbf{k}$ varies,by plotting them as a function of $\mathbf{k}$.When the dimension of the crystal increases,the number of values of $\mathbf{k}$ becomes larger and when $L \to \infty$ it goes to a continuum. Instead of plotting $\omega^2_{p\mathbf{k}}$,it is convenient to plot directly $\omega_{p\mathbf{k}}$ as a function of $\mathbf{k}$,since these are the experimental quantities of interest.We obtain in this way 3ZN "branches",one for each value of p,called "dispersion curves".

Although components of $\mathbf{k}$ are allowed to vary in the range $-\frac{1}{2};+\frac{1}{2}$ (see Eq.1.2.19d),it is unnecessary to consider negative values of $\mathbf{k}$ since,according to 1.2.18a,$\omega^*_{p\mathbf{k}} = \omega_{p-\mathbf{k}}$ and,being the vibrational frequencies real and positive numbers,it follows that $\omega_{p-\mathbf{k}} = \omega_{p\mathbf{k}}$,i.e.that normal modes for $+\mathbf{k}$ or $-\mathbf{k}$ have the same frequency.

An important feature of the dispersion curves is that three of them have the particular property of going to zero when $\mathbf{k} \to 0$.These three branches are called "acoustic" branches because of their connection with ordinary sound waves in an elastic continuum.

The behaviour of the acoustic branches in the region of small values of $\mathbf{k}$ (long wavelength limit) follows directly from the invariance of the crystal potential under a uniform translation.For any given initial configuration of the crystal,not necessarily coinciding with the equilibrium one,the αth component of the total force acting on an atom must remain unchanged when the crystal undergoes a uniform translation. Since the force acting on atom mμi is equal to $-\partial V/\partial U^{m\mu i}_{\alpha}$, the translational invariance requires that the derivative $\partial V/\partial U^{m\mu i}_{\alpha}$ must be unchanged for a uniform translation.

From 1.2.2 we have

$$\partial V/\partial U_\alpha^{m\mu i} = \Phi_\alpha(m\mu i) + \sum_\beta \sum_{n\nu j} \Phi_{\alpha\beta}\binom{m\mu i}{n\nu j} U_\beta^{n\nu j} + \ldots \tag{1.3.1}$$

and for a uniform translation, i.e. for $U_\beta^{n\nu j} = \varepsilon_\beta$ for all atoms, this becomes

$$\partial V/\partial U_\alpha^{m\mu i} = \Phi_\alpha(m\mu i) + \sum_\beta \sum_{n\nu j} \Phi_{\alpha\beta}\binom{m\mu i}{n\nu j} \varepsilon_\beta + \ldots \tag{1.3.2}$$

Since the displacements ε_β of the atoms are completely arbitrary and since the first derivatives for the translated configuration cannot differ from those of the initial configuration, which are given by

$$\partial V/\partial U_\alpha^{m\mu i} = \Phi_\alpha(m\mu i) \tag{1.3.3}$$

we obtain from the coefficients of each power of ε_β in 1.3.2

$$\sum_{n\nu j} \Phi_{\alpha\beta}\binom{m\mu i}{n\nu j} = 0 \tag{1.3.4a}$$

$$\sum_{n\nu j} \Phi_{\alpha\beta\gamma}\begin{pmatrix} m\mu i \\ n\nu j \\ p\lambda k \end{pmatrix} = \sum_{m\mu i} \Phi_{\alpha\beta\gamma}\begin{pmatrix} m\mu i \\ n\nu j \\ p\lambda k \end{pmatrix} = \sum_{p\lambda k} \Phi_{\alpha\beta\gamma}\begin{pmatrix} m\mu i \\ n\nu j \\ p\lambda k \end{pmatrix} = 0 \tag{1.3.4b}$$

etc.

In order to show that these modes have zero frequency at $\mathbf{k} = 0$, we examine the form of Eq. 1.2.16 for $\mathbf{k} = 0$. Using 1.2.14 and 1.2.10, we have

$$\sum_\beta \sum_{n\nu j} \Phi_{\alpha\beta}\binom{m\mu i}{n\nu j} \left(\frac{1}{m_i m_j}\right)^{\frac{1}{2}} e_\beta(\nu j|p0) = \omega_{p0}^2 e_\alpha(\mu i|p0) \tag{1.3.5}$$

If $(1/m_j)^{\frac{1}{2}} e_\beta(\nu j|p0)$ is independent on ν and j, as it is the case when all atoms undergo the same displacement, independently from their nature and position in the crystal, it follows from 1.3.4 that the left-hand side of 1.3.5 is equal to zero and therefore $\omega_{p0}^2 = 0$.

This type of solution occurs for three independent orthogonal directions and corresponds to movements of all atoms in the crystal in the same direction with the same amplitude and phase. The three crystal modes correspond thus to the propagation of the sound in a continuum and for this reason the three phonon branches that go to zero as $\mathbf{k} \to 0$ are called "acoustic" branches.

The remaining 3ZN - 3 branches are called "optical" branches since when $\mathbf{k} \to 0$ the frequency ω_{pk} tends to a finite value ω_{p0} that can be measured in an optical experiment (infrared or Raman spectroscopy).For a molecular crystal the optical branches are further divided into (3N - 6)Z internal and (6Z - 3) external branches.The internal branches involve mainly vibrations of molecular nature in which the relative positions of the atoms in each molecule change without shifting the center of mass and without changing the molecular orientation.The external bands instead correspond to rotational and translational motions of the molecules. At $\mathbf{k} = 0$ all unit cells vibrate in phase. At $\mathbf{k} \neq 0$ molecules in different unit cells vibrate with a phase difference determined by the value of $\mathbf{k}$.

Since the intramolecular forces,i.e. the forces holding together the atoms in a molecule, are much stronger than the forces of intermolecular nature, the internal branches have normally higher frequencies than the external branches.Often a well defined frequency gap separates the two types of branches and in this case the formal classification given above has a real physical meaning since external and internal vibrations are not coupled together and preserve their individuality.In other cases, however,branches of internal origin overlap with the external branches in the low frequency region and the two types of motion can be strongly coupled.More complex motions take place in this case since the molecules combine their internal deformations with a shift of the center of mass and with a change in orientation.

The dispersion curves of crystal modes that develop a dynamic dipole moment (polar phonons) have a peculiar behaviour in the long wavelength limit. A polar phonon is called "transverse" or "longitudinal", when the dipole moment developed is perpendicular or parallel respectively to the propagation direction.More generally, the transverse or longitudinal character is associated to the orientation of the atomic displacements with respect to the direction of $\mathbf{k}$,i.e. with respect to the direction of propagation.

Because of the long-range electrostatic interactions associated with polar modes, the limiting frequencies at $\mathbf{k} = 0$ are different for longitudinal and transverse modes. The frequency difference depends on the magnitude of the dynamic dipole moment.Thus,although this is a general property of all polar modes, a frequency splitting is observed in molecular crystals only for internal or external vibrations that possess strong transition dipoles.

Dispersion curves are best classified according to the irreducible representations of the space groups.This will be discussed in detail in the next Chapter.

A typical example of dispersion curves for a molecular crystal is shown in Fig.1.2 for crystalline iodoform [7]. Iodoform crystallizes in the hexagonal system, with two molecules per unit cell. Since each molecule has five atoms, there are 30 branches altogether. These are clas-

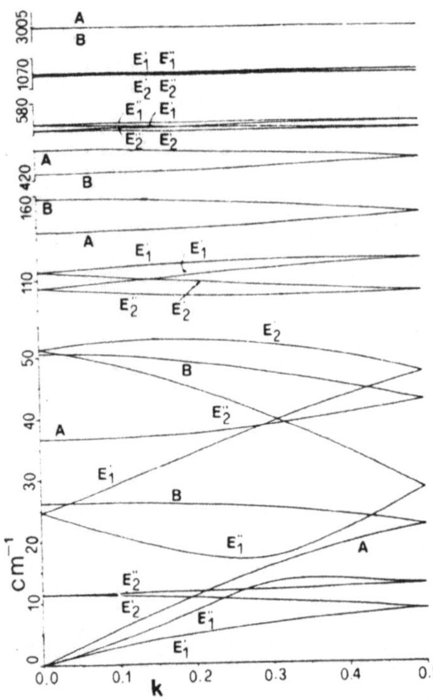

Fig. 1.2 Dispersion curves of crystalline Iodoform for a vector **k** along a C_6 symmetry axis

sified in the figure with the symbols A, B, E'_1, E'_2, E''_1, E''_2 that label the irreducible representations of the C_6 point group. Strictly speaking, this classification is valid only at $\mathbf{k} = 0$ but it is kept through the curves since the $\mathbf{k}$ vector is allowed to vary in the figure along a six fold axis of the crystal. We notice that all the thirty branches are separated in the central part of the figure since they belong to mono-dimensional representations of the space group. Double degenerate frequencies occur, however, at $\mathbf{k} = 0$ and at $\mathbf{k} = (0,0,\frac{1}{2})$ due to a particular symmetry at these points called "time reversal" and discussed in Chapter 2.

The occurrence of three acoustic branches is clearly seen in the figure. These are the three branches that go to zero when $\mathbf{k} \to 0$. It is also seen from the Figure that the long wavelength distinction between acoustic and optical branches looses completely its meaning at the point $\mathbf{k} = (0,0,\frac{1}{2})$, where the acoustic and optical branches stick together.

The figure shows also the separation between the nine external and the fifteen internal optical branches. These are separated by a gap of about 50 cm^{-1} . Finally the figure shows that the external branches have a rather large dispersion whereas the internal branches, as pointed out before, are practically not dispersed.

The dispersion relations, represented grafically by the dispersion curves, allow us to follow the behaviour of individual phonons along specific $\mathbf{k}$ directions. Since the number of frequencies is very large, it can be more convenient for several applications to deal with the frequency distribution or "density of states" $n(\omega)$ of the phonons. These applications include, for instance, the calculation of the specific heat and other thermodynamic properties, the study of two-phonon bands and of incoherent neutron scattering. The quantity $n(\omega)d\omega$ gives the fraction of frequencies in the interval $\omega \div \omega + d\omega$ and the density of phonon states can be defined as

$$n(\omega) = \sum_{j\mathbf{k}} \delta(\omega - \omega_{j\mathbf{k}}) \qquad 1.3.6$$

and can be normalized so that

$$\int_0^{\omega_m} n(\omega)d\omega = 1 \qquad 1.3.7$$

where ω_m is the maximum value of the frequency.

The calculation of the density of phonon states,according to 1.3.6, requires that the dynamical matrix must be diagonalized for a sufficient number of uniformily distributed **k** values.This process is time consuming and the histograms obtained will depend on the density of sampling of the wavevectors **k**. Numerical methods have, however, been devised to produce reliable densities of states without increasing the computation time to an impracticable level.

As an example, the density of states calculated for the ammonia crystal is shown in Fig. 1.3 for the external frequency region.As can be seen from the figure, several gaps may occur in the density of states, whenever some branches of the dispersion curves are isolated from the others. In molecular crystals this is frequently the case for internal vibrations.

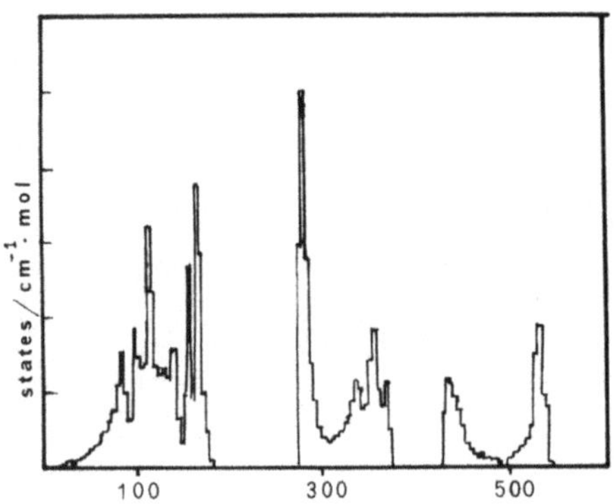

Fig.1.3 Calculated density of states of the ammonia crystal in the region of the lattice frequencies

It may be then useful to consider the density of states for a single branch j

$$n_j(\omega) = \sum_{\mathbf{k}} \delta(\omega - \omega_{j\mathbf{k}}) \qquad 1.3.8$$

1.4 INVARIANCE CONDITIONS

The Cartesian force constants used so far in the lattice dynamics treatment seem at first sight very convenient quantities to use and one has only to regret that their number is exceedingly large for practical purposes. A little thought shows, however, that this is not true and that on the contrary, the use of Cartesian force constants offers rather intriguing difficulties that need to be analyzed for a correct treatment.

We have already seen in the previous Section that the Cartesian force constants must obey the sum rules 1.3.4 that originate from the translational invariance of the crystal. Further restrictions are imposed on them by the fact that the crystal potential must be invariant also under a crystal rotation. This shows that the Cartesian force constants are far from being independent adjustable parameters for the calculation of crystal frequencies but instead are constrained by these invariance conditions.

In order to investigate the rotational invariance, we consider a general rotation of the crystal by an angle ψ about any axis going through a given point P and directed along a unit vector $\boldsymbol{\xi}$. For the sake of simplicity P is taken at the origin of the crystal reference system. To the rotation ψ we can associate a rotation operator $\hat{O}(\psi,\boldsymbol{\xi})$ that, for $\psi \ll 1$, can be written in the form

$$\hat{O}(\psi,\boldsymbol{\xi}) = \hat{E} + \psi\hat{I}_{\xi} \qquad 1.4.1a$$

where $\hat{E}$ is the identity operator and $\hat{I}_{\xi}$ is an infinitesimal rotation operator about the axis $\boldsymbol{\xi}$. The operator $\hat{I}_{\xi}$ can be expressed in terms of infinitesimal rotation operators $\hat{I}_{\gamma}$ about the axes of the crystal reference frame

$$\hat{I}_{\xi} = \sum_{\gamma} \ell_{\gamma} \hat{I}_{\gamma} \qquad 1.4.1b$$

where ℓ_{γ} are the components of $\boldsymbol{\xi}$ in the crystal frame. A rotated vector $\mathbf{X}'$ is then given, in terms of the equilibrium vector $\bar{\mathbf{X}}$, by

$$\mathbf{X}' = (\mathbb{E} + \psi \sum_{\gamma} \ell_{\gamma} \mathbb{M}^{\gamma})\bar{\mathbf{X}} \qquad 1.4.2$$

where $\mathbb{E}$ is the unit matrix and $\mathbb{M}^{\gamma}$ are matrices associated to the operators $\hat{I}_{\gamma}$, with elements $M^{\gamma}_{\alpha\beta} = \delta_{\gamma\beta\alpha}$,expressed in terms of the usual Levi-Civita symbols

$$\delta_{\gamma\beta\alpha} = \begin{cases} 0 & \text{unless } \gamma \neq \beta \neq \alpha \\ +1 & \text{if } \gamma,\beta,\alpha \text{ are in the cyclic order 123} \\ -1 & \text{if } \gamma,\beta,\alpha \text{ are in the cyclic order 132} \end{cases} \qquad 1.4.3$$

Explicitely the matrices $\mathbb{M}^{\gamma}$ are then

$$\mathbb{M}^1 = \begin{vmatrix} 0 & 0 & 0 \\ 0 & 0 & -1 \\ 0 & 1 & 0 \end{vmatrix} \qquad \mathbb{M}^2 = \begin{vmatrix} 0 & 0 & 1 \\ 0 & 0 & 0 \\ -1 & 0 & 0 \end{vmatrix} \qquad \mathbb{M}^3 = \begin{vmatrix} 0 & -1 & 0 \\ 1 & 0 & 0 \\ 0 & 0 & 0 \end{vmatrix} \qquad 1.4.4$$

From 1.4.2 we obtain that the αth Cartesian coordinate $X^{m\mu i}_{\alpha}$ of atom mμi is given, after the rotation, by

$$X^{m\mu i}_{\alpha} = \bar{X}^{m\mu i}_{\alpha} + \Psi\sum_{\gamma}\ell_{\gamma}\sum_{\beta}M^{\gamma}_{\alpha\beta}\bar{X}^{m\mu i}_{\beta} \qquad 1.4.5$$

and thus the atom mμi undergoes displacements $U^{m\mu i}_{\alpha} = X^{m\mu i}_{\alpha} - \bar{X}^{m\mu i}_{\alpha}$, given by

$$U^{m\mu i}_{\alpha} = \Psi\sum_{\gamma}\ell_{\gamma}\sum_{\beta}M^{\gamma}_{\alpha\beta}\bar{X}^{m\mu i}_{\beta} \qquad 1.4.6$$

Applying the transformation 1.4.5 to the components $\Phi_{\alpha}(m\mu i)$ of the force acting on atom mμi in the initial configuration, we obtain that the components of the rotated force are

$$\partial V/\partial U^{m\mu i}_{\alpha} = \Phi_{\alpha}(m\mu i) + \Psi\sum_{\gamma}\ell_{\gamma}\sum_{\beta}M^{\gamma}_{\alpha\beta}\Phi_{\beta}(m\mu i) \qquad 1.4.7$$

By substitution of 1.4.6 in 1.3.1 and by equating the result to 1.4.7, we obtain for each value of γ

$$\sum_{\beta}\sum_{n\nu j}\Phi_{\alpha\beta}\binom{m\mu i}{n\nu j}\sum_{\varepsilon}M^{\gamma}_{\beta\varepsilon}\bar{X}^{n\nu j}_{\varepsilon} = \sum_{\omega}M^{\gamma}_{\alpha\omega}\Phi_{\omega}(m\mu i) \qquad 1.4.8$$

which represents the second set of general invariance conditions,one for each value of γ, first proposed by Born and Huang [3].

The same arguments can be applied to the complete potential expansion 1.2.2 instead of only to the first derivatives.A new set of conditions would be obtained but, as far as the force constants are concerned, these are less restrictive than 1.3.4 and 1.4.8. For this reason these sum rules [3 – 6] will not be discussed here. It is,however, useful to impose the translational and rotational invariance criteria to the potential energy in the initial configuration, limiting the discussion to the conditions involving the first derivatives only.

For a uniform translation of the crystal, i.e. for all $U_{\alpha}^{m\mu i} = \varepsilon_{\alpha}$, we obtain from 1.2.2

$$\sum_{m\mu i} \Phi_{\alpha}(m\mu i) = 0 \qquad 1.4.9a$$

In the same way, for a rotation of the crystal, i.e. by substitution of 1.4.6 in 1.2.2, we obtain

$$\sum_{\alpha}\sum_{m\mu i} \Phi_{\alpha}(m\mu i) \sum_{\beta} M^{\gamma}_{\alpha\beta} \bar{X}^{m\mu i}_{\beta} = 0 \qquad 1.4.9b$$

We recall that the first derivatives of V at the initial configuration were not assumed to be necessarily equal to zero in discussing the invariance properties of the crystal. This was done to ensure the maximum degree of generality to the invariance conditions, that can be thus applied to any scalar defined in terms of crystal Cartesian coordinates such as bond distances, angles, etc. for which no particular set of conditions can be imposed to the first derivatives at equilibrium. Obviously a vibrating crystal must be at equilibrium in the initial configuration. This means that, according to 1.2.3a, each single first order force constant $\Phi_{\alpha}(m\mu i)$ is equal to zero and thus all conditions derived so far in general terms transform into specific sum rules for the force constants. Introducing the condition 1.2.3a in 1.4.8, we can write the translational and rotational invariance conditions in the form

$$\sum_{n\nu j} \Phi_{\alpha\beta}\binom{m\mu i}{n\nu j} = 0 \qquad 1.4.10a$$

$$\sum_{\beta}\sum_{n\nu j} \Phi_{\alpha\beta}\binom{m\mu i}{n\nu j} \sum_{\varepsilon} M^{\gamma}_{\beta\varepsilon} \bar{X}^{n\nu j}_{\varepsilon} = 0 \qquad 1.4.10b$$

which imposes restrictions on the possible values of Cartesian force constants.

Eqs.1.4.10 have been used in dynamical calculations for the evaluation of the so called "self-force constants", defined as the force constants connecting two Cartesian displacements on the same molecule (or on the same atom for atomic crystals). For instance for the low-symmetry atomic crystal of Te, Powell [8] used self-force constants

$$\Phi_{\alpha\beta}\binom{mi}{mi} = -\sum'_{nj}\Phi_{\alpha\beta}\binom{mi}{nj} \qquad 1.4.11$$

where molecular indices have been dropped as unnecessary for atomic crys tals. The symbol $\sum'$ means that the term mi = nj is not counted in the sum. The use of self-terms 1.4.11, determined by all other force constants,ensures that the crystal is invariant under a translation. However, as pointed out by Powell [8] , 1.4.11 does not automatically fulfill Hermiticity of the dynamical matrix. Even if one assumes that, when mi ≠ nj ,

$$\Phi_{\alpha\beta}\binom{mi}{nj} = \Phi_{\beta\alpha}\binom{nj}{mi} \qquad 1.4.12$$

self-terms 1.4.11 are not symmetric (except when so required by particular symmetry properties of the crystal space group) and this gave rise to a debate [9,10] on the Hermitian character of the dynamical matrix. This problem is not limited to the Cartesian force constants but has a counterpart in molecular coordinates, that will be treated in Section 1.6.Here it is sufficient to point out that, as found by Powell [8] ,the relationships 1.4.10 are automatically fulfilled, if a model for V is used in which the potential energy is an explicit function of the interatomic distances r_s

$$V = \sum_s f_s(r_s) \qquad 1.4.13$$

with the sum extended to all possible atom-atom distances in atomic crystals or to all distances between two atoms belonging to different molecules in a molecular crystal.

If f'_s and f''_s are derivatives of V with respect to r_s at equilibrium

$$f'_s = (\partial f_s/\partial r_s)_0 \qquad f''_s = (\partial^2 f_s/\partial^2 r_s)_0 \tag{1.4.14}$$

and if the atom-atom distances are expanded in power series of the displacements

$$r_s = r_s^0 + \sum_\alpha \sum_{mi} B_\alpha^s(mi) U_\alpha^{mi} + \frac{1}{2} \sum_{\alpha\beta} \sum_{mn} \sum_{ij} B_{\alpha\beta}^s\binom{mi}{nj} U_\alpha^{mi} U_\beta^{nj} + \ldots \tag{1.4.15}$$

self-force constants can be explicitely obtained using rules of partial differentiation and the check of translational invariance can be directly made. We note that the expansion 1.4.15, when written including molecular indices, can be used whenever internal displacements are needed as a function of Cartesian displacements in a molecular crystal. All conclusions drawn from 1.4.15 can be thus easily transferred to molecular coordinates. We can apply the conditions 1.4.9a and 1.4.10a to each scalar r_s and find

$$B_\alpha^s(mi) = - \sum'_{nj} B_\alpha^s(nj) \tag{1.4.16a}$$

$$B_{\alpha\beta}^s\binom{mi}{mi} = - \sum'_{nj} B_{\alpha\beta}^s\binom{mi}{nj} \tag{1.4.16b}$$

Hence, the self-terms 1.4.11 are given by

$$\begin{aligned}\Phi_{\alpha\beta}\binom{mi}{mi} &= - \sum_s \left[f'_s \sum'_{nj} B_{\alpha\beta}^s\binom{mi}{nj} + f'' B_\alpha^s(mi) \sum'_{nj} B_\beta^s(nj)\right] = \\ &= \sum_s \left[f'_s B_{\alpha\beta}^s\binom{mi}{mi} + f'' B_\alpha^s(mi) B_\beta^s(mi)\right]\end{aligned} \tag{1.4.17}$$

where 1.4.16 were used in the last step. Self-terms thus correspond to analytical derivatives of a potential function of the kind 1.4.13.They are obviously symmetric and have the property that

$$\sum'_{nj} \Phi_{\alpha\beta}\binom{mi}{nj} = \sum'_{nj} \Phi_{\beta\alpha}\binom{mi}{nj} \tag{1.4.18}$$

a result which is not true in general and derives from the properties of the scalar r_s.

1.5 MOLECULAR COORDINATES

The simple lattice dynamics treatment of Born and Huang which makes use of atomic Cartesian displacements has been extended in the previous sections to the molecular crystals. This choice of coordinates presents, however,several theoretical and practical difficulties that make its use rather inconvenient and call for a better choice, more directly connected to the molecular nature of these crystals.

First of all, we notice that the number of Cartesian force constants for a crystal made of polyatomic molecules is exceedingly high,despite the sum rules constraints discussed in the previous Section. Furthermore, the use of atomic displacements obscures completely the existence of the molecules in the crystal and makes impossible the separation between internal and external crystal modes. This separation is very important for the development of model potentials and for the application of the rigid body approximation.

Molecular spectroscopists have long ago adopted "internal" coordinates [11,13] as a suitable set of variables for the description of the vibrational motions of isolated molecules. These coordinates represent variations of bond lengths,interbond angles,bond torsional angles etc., within each molecule and are connected to the atomic displacements by relations of the type 1.4.15 where the coefficients represent now derivatives at equilibrium of explicit functions defining each scalar.

The internal coordinates are defined [11,13] so that there is no coupling,at least to the first order, between them and the coordinates used to describe the molecular translations and rotations. Their use simplifies considerably the treatment and permits to incorporate in the lattice dynamics of molecular crystals the large body of information accumulated until now on the dynamics of free molecules.

The basic assumption that justifies the choice of molecular coordinates is that the potential energy of the crystal can be expressed as the sum of intra- and intermolecular contributions. In this way we can utilize the results of molecular dynamics and of the whole theory of intermolecular forces to construct suitable crystal potentials of general use. This problem will be discussed in more detail in the next Section.

In order to introduce the internal coordinates of the molecules in the form first proposed by Wilson [11], it is convenient to associate with each molecule a molecule-fixed reference system with the origin in the center of mass and oriented along the axes of the principal inertia tensor of the molecule. In what follows we shall indicate by small letters the Cartesian coordinates of the atoms in the molecule-fixed frame. If $x_\rho^{m\mu i}$ and $\bar{x}_\rho^{m\mu i}$ represent instantaneous and equilibrium coordinates of the atom $m\mu i$, respectively, we define in analogy to 1.2.1, Cartesian displacements in the form

$$u_\rho^{m\mu i} = x_\rho^{m\mu i} - \bar{x}_\rho^{m\mu i} \qquad 1.5.1$$

Furthermore, if $X_\alpha^{m\mu}$ and $\bar{X}_\alpha^{m\mu}$ represent instantaneous and equilibrium coordinates, respectively, of the center of mass of the molecule in the crystal-fixed frame, we define corresponding Cartesian displacements

$$U_\alpha^{m\mu} = X_\alpha^{m\mu} - \bar{X}_\alpha^{m\mu} \qquad 1.5.2$$

Then, if Γ represents the matrix of the instantaneous direction cosines of the molecule-fixed with respect to the crystal-fixed system and Λ the corresponding equilibrium matrix, the relation between coordinates in the two systems are simply

$$X_\alpha^{m\mu i} = X_\alpha^{m\mu} + \sum_\rho \Gamma_{\alpha\rho}^{\mu} x_\rho^{m\mu i} \qquad 1.5.3a$$

$$\bar{X}_\alpha^{m\mu i} = \bar{X}_\alpha^{m\mu} + \sum_\rho \Lambda_{\alpha\rho}^{\mu} \bar{x}_\rho^{m\mu i} \qquad 1.5.3b$$

$$U_\alpha^{m\mu i} = U_\alpha^{m\mu} + \sum_\rho \Gamma_{\alpha\rho}^{\mu} u_\rho^{m\mu i} + \sum_\rho \lambda_{\alpha\rho}^{\mu} \bar{x}_\rho^{m\mu i} \qquad 1.5.3c$$

where $\Gamma_{\alpha\rho}^{\mu}$ represents the direction cosine between the molecule-fixed axis ρ ($x_\rho = x,y,z$) and the crystal-fixed axis α($X_\alpha = X,Y,Z$) and

$$\lambda_{\alpha\rho}^{\mu} = \Gamma_{\alpha\rho}^{\mu} - \Lambda_{\alpha\rho}^{\mu} \qquad 1.5.3d$$

We shall show later that the variation $\lambda_{\alpha\rho}^{\mu}$ in the direction cosines can be expressed in terms of only three independent rotation angles θ_x, θ_y and θ_z about the axes of the molecule-fixed frame.

Equation 1.5.3c is of great importance to us since it shows that the 3N crystal-fixed Cartesian displacements which describe the 3N degrees of freedom of molecule $m\mu$ can be replaced by a new set of coordinates that separate the translations, the rotations and the internal vibrations. In particular, the new set of coordinates is made of the three translational displacements of the center of mass, of the three rotations θ_x, θ_y and θ_z and of the 3N displacements of the atoms in the molecule-fixed reference frame. In this way we have, however, 3N atomic displacements to describe 3N - 6 degrees of internal freedom. We need therefore to impose six constraints on the atomic displacements $u_\rho^{m\mu i}$ and to transform them to a new set of 3N - 6 coordinates that automatically fulfill the constraints.

We define therefore 3N - 6 internal coordinates $S_\ell^{m\mu}$ plus six redundancy conditions (constraints) for each molecule $m\mu$. The definition of the internal coordinates in terms of the atomic displacements $u_\rho^{m\mu i}$ was first given by Wilson [11] in the form of a linear relation

$$S_\ell^{m\mu} = \sum_\rho \sum_i B_\rho^\ell(m\mu i)\, u_\rho^{m\mu i} \tag{1.5.4a}$$

and later extended by Hoy et al. [12] to cover the more general case of curvilinear internal coordinates of interest in anharmonic calculations by adding second-order terms to 1.5.4a

$$S_\ell^{m\mu} = \sum_\rho \sum_i B_\rho^\ell(m\mu i)\, u_\rho^{m\mu i} + \tfrac{1}{2}\sum_{\rho\sigma} \sum_{ij} B_{\rho\sigma}^\ell\binom{m\mu i}{m\mu j} u_\rho^{m\mu i} u_\sigma^{m\mu j} + \ldots \tag{1.5.4b}$$

The coefficients $B_\rho^\ell(m\mu i)$ and $B_{\rho\sigma}^\ell\binom{m\mu i}{m\mu j}$ are defined by the molecular geometry at equilibrium. Their construction represents a standard subject of molecular spectroscopy textbooks [11,13] and will not be discussed here.

The six redundancies among the displacements $u_\rho^{m\mu i}$ were first found by Eckart [14] and Sayvetz [14], imposing that the molecular center of mass is not displaced and that there is no rotation of the molecular frame in an internal vibration. When properly normalized, the translations and rotations of the molecule can be written in the form

$$t_{\rho}^{m\mu} = (\frac{1}{M})^{\frac{1}{2}} \sum_i m_i u_{\rho}^{m\mu i} \qquad 1.5.5a$$

$$r_{\rho}^{m\mu} = \sum_i (\frac{1}{I_\rho})^{\frac{1}{2}} m_i \sum_\sigma u_\sigma^{m\mu i} \sum_\tau M_{\sigma\tau}^{\rho} \bar{x}_\tau^{m\mu i} = \sum_i (\frac{1}{I_\rho})^{\frac{1}{2}} m_i (\bar{x}_\sigma^{m\mu i} u_\tau^{m\mu i} - \bar{x}_\tau^{m\mu i} u_\sigma^{m\mu i}) \qquad 1.5.5b$$

with ρ , σ , τ in cyclic order.

The Eckart conditions are then simply obtained by equating to zero the three rotations and the three translations defined by 1.5.5.

Truly speaking only the three translations are completely separated in 1.5.3c from other coordinates. Internal displacements $u_{\rho}^{m\mu i}$ are instead still coupled to the rotations in the second term of the right hand side of 1.5.3c, since the instantaneous direction cosines $\Gamma_{\alpha\rho}^{\mu}$ are a function of the rotations. This second term can be rewritten in the form

$$\sum_\rho \Gamma_{\alpha\rho}^{\mu} u_{\rho}^{m\mu i} = \sum_\rho \Lambda_{\alpha\rho}^{\mu} u_{\rho}^{m\mu i} + \sum_\rho \lambda_{\alpha\rho}^{\mu} u_{\rho}^{m\mu i} \qquad 1.5.6$$

and thus the rotations are separated from the internal displacements as long as the second term in 1.5.6 can be neglected. We notice that this is a second-order term since it involves the product of two coordinates, one internal displacement and one rotation.We can thus confidently neglect the second term in 1.5.6 and rewrite 1.5.3c in the form

$$U_{\alpha}^{m\mu i} = U_{\alpha}^{m\mu} + \sum_\rho \Lambda_{\alpha\rho}^{\mu} u_{\rho}^{m\mu i} + \sum_\rho \lambda_{\alpha\rho}^{\mu} \bar{x}_{\rho}^{m\mu i} \qquad 1.5.7$$

where,to the first order, all coordinates are uncoupled.

In order to obtain the dependence of $\lambda_{\alpha\rho}^{\mu}$ on the three rotation angles θ_x, θ_y,θ_z about the axes of the molecule-fixed system, we follow a basic treatment due to Oliver and Walmsley [15] that we develop here in a slightly different form.

Let **e** and **e'** represent the molecule-fixed basis at equilibrium and at a generic orientation with respect to the basis **i** of the crystal-fixed frame respectively,such that

$$\mathbf{e} = \mathbf{i}\Lambda \qquad 1.5.8a$$

$$\mathbf{e'} = \mathbf{i}\Gamma \qquad 1.5.8b$$

Let furthermore $\hat{O}(\Psi,\boldsymbol{\xi})$ be a rotation operator, associated with a finite rotation of an angle Ψ about a generic axis directed along a unit vector $\boldsymbol{\xi}$, going through the origin of the molecule-fixed reference system **e**, which brings the basis **e** into the rotated basis e'. The operator $\hat{O}(\Psi,\boldsymbol{\xi})$ acts on the basis **e** according to the relation

$$\hat{O}(\Psi,\boldsymbol{\xi})\mathbf{e} = \mathbf{e}' = \mathbf{e}\mathbb{D}(\Psi,\boldsymbol{\xi}) \qquad 1.5.9$$

where $\mathbb{D}(\Psi,\boldsymbol{\xi})$ is a rotation matrix. The operator $\hat{O}(\Psi,\boldsymbol{\xi})$ can be expressed in terms of an infinitesimal operator $\hat{I}_{\boldsymbol{\xi}}$ about the axis $\boldsymbol{\xi}$, in the form

$$\hat{O}(\Psi,\boldsymbol{\xi}) = \hat{E} + \Psi\hat{I}_{\boldsymbol{\xi}} + \frac{1}{2}\Psi^2\hat{I}^2_{\boldsymbol{\xi}} + \dots \qquad 1.5.10$$

where $\hat{E}$ is the identity operator. In turn, as discussed in the previous Section, the operator $\hat{I}_{\boldsymbol{\xi}}$ can be expressed in terms of infinitesimal rotation operators $\hat{I}_\rho$ about the axes $\mathbf{e}_\rho$ of the molecule-fixed frame

$$\hat{I}_{\boldsymbol{\xi}} = \sum_\rho \ell_\rho \hat{I}_\rho \qquad 1.5.11$$

By substitution of 1.5.11 in 1.5.10 we obtain then

$$\hat{O}(\Psi,\boldsymbol{\xi}) = \hat{E} + \Psi\sum_\rho \ell_\rho \hat{I}_\rho + \frac{1}{2}\Psi^2\sum_{\rho\sigma}\ell_\rho\ell_\sigma\hat{I}_\rho\hat{I}_\sigma + \dots \qquad 1.5.12$$

It follows from 1.5.12 that the rotation matrix $\mathbb{D}(\Psi,\boldsymbol{\xi})$ is given, in terms of the matrices M^ρ associated to the operators $\hat{I}_\rho$, by the relation

$$\mathbb{D}(\Psi,\boldsymbol{\xi}) = \mathbf{E} + \Psi\sum_\rho \ell_\rho \mathbf{M}^\rho + \frac{1}{2}\Psi^2\sum_{\rho\sigma}\ell_\rho\ell_\sigma\mathbf{M}^\rho\mathbf{M}^\sigma + \dots \qquad 1.5.13$$

The form of the matrices $\mathbf{M}^\rho$ is given by 1.4.4.

We define now the rotational variables

$$\theta_\rho = \Psi\ell_\rho \qquad (\rho = x,\ y,\ z) \qquad 1.5.14$$

that we identify with three rotation angles about the axes ρ of the molecular frame. For infinitesimal rotations the angles θ_ρ are independent and can be taken as the three rotational coordinates.

By substitution of 1.5.13 in 1.5.9 and by using 1.5.8, we obtain then the relation between the direction cosines $\Gamma_{\alpha\rho}$ and the rotation angles

$$\mathbf{\Gamma} = \mathbf{\Lambda}[\,\mathbf{E} + \sum_\rho \theta_\rho \mathbf{M}^\rho + \frac{1}{2}\sum_{\rho\sigma}\theta_\rho\theta_\sigma \mathbf{M}^\rho \mathbf{M}^\sigma\,] \qquad 1.5.15$$

or, by performing the matrix multiplication in 1.5.15

$$\Gamma^\mu_{\alpha\rho} = \Lambda^\mu_{\alpha\rho} + \sum_\sigma \theta_\sigma \sum_\tau \Lambda^\mu_{\alpha\tau} M^\sigma_{\tau\rho} + - \sum_{\sigma\chi}\theta_\sigma\theta_\chi \sum_\tau \Lambda^\mu_{\alpha\tau}\sum_\omega M^\sigma_{\tau\omega} M^\chi_{\omega\rho} + \ldots \qquad 1.5.16$$

We obtain therefore

$$\lambda^\mu_{\alpha\rho} = \Gamma^\mu_{\alpha\rho} - \Lambda^\mu_{\alpha\rho} = \sum_\sigma \theta_\sigma \sum_\tau \Lambda^\mu_{\alpha\tau} M^\sigma_{\tau\rho} + \frac{1}{2}\sum_{\sigma\chi}\theta_\sigma\theta_\chi \sum_\tau \Lambda^\mu_{\alpha\tau}\sum_\omega \mathbf{M}^\sigma_{\tau\omega} M^\chi_{\omega\rho} \qquad 1.5.17$$

In order to write 1.5.7 in the most convenient form, we notice that, although the displacements $U^{m\mu}_\alpha$ of the molecular center of mass are perfectly well defined translational coordinates, for simplifying the kinetic energy, it is better to use molecular translations $u^{m\mu}_\rho$ along the axes ρ of the molecular reference frame in the equilibrium position. By projecting the displacements $U^{m\mu}_\alpha$ on these axes we have

$$U^{m\mu}_\alpha = \sum_\rho \Lambda^\mu_{\alpha\rho} u^{m\mu}_\rho \qquad 1.5.18$$

Finally, by defining the inverse transformation of 1.5.4a

$$u^{m\mu i}_\rho = \sum_\ell C^\ell_\rho(m\mu i) S^{m\mu}_\ell \qquad 1.5.19$$

such that

$$\sum_i \sum_\rho C^{\ell'}_\rho(m\mu i) B^\ell_\rho(m\mu i) = \delta_{\ell\ell'} \qquad 1.5.20$$

we can introduce the internal coordinates directly into 1.5.7. By substitution of 1.5.17, 1.5.18 and 1.5.19 in 1.5.7 we obtain then the desired transformation from the set of 3N crystal-fixed coordinates $U^{m\mu i}_\alpha$

to the set of 3N molecular coordinates made of three translations $u_\rho^{m\mu}$ of the center of mass, three rotations $\theta_\rho^{m\mu}$ of the molecular frame and 3N - 6 internal coordinates $S_\ell^{m\mu}$

$$U_\alpha^{m\mu i} = \sum_\rho \Lambda_{\alpha\rho}^\mu u_\rho^{m\mu} + \sum_\rho \sum_\ell \Lambda_{\alpha\rho}^\mu C_\rho^\ell (m\mu i) S_\ell^{m\mu} + \sum_\rho \sum_\sigma \theta_\sigma^{m\mu} \sum_\tau \Lambda_{\alpha\tau}^\mu M_{\tau\rho}^\sigma \bar{x}_\rho^{m\mu i} \qquad 1.5.21$$

where for simplicity only the first order term of 1.5.17 has been used.

1.6 THE DYNAMICAL EQUATIONS IN MOLECULAR COORDINATES

The molecular coordinates defined in the last Section will be now used to set up the dynamical equation for a molecular crystal.

Let us consider first the form of the kinetic energy in molecular coordinates. From 1.2.7 we have

$$T = \sum_{m\mu} T^{m\mu} = \frac{1}{2} \sum_\alpha \sum_{m\mu i} m_i (\dot{U}_\alpha^{m\mu i})^2 \qquad 1.6.1$$

and thus, using 1.5.21, we obtain

$$T^{m\mu} = \frac{1}{2} \sum_\alpha \sum_i m_i \{\sum_\rho [\Lambda_{\alpha\rho}^\mu \dot{u}_\rho^{m\mu} + \sum_\ell \Lambda_{\alpha\rho}^\mu C_\rho^\ell (m\mu i) \dot{S}_\ell^{m\mu} + \sum_\sigma \dot{\theta}_\sigma^{m\mu} \sum_\tau \Lambda_{\alpha\tau}^\mu M_{\tau\rho}^\sigma \bar{x}_\rho^{m\mu i}]\}^2 \qquad 1.6.2$$

Using the time derivatives of the Eckart conditions 1.5.5 and the orthogonality of the direction cosines

$$\sum_\alpha \Lambda_{\alpha\rho}^\mu \Lambda_{\alpha\sigma}^\mu = \delta_{\rho\sigma} \qquad 1.6.3$$

we obtain

$$T^{m\mu} = \frac{1}{2} \sum_\rho [M(\dot{u}_\rho^{m\mu})^2 + \sum_i m_i \sum_{\ell'\ell} C_\rho^\ell (m\mu i) C_\rho^{\ell'} (m\mu i) \dot{S}_{\ell'}^{m\mu} \dot{S}_\ell^{m\mu} + \qquad 1.6.4$$

$$+ \sum_i \sum_{\sigma\chi} \sum_{\tau\xi} (m_i \bar{x}_\tau^{m\mu i} \bar{x}_\xi^{m\mu i} M_{\rho\tau}^\sigma M_{\rho\xi}^\chi) \dot{\theta}_\sigma^{m\mu} \dot{\theta}_\chi^{m\mu}]$$

where M is the molecular mass.

From the definition 1.4.4 of the rotation matrices $M_{\rho\tau}^\sigma$, it is easily seen that the terms in the last sum of 1.6.4 are of two types only and, owing to the fact that the axes of the molecule-fixed frame coincide with those of the principal inertia tensor, are given by

$$\sum_i m_i (\bar{x}_\sigma^2 + \bar{x}_\tau^2)\dot{\theta}_\rho^2 = I_\rho \dot{\theta}_\rho^2 \qquad 1.6.5a$$

$$\sum_i m_i \bar{x}_\sigma \bar{x}_\tau \dot{\theta}_\sigma \dot{\theta}_\tau = I_{\sigma\tau} \dot{\theta}_\sigma \dot{\theta}_\tau = 0 \qquad 1.6.5b$$

where for simplicity the molecular label $m\mu$ has been omitted. We have therefore

$$T^{m\mu} = \frac{1}{2} \sum_\rho \left[\mathbf{M}(\dot{u}_\rho^{m\mu})^2 + I_\rho^{m\mu}(\dot{\theta}_\rho^{m\mu})^2 + \sum_i \sum_{\ell'\ell} C_\rho^\ell (m\mu i) C_\rho^{\ell'}(m\mu i) m_i \dot{S}_\ell^{m\mu} \dot{S}_{\ell'}^{m\mu}\right] \qquad 1.6.6$$

that can be rewritten in the form

$$T^{m\mu} = \frac{1}{2} \left[\sum_\rho \mathbf{M}(\dot{u}_\rho^{m\mu})^2 + \sum_\rho I_\rho^{m\mu}(\dot{\theta}_\rho^{m\mu})^2 + \sum_\ell \sum_{\ell'} (g_{\ell\ell'}^{m\mu})^{-1} \dot{S}_\ell^{m\mu} \dot{S}_{\ell'}^{m\mu}\right] \qquad 1.6.7$$

where

$$(g_{\ell\ell'}^{m\mu})^{-1} = \sum_i \sum_\rho m_i C_\rho^\ell (m\mu i) C_\rho^{\ell'}(m\mu i) \qquad 1.6.8$$

The elements $(g_{\ell\ell'}^{m\mu})^{-1}$ can be viewed as covariant components of the metric tensor $\mathbb{G}^{-1}$ of the subspace spanned by the coordinates $S_\ell^{m\mu}$. The inverse of the kinetic energy matrix $\mathbb{G}^{-1}$, i.e. the matrix $\mathbb{G}$ whose elements are contravariant components of the same metric tensor, is a well known matrix in Wilson treatment of molecular vibrations [11]. Standard techniques of construction of the elements of $\mathbb{G}$ are given in all textbooks of molecular dynamics [11-13].

Equation 1.6.7 can be further simplified by introducing normalized external coordinates in the form

$$u_\rho^{m\mu} = M^{-\frac{1}{2}} t_\rho^{m\mu} \qquad 1.6.9a$$

$$\theta_\rho^{m\mu} = I_\rho^{-\frac{1}{2}} r_\rho^{m\mu} \qquad 1.6.9b$$

By substitution of 1.6.9 in 1.6.7 the kinetic energy becomes

$$T = \frac{1}{2}\sum_{m\mu}[\sum_\rho(\dot{t}_\rho^{m\mu})^2 + \sum_\rho(\dot{r}_\rho^{m\mu})^2 + \sum_{\ell\ell'}(g_{\ell\ell'}^{m\mu})^{-1}\dot{S}_\ell^{m\mu}\dot{S}_{\ell'}^{m\mu}] \qquad 1.6.10$$

The metric tensor of the subspace spanned by the external coordinates $t_\rho^{m\mu}$ and $r_\rho^{m\mu}$ is thus the unit matrix.This simplification in the form of the kinetic energy has obviously a counterpart in the fact that the molecular mass **M** and the inertia moments I_ρ will now appear in the potential energy expanded in terms of the normalized coordinates.

For simplicity and compactness of the expressions we shall use, whenever convenient, the comprehensive symbol $R_\ell^{m\mu}$ for all molecular coordinates $t_\rho^{m\mu}, r_\rho^{m\mu}$ and $S_\ell^{m\mu}$ with the convention that for $\ell = 1,2,3$ $R_\ell^{m\mu} = t_\rho^m$ for $\ell = 4,5,6$ $R_\ell^{m\mu} = r_\rho^{m\mu}$ and for $\ell = 7,8,..,3N$ $R_\ell^{m\mu} = S_\ell^{m\mu}$.

In this way we can write the kinetic energy in the compact form

$$T = \frac{1}{2}\sum_{m\mu}\sum_{\ell\ell'}(g_{\ell\ell'}^{m\mu})^{-1}\dot{R}_\ell^{m\mu}\dot{R}_{\ell'}^{m\mu} \qquad 1.6.11$$

where, for $\ell,\ell' = 7,8,...,3N$, $(g_{\ell\ell'}^{m\mu})^{-1}$ is given by 1.6.8 and

$$(g_{\ell\ell'}^{m\mu})^{-1} = \delta_{\ell\ell'} \qquad \text{for} \quad \ell,\ell' = 1,2,...,6$$

$$(g_{\ell\ell'}^{m\mu})^{-1} = 0 \qquad \text{for} \quad \begin{matrix}\ell = 1,2,...,6\\ \ell' = 7,8,...,3N\end{matrix} \quad \text{or viceversa} \qquad 1.6.12$$

The potential energy, expanded in power series of the molecular coordinates, is given by

$$\begin{aligned}V = &\sum_{m\mu}\sum_\ell F_\ell(m\mu)R_\ell^{m\mu} + \frac{1}{2}\sum_{mn}\sum_{\mu\nu}\sum_{\ell\ell'}F_{\ell\ell'}\binom{m\mu}{n\nu}R_\ell^{m\mu}R_{\ell'}^{n\nu} + \\ &+ \frac{1}{3!}\sum_{mnp}\sum_{\mu\nu\pi}\sum_{\ell\ell'\ell''}F_{\ell\ell'\ell''}\begin{pmatrix}m\mu\\n\nu\\p\pi\end{pmatrix}R_\ell^{m\mu}R_{\ell'}^{n\nu}R_{\ell''}^{p\pi} + \\ &+ \frac{1}{4!}\sum_{mnpr}\sum_{\mu\nu\pi\rho}\sum_{\ell\ell'\ell''\ell'''}F_{\ell\ell'\ell''\ell'''}\begin{pmatrix}m\mu & p\pi\\ n\nu & r\rho\end{pmatrix}R_\ell^{m\mu}R_{\ell'}^{n\nu}R_{\ell''}^{p\pi}R_{\ell'''}^{r\rho} + \ldots\end{aligned} \qquad 1.6.13$$

where

$$F_\ell(m\mu) = (\partial V/\partial R_\ell^{m\mu})_0 \qquad F_{\ell\ell'}\binom{m\mu}{n\nu} = (\partial^2 V/\partial R_\ell^{m\mu}\partial R_{\ell'}^{n\nu})_0 \qquad 1.6.14$$

$$F_{\ell\ell'\ell''}\begin{pmatrix}m\mu\\n\nu\\p\pi\end{pmatrix} = (\partial^3 V/\partial R_\ell^{m\mu}\partial R_{\ell'}^{n\nu}\partial R_{\ell''}^{p\pi})_0 \qquad \text{etc.}$$

As discussed in Section 2, where the potential was expanded in terms of the Cartesian displacements, the equilibrium conditions of the crystal require that all the first derivatives of V at equilibrium must vanish. The condition 1.2.3a becomes then

$$F_{\ell}(m\mu) = 0 \qquad 1.6.15$$

In the harmonic approximation we neglect cubic and higher terms in the expansion of the potential. Eq.1.6.13 reduces then to

$$V = \frac{1}{2} \sum_{mn}\sum_{\mu\nu}\sum_{\ell\ell'} F_{\ell\ell'}\binom{m\mu}{n\nu} R_{\ell}^{m\mu} R_{\ell'}^{n\nu} \qquad 1.6.16$$

The force constants $F_{\ell\ell'}\binom{m\mu}{n\nu}$ must obey the same symmetry conditions discussed in Section 1.2 for the Cartesian force constants and arising from the lattice periodicity. We have thus

$$F_{\ell\ell'}\binom{m\mu}{n\nu} = F_{\ell\ell'}\binom{0 \quad \mu}{n-m \quad \nu} = F_{\ell\ell'}\binom{m-n \quad \mu}{0 \quad \nu} \qquad 1.6.17$$

Furthermore condition 1.2.6 becomes now

$$F_{\ell\ell'}\binom{m\mu}{n\nu} = F_{\ell'\ell}\binom{n\nu}{m\mu} \qquad 1.6.18$$

The molecular force constants must also obey the translational and rotational invariance conditions. These can be introduced in the simplest way as a set of constraints on the force constants. From the definition 1.6.14 and using the linear transformation

$$U_{\alpha}^{m\mu i} = \sum_{\ell} (\partial U_{\alpha}^{m\mu i}/\partial R_{\ell}^{m\mu}) R_{\ell}^{m\mu} \qquad 1.6.19a$$

which is a generalization of 1.5.21,we obtain

$$F_{\ell\ell'}\binom{m\mu}{n\nu} = \sum_{ij}\sum_{\alpha\beta}\Phi_{\alpha\beta}\binom{m\mu i}{n\nu j}(\partial U_{\alpha}^{m\mu i}/\partial R_{\ell}^{m\mu})(\partial U_{\beta}^{n\nu j}/\partial R_{\ell'}^{n\nu}) \qquad 1.6.19b$$

For $R_{\ell}^{m\mu} = t_{\rho}^{m\mu}$ (ρ = 1,2,3) it is easily shown, using 1.5.21, that the translational invariance condition 1.4.10a assumes the form

$$\sum_{n\nu}\sum_{\rho}F_{\ell\rho}\binom{m\mu}{n\nu}\Lambda^{\nu}_{\beta\rho} = 0 \qquad 1.6.20a$$

where

$$F_{\ell\rho}\binom{m\mu}{n\nu} = (\partial^2 V/\partial R^{m\mu}_{\ell}\partial t^{n\nu}_{\rho})_0 \qquad 1.6.20b$$

In the same way, for $R^{m\mu}_{\ell} = r^{m\mu}_{\rho}$, using 1.4.10b and the translational sum rule 1.6.20a, it is possible to show that

$$\sum_{n\nu}\sum_{\rho}[F'_{\ell\rho}\binom{m\mu}{n\nu}\Lambda^{\nu}_{\alpha\rho} + F_{\ell\rho}\binom{m\mu}{n\nu}\sum_{\beta}\Lambda^{\nu}_{\beta\rho}\sum_{\delta}M^{\alpha}_{\beta\delta}\bar{X}^{m\mu}_{\delta}] = 0 \qquad 1.6.21a$$

where

$$F'_{\ell\rho}\binom{m\mu}{n\nu} = (\partial^2 V/\partial R^{m\mu}_{\ell}\partial r^{n\nu}_{\rho})_0 \qquad 1.6.21b$$

Of fundamental importance for the theory that we shall develop in the next Chapters is the possibility of using suitable models for the crystal potential in which the existence of molecular units, connected by weak intermolecular forces, is clearly taken into account. For this it is very convenient to write the total crystal potential as the sum of two terms

$$V = V_M + V_I \qquad 1.6.22$$

where V_M collects all the intramolecular potentials of the molecules in the crystal and V_I includes all the intermolecular interactions.

If $V(m\mu)$ represents the internal potential of molecule $m\mu$, taken with the effective structure and symmetry that it possesses at the crystal site μ, the intramolecular potential is

$$V_M = \sum_{m\mu} V(m\mu) \qquad 1.6.23$$

The internal potential $V(m\mu)$ is a function of the internal coordinates of molecule $m\mu$ only and is thus completely decoupled from all other molecules in the crystal.

In the same way, if $V(^{m\mu}_{n\nu})$ represents the pairwise interaction potential between molecule $m\mu$ and molecule $n\nu$, the intermolecular potential assumes the form

$$V_I = \frac{1}{2} \sum_{mn} \sum_{\mu\nu}{}' V(^{m\mu}_{n\nu}) \quad + \ldots \qquad 1.6.24$$

where the symbol $\sum'$ means that $\mu \neq \nu$ if $m = n$. Only pairwise interactions have been explicitly written in 1.6.24. Three-body forces are normally very weak in molecular crystals and are thus generally neglected. In some few cases, however, they can be of some importance and must be taken into account. We shall discuss three-body forces in Chapter 3.

By substitution of 1.6.22 in the equilibrium conditions 1.6.15, we have at the initial configuration

$$\frac{\partial V(m\mu)}{\partial s_\ell^{m\mu}} + \sum_{n\nu} \frac{\partial V(^{m\mu}_{n\nu})}{\partial s_\ell^{m\mu}} = 0$$

$$\sum_{n\nu} \frac{\partial V(^{m\mu}_{n\nu})}{\partial t_\rho^{m\mu}} = 0 \qquad 1.6.25$$

$$\sum_{n\nu} \frac{\partial V(^{m\mu}_{n\nu})}{\partial r_\rho^{m\mu}} = 0$$

The intramolecular potential is an unknown function of the internal coordinates of the molecules. For this reason $V(m\mu)$ is normally expanded in a power series of the internal coordinates and the force constants defined in this way are used as adjustable parameters to fit the frequencies of the internal modes.

Analytical forms are instead available for the intermolecular potential $V(^{m\mu}_{n\nu})$. These will be discussed in Chapter 3. It is, however, important to notice that, if the force constants are computed analytically [16,17], no particular care must be payed to the invariance properties since any intermolecular potential which depends on the intermolecular distances is automatically invariant under a translation or a rotation of the whole crystal [18]. If instead, the intermolecular force constants are used as parameters [19], the "self-term" formalism, previously discussed for the Cartesian force constants, and properly adapted [20] to translational and rotational coordinates, can be utilized. In this approach emphasis is placed on the invariance conditions and thus force constants involving two molecular coordinates on the same molecule are computed from terms relating different molecules. As a consequence of this procedure, the Hermiticity of the resulting potential energy matrix comes under discussion, as self-terms are symmetric only if the equilibrium conditions 1.6.25 are strictly obeyed. It has been shown [18] that, if the potential fulfills the equilibrium conditions, the self-term and the analytical approaches produce equivalent results. In this case the self-term formalism amounts to unnecessary complications, although it stresses the importance of equilibrium conditions. If these conditions are not obeyed, the self-term method produces a non-Hermitian force constant matrix. On the other hand the analytical procedure, which necessarily yields symmetric force constants, may lead in this case to an overall potential which is not invariant under rotation.

In Chapter 3 we shall discuss in detail the problem of constructing the force constants from specific models of intermolecular potentials. Assuming for the moment that the force constants are known, we can solve the dynamical problem in terms of molecular coordinates, using the kinetic and potential energy expressions 1.6.11 and 1.6.16. We notice that the kinetic energy matrix is diagonal and equal to the unit matrix for the translational and rotational coordinates (see Eq.1.6.10) but is non-diagonal for the block involving the internal coordinates. For this reason it is convenient to introduce molecular normal coordinates defined by the relations

$$S_{\ell}^{m\mu} = \sum_t L_t^{\ell}(m\mu) q_t^{m\mu} \qquad \text{for } t,\ell = 7,8,\dots,3N \qquad 1.6.26a$$
$$t_{\rho}^{m\mu} = q_t^{m\mu} \qquad \text{for } t,\rho = 1,2,3 \qquad 1.6.26b$$
$$r_{\rho}^{m\mu} = q_t^{m\mu} \qquad \text{for } t = 4,5,6 \;;\; \rho = 1,2,3 \qquad 1.6.26c$$

so that the kinetic and the potential energy of molecule $m\mu$ have both the complete diagonal form

$$T^{m\mu} = \frac{1}{2}\sum_t \dot{q}_t^{m\mu}\dot{q}_t^{m\mu} \qquad 1.6.27a$$

$$V(m\mu) = \frac{1}{2}\sum_t \omega_t^2 q_t^{m\mu} q_t^{m\mu} \qquad 1.6.27b$$

We noticc that in 1.6.27b the frequencies of the isolated molecule are equal to zero for t = 1,2,...,6.

By comparison of eqs.1.6.27 with 1.6.10 and 1.6.16 it is easily seen that the coefficients $L_t^{\ell}(m\mu)$ obey the relations

$$\sum_{\ell\ell'} L_t^{\ell}(m\mu) L_{t'}^{\ell'}(m\mu) (g_{\ell\ell'}^{m\mu})^{-1} = \delta_{tt'} \qquad 1.6.28a$$

$$\sum_{\ell\ell'} L_t^{\ell}(m\mu) L_{t'}^{\ell'}(m\mu) F_{\ell\ell'}\binom{m\mu}{m\mu} = \omega_t^2 \delta_{tt'} \qquad 1.6.28b$$

In terms of normal coordinates of the isolated molecules the total crystal potential 1.6.22 assumes then the form

$$V = \frac{1}{2} \sum_{mn} \sum_{\mu\nu} \sum_{tt'} \left[F_{tt'}\binom{m\mu}{n\nu} + \delta_{tt'} \delta_{mn} \delta_{\mu\nu} \omega_t^2 \right] q_t^{m\mu} q_{t'}^{n\nu} \qquad 1.6.29$$

where now

$$F_{tt'}\binom{m\mu}{n\nu} = \left[\partial^2 V\binom{m\mu}{n\nu} / \partial q_t^{m\mu} \partial q_{t'}^{n\nu} \right]_0 \qquad 1.6.30$$

We construct now, as done for the Cartesian coordinates in section 2, site symmetrized normal coordinates of the isolated molecules, through the transformation

$$q_t^{\mu}(\mathbf{k}) = L^{-\frac{1}{2}} \sum_m q_t^{m\mu} e^{-i\mathbf{k}\cdot\mathbf{R}_m} \qquad 1.6.31$$

In terms of the symmetrized coordinates 1.6.31 we obtain then the dynamical equations

$$\sum_{t'} \sum_{\nu} D_{tt'}\left(\begin{smallmatrix}\mu\\ \nu\end{smallmatrix}\middle|\mathbf{k}\right) e(t'\nu|p\mathbf{k}) = \omega_{p\mathbf{k}}^2 e(t\mu|p\mathbf{k}) \qquad 1.6.32$$

where

$$D_{tt'}\left(\begin{smallmatrix}\mu\\ \nu\end{smallmatrix}\middle|\mathbf{k}\right) = \sum_n \left[F_{tt'}\binom{o\mu}{n\nu} + \delta_{tt'} \delta_{on} \delta_{\mu\nu} \omega_t^2 \right] e^{i\mathbf{k}\cdot\mathbf{R}_n} \qquad 1.6.33$$

The normalized eigenvectors $e(t\mu|p\mathbf{k})$, that obviously obey relations of the type 1.2.17, define a transformation to crystal normal coordinates

$$q_t^{\mu}(\mathbf{k}) = \sum_p e(t\mu|p\mathbf{k})Q_{p\mathbf{k}} \qquad 1.6.34$$

with inverse transformation

$$Q_{p\mathbf{k}} = \sum_{t\mu} e^*(t\mu|p\mathbf{k})q_t^{\mu}(\mathbf{k}) \qquad 1.6.35$$

In terms of the localized coordinates $q_t^{m\mu}$,these expressions become

$$q_t^{m\mu} = L^{-\frac{1}{2}}\sum_{p\mathbf{k}} e(t\mu|p\mathbf{k})Q_{p\mathbf{k}}e^{i\mathbf{k}\cdot\mathbf{R}_m} \qquad 1.6.36$$

$$Q_{p\mathbf{k}} = L^{-\frac{1}{2}}\sum_{tm\mu} e^*(t\mu|p\mathbf{k})q_t^{m\mu}e^{-i\mathbf{k}\cdot\mathbf{R}_m} \qquad 1.6.37$$

In the rigid body approximation the formulation from 1.6.27 to 1.6.37 remains the same with the obvious limitation that the indices t and t' run only from 1 to 6 and the corresponding frequencies ω_t are equal to zero.

CHAPTER 2

SYMMETRY

2.1 SPACE GROUP SYMMETRY

Symmetry plays an important role in the description of molecular crystals since symmetry considerations permit the computation of bulk properties of the entire crystal from a knowledge of the behavior of a very small region. Group theoretical procedures can be used as is customary for isolated molecules, although there is a relevant difference between the collection of all symmetry elements which leave a given crystal invariant and those which leave a given molecule invariant. For isolated molecules, the set of all possible symmetry elements defines the molecular point group such that at least one point is left fixed in space. The corresponding coordinate transformation can be given in terms of orthogonal matrices representing rotations about some axis, reflections in some plane or combinations of the two. These transformations are usually referred to as (proper or improper) rotations. Rotational symmetry is also present in crystals which are, however, characterized by a structure made up of identical elementary cells, each containing a set of atoms which form the "basis" of the crystal structure and is repeated through space. Translations are thus symmetry operations for crystals. They transform every point into an equivalent one and the collection of all operations (translations, rotations and combinations of the two) which leave the crystal invariant forms a group, the "space group". Obviously a given crystal cannot be reproduced by any translation, i.e. the elements of the translational subgroup of the space group are only those associated with translational vectors which interconnect points of a three-dimensional lattice.

As it is known, there are only 14 possible lattice types, usually referred to as Bravais lattices grouped into 7 crystal systems, according to the point group symmetry which leaves the lattice invariant. A given lattice is uniquely identified once a set of non planar basic translations t_1, t_2, t_3 is defined such that any lattice translation can be written as

$$\mathbf{R}_n = n_1\mathbf{t}_1 + n_2\mathbf{t}_2 + n_3\mathbf{t}_3 \qquad 2.1.1$$

with n_1, n_2, n_3 = 0, ±1, ±2,... However, the same lattice can be generated by an infinite number of sets of three independent basis vectors. Among all possible unit cells, which can be defined in this way for a given lattice, one can select a symmetrical unit cell known to physicists as the "Wigner-Seitz primitive cell". The construction of such unit cell starts by joining a given lattice point to all nearby lattice points and by considering planes perpendicular to each line , from the given lattice point to neighboring points , midway between lattice points. At the given

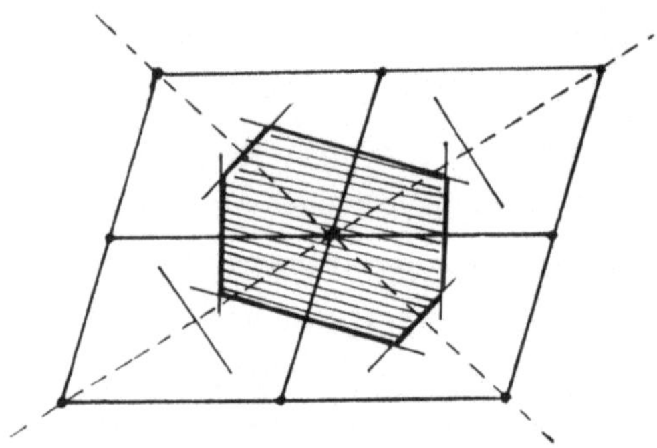

Fig. 2.1 Wigner-Seitz symmetrical primitive cell for a bidimensional lattice.

lattice point the smallest volume enclosed in this way is the Wigner-Seitz primitive cell. Boundary planes occur in pairs and the cell is invariant with respect to all rotations which are symmetry elements of the lattice. A simple example of such symmetrical cell is given in Fig. 2.1 for a planar lattice.

Pure lattice translations $\mathbf{R}_n$ as defined in 2.1.1 are a particular example of space group elements whose more general form is given by

$$\{\alpha|\mathbf{a}\} \qquad 2.1.2$$

using the customary Seitz [21] notation, where α represents the rotational part and $\mathbf{a}$ the translational part. Thus a lattice point $\mathbf{r}$ is transformed into $\mathbf{r}'$ in the active sense

$$\mathbf{r}' \equiv \{\alpha|\mathbf{a}\}\mathbf{r} = \alpha\mathbf{r} + \mathbf{a} \qquad 2.1.3$$

Hence if E represents the identity rotation, lattice translations are written as $\{E|\mathbf{R}_n\}$ while the identity, which transforms a vector $\mathbf{r}$ into itself without rotation and translation, is denoted by $\{E|\mathbf{0}\}$. Obviously transformations $\{\alpha|\mathbf{a}\}$ are not linear if the translational part is non--zero. By repeated application of 2.1.3 it is simple to derive a general multiplication rule for space group elements. Thus, by applying $\{\alpha|\mathbf{a}\}$ to a vector $\mathbf{r}'=\{\beta|\mathbf{b}\}\mathbf{r}$ we have

$$\mathbf{r}'' = \alpha(\beta\mathbf{r} + \mathbf{b}) + \mathbf{a} = \alpha\beta\mathbf{r} + \alpha\mathbf{b} + \mathbf{a} \qquad 2.1.4$$

Therefore, the formal definition of the product of two space group elements is given as

$$\{\alpha|\mathbf{a}\}\ \{\beta|\mathbf{b}\} = \{\alpha\beta|\alpha\mathbf{b} + \mathbf{a}\} \qquad 2.1.5$$

It follows that the inverse of an element $\{\alpha|\mathbf{a}\}$ is given by

$$\{\alpha|\mathbf{a}\}^{-1} = \{\alpha^{-1}|-\alpha^{-1}\mathbf{a}\} \qquad 2.1.6$$

and thus the transformations $\{\alpha|\mathbf{a}\}$ form a group in the usual sense. The pure translations form an invariant subgroup of the whole space group G : this subgroup will be called T. It is invariant since, for every $\{\alpha|\mathbf{a}\}$, if $\{E|\mathbf{R}_n\}$ is an element of T then

$$\{\alpha|\mathbf{a}\}^{-1}\{E|\mathbf{R}_n\}\{\alpha|\mathbf{a}\} = \{E|\alpha^{-1}\mathbf{R}_n\} \qquad 2.1.7$$

is again a member of the translational subgroup, as the lattice of a space group is unchanged by any element of the point group G_0 formed by all rotations α present in the space group G.

Since T is an invariant subgroup of the space group G, the latter can be decomposed into g cosets with respect to T as

$$G = T + \{\alpha_2|\mathbf{v}(\alpha_2)\}T + \{\alpha_3|\mathbf{v}(\alpha_3)\}T + \dots + \{\alpha_g|\mathbf{v}(\alpha_g)\}T \qquad 2.1.8$$

where elements of the same coset have the same rotational part α while elements of different cosets have different rotational parts. In this way we define the factor group G/T isomorphic with the point group G_0 and $\{\alpha_i|\mathbf{v}(\alpha_i)\}$, a representative of the ith coset, contains a translatio $\mathbf{v}(\alpha_i)$ of a minimum length correlated with α_i. Then, any element of the space group can be specified by

$$\{\alpha|\mathbf{a}\} = \{E|\mathbf{R}_n\}\{\alpha|\mathbf{v}(\alpha)\} = \{\alpha|\mathbf{v}(\alpha)+\mathbf{R}_n\} \qquad 2.1.9$$

where $\mathbf{v}(\alpha)$ is a fractional, non-primitive translation associated with α and depends on the choice of the origin through which the rotation α is defined. In fact, if $\mathbf{x}$ and $\mathbf{x}'$ are coordinates of vectors $\mathbf{r}$ and $\mathbf{r}'$ referred to a given origin of the coordinate system such that

$$\mathbf{x}' = \{\alpha|\mathbf{v}(\alpha)\}\mathbf{x} \qquad 2.1.10$$

and if the origin of the coordinate system is displaced by any vector $\mathbf{a}$ such that $\mathbf{y}$ and $\mathbf{y}'$ are the new coordinates

$$\mathbf{y} = \{E|\mathbf{a}\}\mathbf{x} \quad ; \quad \mathbf{y}' = \{E|\mathbf{a}\}\mathbf{x}' \qquad 2.1.11$$

then the transformation 2.1.10 in the new coordinate system is given by

$$\mathbf{y}' = \{\alpha'|\mathbf{v}(\alpha')\}\mathbf{y} \qquad 2.1.12$$

Introducing 2.1.11 in this equation and comparing with 2.1.10 we obtain the following similarity transformation

$$\{\alpha'|\mathbf{v}(\alpha')\} = \{E|\mathbf{a}\}\{\alpha|\mathbf{v}(\alpha)\}\{E|\mathbf{a}\}^{-1} \qquad 2.1.13$$

where $\{\alpha'|\mathbf{v}(\alpha')\}$, an operator in the 'new' space group referred to the new origin, differs from $\{\alpha|\mathbf{v}(\alpha)\}$ only because a change has occurred in the non-primitive translation

$$\mathbf{v}(\alpha') \equiv \mathbf{v}'(\alpha) = \mathbf{v}(\alpha) - \alpha\mathbf{a} + \mathbf{a} \qquad 2.1.14$$

For any space group, it is always possible to select a proper origin such that a specific coset representative is expressed by the simple form $\{\alpha|\mathbf{0}\}$. However, for most space groups, it is not possible to set $\mathbf{v}(\alpha) = \mathbf{0}$ for all $\alpha \in G_0$ for some origin of the coordinate system. There will be some $\mathbf{v}(\alpha) \neq 0$ which can be usually written as fractions of primitive translations and are thus referred to as fractional translations. This means that pure rotations $\{\alpha|\mathbf{0}\}$ do not all belong to the space group and fractional translations may occur in combination with rotations (screw axes) or with reflection planes (glide planes). Only for 73 space groups out of the total 230 can all fractional translations be taken as zero. These space groups are called "symmorphic" and the representatives $\{\alpha|\mathbf{0}\}$ themselves form a group isomorphic with the point group G_0. All other 157 space groups are non-symmorphic and the multiplication of two coset representatives

$$\{\alpha_i|\mathbf{v}(\alpha_i)\}\{\alpha_j|\mathbf{v}(\alpha_j) = \{\alpha_i\alpha_j|\alpha_i\mathbf{v}(\alpha_j)+\mathbf{v}(\alpha_i)\} \qquad 2.1.15$$

may produce an element which differs from the other coset representatives due to a different lattice translation. This implies that the relation

$$\{\alpha_i|\mathbf{v}(\alpha_i)\}\{\alpha_j|\mathbf{v}(\alpha_j)\} = \{\alpha_k|\mathbf{v}(\alpha_k)+\mathbf{R}_n\} \qquad 2.1.16$$

may hold where

$$\alpha_k = \alpha_i\alpha_j \quad \text{and } \mathbf{R}_n = \alpha_i\mathbf{v}(\alpha_j) + \mathbf{v}(\alpha_i) - \mathbf{v}(\alpha_k) \qquad 2.1.17$$

Hence, the product of two coset representatives is a coset representative modulo an element of the invariant subgroup of pure translations. These multiplication properties of the coset representatives of a space group are of fundamental importance for obtaining the irreducible representations of space groups on which all possible applications of symmetry to physical problems are based. For symmorphic space groups, the product of coset representatives is simplified since in it all possible $\mathbf{R}_n$ values (see 2.1.16 and 2.1.17) are zero. The irreducible representations are then simply related to those of ordinary point groups, which are well known and tabulated in many books (see for instance Ref.22). This is generally not the case with non-symmorphic space groups as we will see in the next Sections.

Before entering into a discussion on the structure of space group representations, we will consider the transformation properties of functions, rather than coordinates, as introduced by Wigner [23]. If a point $\mathbf{r}$ is rotated into $\mathbf{r}'$ by a space group operation $\{\alpha|\mathbf{a}\}$ as given by 2.1.3, we obtain in terms of the components in a given fixed orthonormal basis

$$x'_s = \sum_t \mathbf{D}(\alpha)_{st}\, x_t + a_s \qquad 2.1.18$$

where x_t (t=1,2,3) are components of $\mathbf{r}$ and x'_s corresponding components of the rotated vector $\mathbf{r}'$; $\mathbf{D}$ is an orthogonal matrix associated with the rotation and a_s are components of the translation $\mathbf{a}$. Thus, a given function of the coordinates

$$f(x_1,\ x_2,\ x_3) \qquad 2.1.19$$

becomes, after transformation due to $\{\alpha|\mathbf{a}\}$

$$f[\sum_s \mathbf{D}(\alpha)_{s1}(x'_s - a_s), \sum_s \mathbf{D}(\alpha)_{s2}(x'_s - a_s),\ \sum_s \mathbf{D}(\alpha)_{s3}(x'_s - a_s)] \qquad 2.1.20$$

Comparing 2.1.20 with 2.1.19 we may conclude that the function f in the coordinates x was changed into a new function F in the new coordinates x' by the application of an operator $\hat{O}(\{\alpha|\mathbf{a}\})$ associated with $\{\alpha|\mathbf{a}\}$ whic acts on functions rather than on coordinates, i.e.

$$f(\ \dots\ x\ \dots\ )\ \xrightarrow{\hat{O}(\{\alpha|\mathbf{a}\})}\ F(\ \dots\ x'\ \dots\ ) \qquad 2.1.21$$

Introducing instead of F a new symbol $\hat{O}(\{\alpha|\mathbf{a}\})f$ to recall explicitly the dependence of the new function on the operator $\hat{O}(\{\alpha|\mathbf{a}\})$, an obvious requirement is that the new function in a point identified by x_1', x_2', x_3' has the same numerical value as the old function in the corresponding non-rotated point x_1, x_2, x_3, that is

$$\hat{O}(\{\alpha|\mathbf{a}\})\ f(\mathbf{r}') \quad = \quad f(\mathbf{r}) \qquad 2.1.22$$

or, when using 2.1.3 to rewrite r on the right-hand side and putting $\mathbf{r}' \rightarrow \mathbf{r}$

$$\hat{O}(\{\alpha|\mathbf{a}\})\ f(\mathbf{r}) \quad = \quad f(\{\alpha|\mathbf{a}\}^{-1}\mathbf{r}) \qquad 2.1.23$$

Operators associated to space group elements which leave the crystal invariant commute with the potential and kinetic energy operators and are thus called symmetry operators of the Hamiltonian. Using the definition 2.1.23, it is possible to establish an isomorphism between space group elements and symmetry operators and to define a symmetry group of the Hamiltonian as the group of its symmetry operators. If E_j is an eigenvalue of H with eigenfunction ψ

$$H\ \psi \quad = \quad E_j\ \psi \qquad 2.1.24$$

and if $\hat{O}(\{\alpha|\mathbf{a}\})$ is a symmetry operator of H, then

$$H\ \hat{O}(\{\alpha|\mathbf{a}\})\psi \quad = \quad E_j\ \hat{O}(\{\alpha|\mathbf{a}\})\psi \qquad 2.1.25$$

since $\hat{O}(\{\alpha|\mathbf{a}\})$ commutes with H and, of course, with the eigenvalue E_j. It follows that any function $\hat{O}(\{\alpha|\mathbf{a}\})\psi$, obtained from an eigenfunction ψ by a symmetry operator, will also be eigenfunction having the same eigenvalue as the original one. Among the set of functions $\hat{O}(\{\alpha|\mathbf{a}\})\psi$, one for each element of the symmetry group of the Hamiltonian, there are l_j linearly independent functions

$$\psi_1^j\ \ \psi_2^j,\ \ldots\ ,\ \psi_{l_j}^j \qquad 2.1.26$$

which form a basis for the jth irreducible representation of the group, Γ^j. Explicitly

$$\hat{O}(\{\alpha|\mathbf{a}\})\psi_\mu^j = \sum_\nu \Gamma^j(\{\alpha|\mathbf{a}\})_{\nu\mu}\psi_\nu^j \qquad 2.1.27$$

Thus E_j is an l_j-fold degenerate eigenvalue belonging to the jth irreducible representation of the symmetry group of the Hamiltonian.

2.2 IRREDUCIBLE REPRESENTATIONS OF THE TRANSLATIONAL GROUP

As briefly described in the preceding Section, the irreducible representations of space groups play a fundamental role in lattice dynamic as they allow a classification of states of the system which is of great importance for the determination of the probability of transition from one state to the other. Before discussing the general method for obtaining the irreducible representations of the space groups, it will be shown how they may be classified using the property that the translational group T is an invariant subgroup of any space group. Therefore, the discussion will be initiated by considering a pure tridimensional translational group which is also the simplest of all space groups. In Section 2.3 more complicated space groups will be examined, namely those for which the order g is greater than unity in the expansion 2.1.8.

The translational group T is an infinite group but, as anticipated in Section 1.2, it is made finite by the use of periodic boundary conditions such that

$$\{E|\mathbf{t}_i\}^{L_i} = \{E|\mathbf{0}\} \qquad (i=1,2,3) \qquad 2.2.1$$

where L_i is a large number. This implies that the infinite crystal is divided into identical parallelepipeds with sides $L_i\mathbf{t}_i$ so that the physical properties in one point of a given region are identical to those of the corresponding points in every other region. In this way, the translational group T becomes of finite order $L = L_1L_2L_3$, it is an Abelian group and may be regarded as the direct product of three cyclic translational subgroups in one dimension with

$$\{E|\mathbf{R}_n\} = \{E|n_1\mathbf{t}_1\}\{E|n_2\mathbf{t}_2\}\{E|n_3\mathbf{t}_3\} \qquad 2.2.2$$

The irreducible representations of T are then given by forming the outer Kronecker product of the one dimensional groups and the problem reduces to find the irreducible representations of the translational group

$$a_0 = \{E|0\}, \quad a_1 = \{E|\mathbf{t}_i\}, \ldots, \; a_n = \{E|n\mathbf{t}_i\}, \; \ldots, \; a_l = \{E|L_i\mathbf{t}_i\} \qquad 2.2.3$$

This is a cyclic group for which, because of 2.2.1, it must be

$$(a_1)^{L_i} = a_l = a_0 \qquad 2.2.4$$

and L_i irreducible representations of dimensionality 1 are derived. If r is a scalar which represents a_1 for a given irreducible representation of the one dimensional group of translations such that

$$\Gamma(a_0) = 1 \quad \text{and} \; \Gamma(a_1) = r \qquad 2.2.5$$

then

$$\Gamma(a_n) = \Gamma(a_1)^n = r^n \qquad 2.2.6$$

Because of the cyclic conditions 2.2.4 it follows that $r^{L_i} = 1$. Hence r assumes L_i possible values which are the L_i roots of unity given by

$$r = e^{2\pi i \nu_i / L_i} \qquad (\nu_i = 1, 2, \ldots, L_i) \qquad 2.2.7$$

This means that $\{E|\mathbf{t}_i\}$is represented by L_i possible numbers, one for each irreducible representation of the group, and in general we have

$$\Gamma_{\nu_i}(\{E|n_i\mathbf{t}_i\}) = e^{2\pi i n_i \nu_i / L_i} \qquad 2.2.8$$

for the scalar which represents the element $\{E|n_i\mathbf{t}_i\}$of the group of the translations along $\mathbf{t}_i$, in the ν_i-th irreducible representation. As pointed out in Ref. 24 a negative sign in the exponential 2.2.8 could be used in order to be fully consistent with the definition of active Seitz operators given in 2.1.3. A positive sign gives, of course, precisely the same results in all applications and is adopted here in agreement with the choice established in most textbooks on lattice dynamics. It is possible now to obtain the irreducible representations of the group T by forming the Kronecker product

$$\boldsymbol{\Gamma}_{\nu_1\nu_2\nu_3}(\{E|\mathbf{R}_n\}) = \boldsymbol{\Gamma}_{\nu_1}(\{E|n_1\mathbf{t}_1\})\,\boldsymbol{\Gamma}_{\nu_2}(\{E|n_2\mathbf{t}_2\})\,\boldsymbol{\Gamma}_{\nu_3}(\{E|n_3\mathbf{t}_3\}) =$$

$$= e^{2\pi i(n_1\nu_1/L_1 + n_2\nu_2/L_2 + n_3\nu_3/L_3)} \qquad 2.2.9$$

The scalar in the square bracket of 2.2.9 may be considered as being formed by the scalar product of a vector $\mathbf{R}_n$ times a vector $\mathbf{k}$

$$\mathbf{k} = \frac{\nu_1}{L_1}\mathbf{b}_1 + \frac{\nu_2}{L_2}\mathbf{b}_2 + \frac{\nu_3}{L_3}\mathbf{b}_3 \qquad 2.2.10$$

Here $\mathbf{b}_1$, b_2, $\mathbf{b}_3$ are basic translations of the reciprocal lattice so that

$$\mathbf{b}_i \cdot \mathbf{t}_j = 2\pi\,\delta_{ij} \qquad 2.2.11a$$

They are then given in terms of basis vectors in direct space by

$$\mathbf{b}_i = \frac{2\pi\mathbf{t}_j\times\mathbf{t}_k}{\mathbf{t}_i\cdot(\mathbf{t}_j\times\mathbf{t}_k)} \qquad (i=1,2,3) \qquad 2.2.11b$$

with i, j, k arranged in cyclic order. In conclusion a vector $\mathbf{k}$ identifies an irreducible representation of the translational space group as

$$\boldsymbol{\Gamma}_{\mathbf{k}}(\{E|\mathbf{R}_n\}) = e^{i\mathbf{k}\cdot\mathbf{R}_n} \qquad 2.2.12$$

where $\mathbf{k}$ assumes fractional components in terms of basis vectors of the reciprocal lattice. This lattice is left invariant by certain rotational operations as it is for the direct lattice, that is, the direct and re-

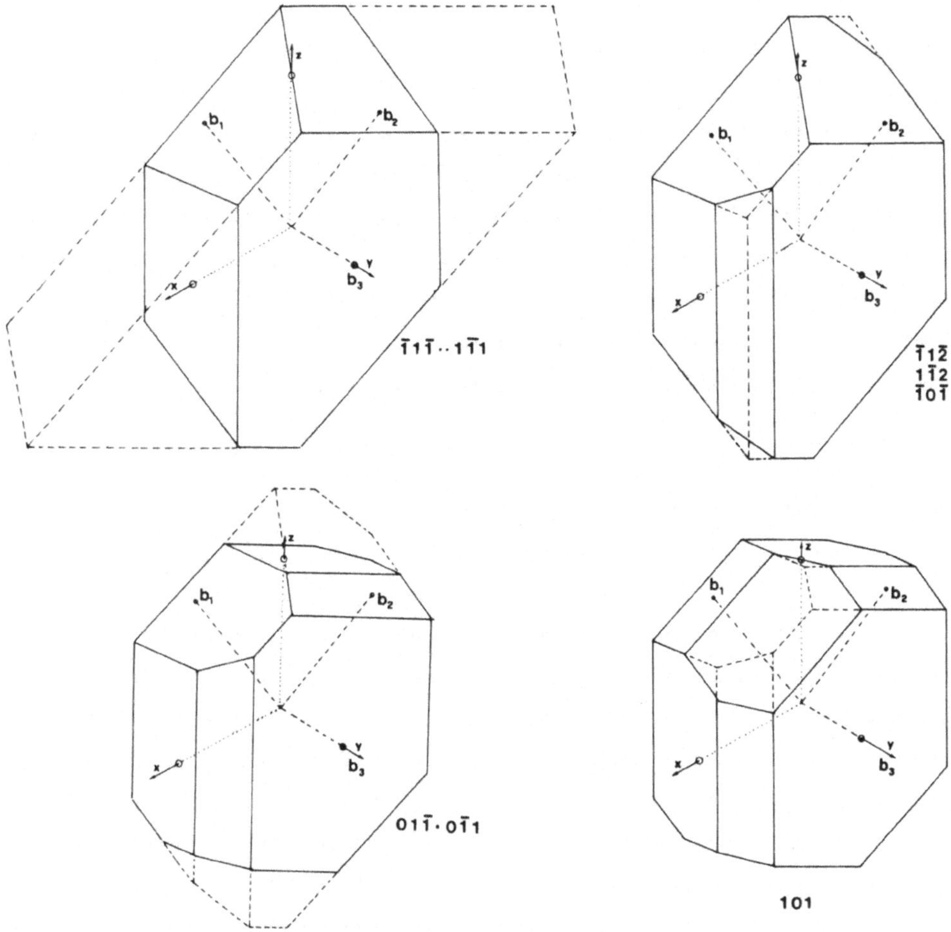

Fig. 2.2 Brillouin zone for face centered (B) monoclinic lattice. Basis vectors in direct and reciprocal space are

$\mathbf{t}_1 = (a/2)\mathbf{i} + (c/2)\mathbf{k}$ $\qquad \mathbf{b}_1 = (2\pi/a)\mathbf{i} - (2\pi b/ad)\mathbf{j} + (2\pi/c)\mathbf{k}$

$\mathbf{t}_2 = -(a/2)\mathbf{i} + (c/2)\mathbf{k}$ $\qquad \mathbf{b}_2 = -(2\pi/a)\mathbf{i} + (2\pi b/ad)\mathbf{j} + (2\pi/c)\mathbf{k}$

$\mathbf{t}_3 = b\mathbf{i} + d\mathbf{j}$ $\qquad \mathbf{b}_3 = (2\pi/d)\mathbf{j}$

with $a>d\gg b>c$. Lattice points, to which boundary planes are referred, are indicated. Reciprocal lattice vectors from the origin to points ($\frac{1}{2}00$), ($0\frac{1}{2}0$) and ($00\frac{1}{2}$) are given.

ciprocal lattices have the same holohedry and belong to the same lattice system although the reciprocal lattice may belong to a Bravais class which is different from that of the direct lattice. For instance, it may be verified that the lattice corresponding to the face centered cubic lattice is the body centered cubic lattice in reciprocal space.

As far as the irreducible representations of T are concerned, two vectors **k** and **k'** = **k** + $\mathbf{K}_m$ are considered to be equivalent, that is $\mathbf{k}\equiv\mathbf{k}'$, if K_m is any lattice translation in reciprocal space, analogous to R_n in direct space. In fact **k** and **k'** give rise to the same exponential in 2.2.12 , thus they label the same irreducible representation of T and this is decribed by

$$\mathbf{k} \equiv \mathbf{k} + \mathbf{K}_m \quad \text{with } \mathbf{K}_m = m_1\mathbf{b}_1 + m_2\mathbf{b}_2 + m_3\mathbf{b}_3 \qquad 2.2.13$$
$$(m_1, m_2, m_3 = 0,\ 1,\ 2, ..\ )$$

It is customary to define **k** within a symmetrical unit cell of the reciprocal lattice, called the "first Brillouin zone" or simply "Brillouin zone", corresponding to the Wigner-Seitz cell in the direct lattice. This definition is made as the symmetry properties of the **k** vectors are of relevant importance in the discussion of the irreducible representations of the full space group. Then, it is convenient to locate **k** vectors in the Brillouin zone which shows the full symmetry of the reciprocal lattice. The construction of the first Brillouin zone is carried out as for the Wigner-Seitz cell and the process, in which the final shape is obtained through successive analysis of boundary planes for different lattice points, is shown in Fig. 2.2 for the Brillouin zone of the base centered monoclinic lattice.

Having defined the irreducible representations of the translational group we will now consider eigenfunctions, of the Hamiltonian operator which is invariant under lattice translations, for a system corresponding to electrons moving in a periodic potential field. As eigenfunctions belong to irreducible representations of T, they are labelled by a wave vector **k** and written as $\psi_{\mathbf{k}}(r)$. According to the definition 2.1.23, if an eigenfunction is subjected to a translational operator of the group of the Hamiltonian we have

$$\hat{O}(\{E|\mathbf{R}_n\})\,\psi_{\mathbf{k}}(\mathbf{r}) = \psi_{\mathbf{k}}(\mathbf{r} - \mathbf{R}_n) \qquad 2.2.14$$

On the other hand since $\psi_{\mathbf{k}}(\mathbf{r})$ belongs to the **k**-th irreducible representation of T it must be according to 2.1.27

$$\hat{O}(\{E|\mathbf{R}_n\})\,\psi_{\mathbf{k}}(\mathbf{r}) = e^{i\mathbf{k}\cdot\mathbf{R}_n}\,\psi_{\mathbf{k}}(\mathbf{r}) \qquad 2.2.15$$

If this relation is compared with 2.2.14 we have

$$\psi_{\mathbf{k}}(\mathbf{r} - \mathbf{R}_n) = e^{i\mathbf{k}\cdot\mathbf{R}_n}\,\psi_{\mathbf{k}}(\mathbf{r}) \qquad 2.2.16$$

or, after multiplication of both sides by $e^{i\mathbf{k}\cdot(\mathbf{r} - \mathbf{R}_n)}$,

$$e^{i\mathbf{k}\cdot(\mathbf{r} - \mathbf{R}_n)}\,\psi_{\mathbf{k}}(\mathbf{r} - \mathbf{R}_n) = e^{i\mathbf{k}\cdot\mathbf{r}}\,\psi_{\mathbf{k}}(\mathbf{r}) \qquad 2.2.17$$

The last equality holds for any value of $\mathbf{R}_n$ and means that the combination $e^{i\mathbf{k}\cdot\mathbf{r}}\psi_{\mathbf{k}}(\mathbf{r})$ is unchanged by the translational operator, exibits the periodicity of the lattice and will be referred to as $u_{\mathbf{k}}(\mathbf{r})$. It follows that eigenfunctions of the Hamiltonian, which includes a periodic potential, have the general form stated by the Block theorem

$$\psi_{\mathbf{k}}(\mathbf{r}) = e^{-i\mathbf{k}\cdot\mathbf{r}}\,u_{\mathbf{k}}(\mathbf{r}) \qquad 2.2.18$$

and can be used as a basis for a generally reducible representation of the whole space group. Eigenfunctions for each value of **k**

$$\psi_{\mathbf{k}_1}(\mathbf{r}),\ \psi_{\mathbf{k}_2}(\mathbf{r}),\ \ldots,\ \psi_{\mathbf{k}_n}(\mathbf{r}),\ \ldots \qquad 2.2.19$$

are uncoupled and form a basis for a completely reduced representation if the space group does not contain any rotational symmetry. Using the periodicity of $u_{\mathbf{k}}(\mathbf{r})$, it can be proved that each eigenfunction belongs indeed to an irreducible representation of the translational group since

$$\hat{O}(\{E|\mathbf{R}_n\})\psi_{\mathbf{k}_i}(\mathbf{r}) = e^{-i\mathbf{k}_i\cdot(\mathbf{r} - \mathbf{R}_n)}\,u_{\mathbf{k}_i}(\mathbf{r} - \mathbf{R}_n) =$$

$$= e^{-i\mathbf{k}_i\cdot(\mathbf{r} - \mathbf{R}_n)} u_{\mathbf{k}_i}(\mathbf{r}) = e^{i\mathbf{k}_i\cdot\mathbf{R}_n}\psi_{\mathbf{k}_i}(\mathbf{r}) \qquad 2.2.20$$

If $\{\alpha|\mathbf{a}\}$ is a symmetry element of the space group and if $\psi_{\mathbf{k}}(\mathbf{r})$ belongs to the k-th irreducible representation of T, it can be readily shown that the Block function $\hat{O}(\{\alpha|\mathbf{a}\})\psi_{\mathbf{k}}(\mathbf{r})$ belongs to the $\alpha\mathbf{k}$ irreducible representation of the same group. Using a procedure similar to that leading to 2.2.19 one has

$$\hat{O}(\{E|\mathbf{R}_n\})[\hat{O}(\{\alpha|\mathbf{a}\})\psi_{\mathbf{k}}(\mathbf{r})] = \hat{O}(\{\alpha|\mathbf{a}\})\psi_{\mathbf{k}}(\mathbf{r} - \mathbf{R}_n) =$$

$$= \psi_{\mathbf{k}}(\alpha^{-1}\mathbf{r} - \alpha^{-1}\mathbf{R}_n - \alpha^{-1}\mathbf{a}) =$$

$$= e^{-i\mathbf{k}\cdot(\alpha^{-1}\mathbf{r} - \alpha^{-1}\mathbf{R}_n - \alpha^{-1}\mathbf{a})} u_{\mathbf{k}}(\alpha^{-1}\mathbf{r} - \alpha^{-1}\mathbf{R}_n - \alpha^{-1}\mathbf{a}) =$$

$$= e^{i\mathbf{k}\cdot\alpha^{-1}\mathbf{R}_n} e^{-i\mathbf{k}\cdot(\alpha^{-1}\mathbf{r} - \alpha^{-1}\mathbf{a})} u_{\mathbf{k}}(\alpha^{-1}\mathbf{r} - \alpha^{-1}\mathbf{a}) =$$

$$= e^{i\alpha\mathbf{k}\cdot\mathbf{R}_n}[\hat{O}(\{\alpha|\mathbf{a}\})\psi_{\mathbf{k}}(\mathbf{r})] \qquad 2.2.21$$

Here $\alpha\mathbf{k}$ is the vector obtained from k through a rotation α, which is an element of the point group G_0 associated with the space group G. If $\alpha\mathbf{k}$ coincides with k or it differs by a translation $\mathbf{K}_m$ of the reciprocal lat tice, then α is an element of a point group, that is denoted by $G_0(\mathbf{k})$ and is a subgroup of G_0, which leaves that particular $\mathbf{k}$ vector unchanged The order q of $G_0(\mathbf{k})$ is a divisor of g, the order of the point group G_0, such that

$$g/q = \text{integer} = d \qquad 2.2.22$$

Those q coset representatives of G in the decomposition 2.1.8, whose rotational parts leave a given wave vector $\mathbf{k}$ invariant, define the space group of k, G(k), which is very important in physical applications and determines the irreducible representations of the whole space group G. On the other hand starting with a given vector $\mathbf{k}_1$, say, the set of all distinct vectors $\mathbf{k}_2$, $\mathbf{k}_3$, ... , $\mathbf{k}_d$, which are obtained from $\mathbf{k}_1$ through application of all elements of G_0, is called the "star" of **k**. The star of **k** has d "points" and d is determined by 2.2.22 from the orders of the point groups G_0 and $G_0(\mathbf{k})$.

Before concluding this Section the irreducible representations of T will be used to prove the fundamental results

$$\sum_n e^{i\mathbf{k}\cdot\mathbf{R}_n} = \begin{matrix} 0 \text{ if } \mathbf{k} \neq \mathbf{K}_m \\ L \text{ if } \mathbf{k} = \mathbf{K}_m \end{matrix} \qquad 2.2.23a$$

$$\sum_{\mathbf{k}} e^{i\mathbf{k}\cdot\mathbf{R}_n} = \begin{matrix} 0 \text{ if } \mathbf{R}_n \neq \mathbf{0} \\ L \text{ if } \mathbf{R}_n = \mathbf{0} \end{matrix} \qquad 2.2.23b$$

where $\mathbf{R}_n$ and $\mathbf{K}_m$ are lattice translations in direct and reciprocal space, respectively, and the sum is extended over all L possible vectors. These results follow from a direct application of the great orthogonality theorem due to Wigner [23] for finite groups. This theorem states that

$$\sum_a \Gamma^i(a)^*_{\mu\nu}\Gamma^j(a)_{\mu'\nu'} = h/l_i\ \delta_{ij}\delta_{\mu\mu'}\delta_{\nu\nu'} \qquad 2.2.24$$

where the sum is over the h elements a of the group, and l_i is the dimensionality of the irreducible representation whose matrices are $\Gamma^i(a)$. Using 2.2.24 for the irreducible representations of T given by Eq.2.2.12 and summing over all L elements of the group, we obtain for two different representations labelled by **k** and k'

$$\sum_n e^{i(\mathbf{k} - \mathbf{k}')\cdot\mathbf{R}_n} = L\ \delta_{\mathbf{k}\mathbf{k}'} \qquad 2.2.25$$

from which 2.2.23a results when $\mathbf{k}' \equiv 0$. On the other hand since the irreducible representations of T are monodimensional and each element forms a class by itself, the columns of the irreducible representation table must, as a consequence of 2.2.24, also be orthogonal each other, i.e.

$$\sum_{\mathbf{k}} e^{i\mathbf{k}\cdot(\mathbf{R}_n - \mathbf{R}_{n'})} = L\ \delta_{nn'} \qquad 2.2.26$$

From 2.2.26 expression 2.2.23b is obtained.

2.3 IRREDUCIBLE REPRESENTATIONS OF THE SPACE GROUPS

In this paragraph the analysis previously initiated will be extened to space groups G which contain rotational symmetry. Since for a given wave vector $\mathbf{k}_1$ we need to distinguish between elements that leave it invariant and elements which transform k_1 into a different point of the star, let the elements of the point group $G_0(\mathbf{k}_1)$ be specified as

$$\alpha_1^1, \alpha_2^1, \ldots, \alpha_i^1, \ldots, \alpha_q^1 \qquad \text{such that } \alpha_i^1\mathbf{k}_1 \equiv \mathbf{k}_1 \qquad 2.3.1$$

This list of elements depends, of course, on the particular selection of the vector $\mathbf{k}_1$ from the star of $\mathbf{k}$. However, the symmetry properties of the group of $\mathbf{k}$ and its relationship with the space group G can be discussed without the need to specify a particular vector of a given star. In fact if by

$$\alpha_1^1, \alpha_1^2, \ldots, \alpha_1^i, \ldots, \alpha_1^d \qquad \text{such that } \alpha_1^i\mathbf{k}_1 = \mathbf{k}_i \qquad 2.3.2$$

we define elements of G_0 which send $\mathbf{k}_1$ into a different vector of the star, then $\alpha_1^i\ \alpha_j^1\ (\alpha_1^i)^{-1}$ is an element which leaves $\mathbf{k}_i$ invariant since

$$\alpha_1^i\ \alpha_j^1\ (\alpha_1^i)^{-1}\ \mathbf{k}_i = \alpha_1^i\ \alpha_j^1\ \mathbf{k}_1 = \alpha_1^i\ \mathbf{k}_1 = \mathbf{k}_i \qquad 2.3.3$$

Therefore $G_0(k_1)$ and $G_0(k_i)$ are isomorphic point groups, their elements being related by a similarity transformation. The same multiplication table applies and it is immaterial which vector of the star we start wit in discussing the irreducible representations of G.

The space group G is decomposed into left cosets of its subgroup $G(\mathbf{k}_1)$ as

$$G = \{\alpha_1^1|\mathbf{v}(\alpha_1^1)\}G(k_1) + \{\alpha_1^2|\mathbf{v}(\alpha_1^2)\}G(\mathbf{k}_1) + \ldots + \{\alpha_1^d|\mathbf{v}(\alpha_1^d)\}G(\mathbf{k}_1) \qquad 2.3.4$$

where the coset representatives will be assumed to be chosen once for all.For instance $\{\alpha_1^1|\mathbf{v}(\alpha_1^1)\}$ will always be the identity and by $\{\alpha_1^i|\mathbf{v}(\alpha_1^i)$

we mean a particular representative of the ith coset and not any element of the coset. Let then assume that a given irreducible representation $\Gamma_{\mathbf{k}_1}$ of the group $G(\mathbf{k}_1)$ is known, of dimensionality p, spanned by basis functions

$$\Psi^1_{\mathbf{k}_1} = e^{-i\mathbf{k}_1\cdot\mathbf{r}} u^1_{\mathbf{k}_1}(\mathbf{r})$$

$$\Psi^2_{\mathbf{k}_1} = e^{-i\mathbf{k}_1\cdot\mathbf{r}} u^2_{\mathbf{k}_1}(\mathbf{r}) \qquad 2.3.5$$

....

$$\Psi^p_{\mathbf{k}_1} = e^{-i\mathbf{k}_1\cdot\mathbf{r}} u^p_{\mathbf{k}_1}(\mathbf{r})$$

which are all classified according to the $\mathbf{k}_1$-th irreducible representation of T whose elements are obviously contained in $G(\mathbf{k}_1)$. In $\Gamma_{\mathbf{k}_1}$, all translations $\{E|\mathbf{R}_n\}$ are represented, according to 2.2.12, by $e^{i\mathbf{k}_1\cdot\mathbf{R}_n}\,\mathbb{E}$, where $\mathbb{E}$ is the identity matrix of dimensionality p. Thus each element of $G(\mathbf{k}_1)$ of the kind $\{\alpha^1_j|\mathbf{v}(\alpha^1_j) + \mathbf{R}_n\}$ is represented by

$$\Gamma_{\mathbf{k}_1}(\{\alpha^1_j|\mathbf{v}(\alpha^1_j) + \mathbf{R}_n\}) = e^{i\mathbf{k}_1\cdot\mathbf{R}_n}\,\Gamma_{\mathbf{k}_1}(\{\alpha^1_j|\mathbf{v}(\alpha^1_j)\}) \qquad 2.3.6$$

where the matrices on the right-hand side are assumed to be known and each coset representative in $G(\mathbf{k}_1)$ acts on functions 2.3.5 to give

$$\hat{O}(\{\alpha^1_j|\mathbf{v}(\alpha^1_j)\})\,\Psi^\mu_{\mathbf{k}_1} = \sum_\nu \Psi^\nu_{\mathbf{k}_1}\Gamma_{\mathbf{k}_1}(\{\alpha^1_j|\mathbf{v}(\alpha^1_j)\})_{\nu\mu} \qquad 2.3.7$$

The p functions 2.3.5 and the following functions for j=2 to d

$$\Psi^1_{\mathbf{k}_j} = \hat{O}(\{\alpha^j_1|\mathbf{v}(\alpha^j_1)\})\Psi^1_{\mathbf{k}_1}$$

$$\Psi^2_{\mathbf{k}_j} = \hat{O}(\{\alpha^j_1|\mathbf{v}(\alpha^j_1)\})\Psi^2_{\mathbf{k}_1} \qquad 2.3.8$$

....

$$\Psi^p_{\mathbf{k}_j} = \hat{O}(\{\alpha^j_1|\mathbf{v}(\alpha^j_1)\})\Psi^p_{\mathbf{k}_1}$$

will be taken as a basis of a d×p dimensional irreducible representation of the whole space group G whose structure is obtained simply by examining the effect of a general space group element $\{\alpha|a\}$ on a basis function $\Psi_{\mathbf{k}_r}$. If the rotational part α is such that it moves a vector $\mathbf{k}_r$ into a vector k_s of the star

$$\alpha \mathbf{k}_r = \mathbf{k}_s \tag{2.3.9}$$

then two elements α_1^s and α_1^r can be defined so that

$$(\alpha_1^s)^{-1} \alpha \, \alpha_1^r = \alpha_f^1 = \text{element of } G_0(\mathbf{k}_1) \tag{2.3.10}$$

In terms of the corresponding coset representatives we have

$$\{\alpha_1^s|\mathbf{v}(\alpha_1^s)\}^{-1} \{\alpha|\mathbf{a}\} \{\alpha_1^r|\mathbf{v}(\alpha_1^r)\} = \{E|\mathbf{R}_n\} \{\alpha_f^1|\mathbf{v}(\alpha_f^1)\} \tag{2.3.11a}$$

where

$$\mathbf{R}_n = (\alpha_1^s)^{-1} [\alpha \mathbf{v}(\alpha_1^r) + \mathbf{a} - \mathbf{v}(\alpha_1^s)] - \mathbf{v}(\alpha_f^1) \tag{2.3.11b}$$

Therefore, the matrices of the representation based on functions 2.3.5 and 2.3.8 are defined by

$$\begin{aligned}
\hat{O}(\{\alpha|\mathbf{a}\})\Psi^{\mu}_{\mathbf{k}_r} &= \hat{O}(\{\alpha_1^s|\mathbf{v}(\alpha_1^s)\})\ \hat{O}(\{\alpha_f^1|\mathbf{v}(\alpha_f^1) + \mathbf{R}_n\})\ \hat{O}(\{\alpha_1^r|\mathbf{v}(\alpha_1^r)\}^{-1})\Psi^{\mu}_{\mathbf{k}_r} = \\
&= \hat{O}(\{\alpha_1^s|\mathbf{v}(\alpha_1^s)\})\ \hat{O}(\{\alpha_f^1|\mathbf{v}(\alpha_f^1) + \mathbf{R}_n\})\Psi^{\mu}_{\mathbf{k}_1} = \\
&= \hat{O}(\{\alpha_1^s|\mathbf{v}(\alpha_1^s)\})\ e^{i\mathbf{k}_1\cdot\mathbf{R}_n}\sum_{\nu}\Psi^{\nu}_{\mathbf{k}_1}\Gamma_{\mathbf{k}_1}(\{\alpha_f^1|\mathbf{v}(\alpha_f^1)\})_{\nu\mu} = \\
&= e^{i\mathbf{k}_1\cdot\mathbf{R}_n}\sum_{\nu}\Psi^{\nu}_{\mathbf{k}_s}\Gamma_{\mathbf{k}_1}(\{\alpha_f^1|\mathbf{v}(\alpha_f^1)\})_{\nu\mu}
\end{aligned} \tag{2.3.12}$$

The matrix of the representation 2.3.12 has the dimensionality d×p and consists of p×p blocks with only one non-zero block in each row or column. Thus the irreducible representations of the whole space group G are determined according to 2.3.12 if those of the groups of **k** are known.

It is interesting to apply this procedure when $\{\alpha|\mathbf{a}\} \equiv \{E|\mathbf{R}_m\}$. In this case using the prescription 2.3.11 one obtains

$$\{E|\mathbf{R}_n\}\ \{\alpha_f^1|v(\alpha_f^1)\} = \{E|\ (\alpha_1^r)^{-1}\ \mathbf{R}_m\} \qquad 2.3.13$$

and from Eq. 2.3.12

$$\hat{O}(\{E|\mathbf{R}_m\})\Psi^{\mu}_{\mathbf{k}_r} = \Psi^{\mu}_{\mathbf{k}_r}\ e^{ik_r\cdot R_m} \qquad 2.3.14$$

which confirms that $\Psi^{\mu}_{\mathbf{k}_r}$ belongs to the $\mathbf{k}_r$-th irreducible representation of the subgroup of the pure translations in agreement with the definition 2.3.8 and the transformation properties 2.2.21.

The representation 2.3.12 is irreducible since it was assumed that the set of functions 2.3.5 form a basis for an irreducible representation of G(**k**). That is, the p functions 2.3.5 cannot split into subsets which do not mix under the space group of **k**. On the other hand, the star of **k** cannot be split and thus the set of all the functions 2.3.8 cannot be divided into non mixing subspaces, as we need the whole star of **k** for a representation of G. The irreducibility of the representation 2.3.12 could also be demonstrated using standard criteria of group theory and, accordingly, it is possible to show that all irreducible representations of G are obtained from those of the groups G(**k**). For a rigorous mathematical treatment of the subject, the reader is referred to several excellent books on the subject[25-27]. We only want to show that all symmetry properties and applications of space group symmetry are determined by G(**k**), which is still a space group whose representations are not yet known but has particular properties which greatly simplify the derivation of its irreducible representations.

A discussion of methods for obtaining the irreducible representations of the groups of **k** starts by noting that, as any other space group, G(**k**) can be factorized into left cosets with respect to the translational group as

$$G(\mathbf{k}) = \{\alpha_1|\mathbf{v}(\alpha_1)\}T + \{\alpha_2|\mathbf{v}(\alpha_2)\}T + \dots + \{\alpha_q|\mathbf{v}(\alpha_q)\}T \qquad 2.3.15$$

Here a particular vector, **k**, of the star is selected and the coset representatives,i.e. their fractional translations, are fixed once for all The rotations α define a point group $G_0(k)$ and are such that $\alpha_i\mathbf{k}=\mathbf{k}+\mathbf{K}_m$. If two coset representatives are multiplied,a third coset representativ may not be obtained as coset representatives do not in general form a group. As explicitly pointed out in 2.1.16 and 2.1.17, if Γ_k^p is the pth irreducible representation of G(k) then it must be

$$\Gamma_{\mathbf{k}}^p(\{\alpha_i|\mathbf{v}(\alpha_i)\})\ \Gamma_{\mathbf{k}}^p(\{\alpha_j|\mathbf{v}(\alpha_j)\}) = e^{i\mathbf{k}\cdot[\alpha_i\mathbf{v}(\alpha_j)+\mathbf{v}(\alpha_i)-\mathbf{v}(\alpha_k)]}\Gamma_{\mathbf{k}}^p(\{\alpha_k|\mathbf{v}(\alpha_k)\}) \tag{2.3.16}$$

using 2.3.6 and considering that $\alpha_i\alpha_j = \alpha_k$. Because of the exponential function which appears on the right-hand side of 2.3.16, the matrices $\Gamma_{\mathbf{k}}^p$ do not form a group in the usual sense. If new matrices $\mathbf{D}_{\mathbf{k}}^p$ are defined as

$$\Gamma_{\mathbf{k}}^p(\{\alpha_i|\mathbf{a}_i\}) = e^{i\mathbf{k}\cdot\mathbf{a}_i}\ \mathbf{D}_{\mathbf{k}}^p(\{\alpha_i|\mathbf{a}_i\}) \tag{2.3.17}$$

Eq. 2.3.16 can be rewritten as

$$\mathbf{D}_{\mathbf{k}}^p(\{\alpha_i|\mathbf{v}(\alpha_i)\})\ \mathbf{D}_{\mathbf{k}}^p(\{\alpha_j|\mathbf{v}(\alpha_j)\}) = e^{i\mathbf{k}\cdot[\alpha_i\mathbf{v}(\alpha_j)-\mathbf{v}(\alpha_j)]}\ \mathbf{D}_{\mathbf{k}}^p(\{\alpha_k|\mathbf{v}(\alpha_k)\}) \tag{2.3.18}$$

The exponent in this equation can be rewritten, using the property that a scalar product does not change if both vectors are rotated by α_i^{-1}, as $e^{-i\mathbf{k}\cdot\mathbf{v}(\alpha_j)}e^{i\alpha^{-1}\mathbf{k}\cdot\mathbf{v}(\alpha_j)}$. On the other hand α_i^{-1} is also a member of $G_0(k)$ and then $\alpha_i^{-1}\mathbf{k} = \mathbf{k} + \mathbf{K}_m$. Thus we obtain from Eq. 2.3.18

$$\mathbf{D}_{\mathbf{k}}^p(\{\alpha_i|\mathbf{v}(\alpha_i)\})\ \mathbf{D}_{\mathbf{k}}^p(\{\alpha_j|\mathbf{v}(\alpha_j)\}) = e^{i\mathbf{K}_m\cdot\mathbf{v}(\alpha_j)}\ \mathbf{D}_{\mathbf{k}}^p(\{\alpha_k|\mathbf{v}(\alpha_k)\}) \tag{2.3.19}$$

According to the definition 2.3.17 the matrices $\mathbf{D}_{\mathbf{k}}^p$ are the same for all members of a given coset in the expansion 2.3.15, i.e.

$$\mathbf{D}_{\mathbf{k}}^p(\{\alpha_i|\mathbf{v}(\alpha_i)\}) = \mathbf{D}_{\mathbf{k}}^p(\{\alpha_i|\mathbf{v}(\alpha_i) + \mathbf{R}_n\}) \tag{2.3.20}$$

This means that they define a matrix-valued function of the factor grou G(k)/T or, since the factor group is isomorphic with the point group of **k**, of the appropriate crystallographic point group. In terms of matrice

$\mathbf{D}^p_{\mathbf{k}}$ pure translational elements thus become unimportant for discussing the symmetry properties of G(**k**) : rather one has to discuss properties of point groups for which

$$\mathbf{D}^p_{\mathbf{k}}(\alpha_i)\,\mathbf{D}(\alpha_j) = e^{i\mathbf{K}_m\cdot\mathbf{v}(\alpha_j)}\,\mathbf{D}^p_{\mathbf{k}}(\alpha_k) \qquad 2.3.21$$

where $\alpha_i\alpha_j = \alpha_k$ and $\alpha_i^{-1}\,\mathbf{k} = \mathbf{k} + K_m$. It is evident that if $\mathbf{v}(\alpha_j) = 0$ for all elements of G(**k**), that is, if the group of **k** is a symmorphic space group, Eq. 2.3.21 reduces to the usual matrix multiplication rule and the matrices $\mathbf{D}^p_{\mathbf{k}}$ produce a representation of the point group G_0(k). The irreducible representations of all crystallographic point groups are well known and then the irreducible representations of symmorphic G(k)'s are given ,according to 2.3.17, by matrices

$$\mathbf{\Gamma}^p_{\mathbf{k}}(\{\alpha_i|\mathbf{a}_i\}) = e^{i\mathbf{k}\cdot\mathbf{a}_i}\,\mathbf{\Gamma}^p(\alpha_i) \qquad 2.3.22$$

where Γ^p is an irreducible representation of the appropriate point group. The same is true when k = **0** or when it is a point in the interior of the Brillouin zone, regardless of whether the space group is symmorphic or not. If indeed **k** is inside the Brillouin zone, the vector $\alpha_i^{-1}\,\mathbf{k}$ is also inside the zone and by definition of G_0(**k**) must be equivalent to **k**. Hence all reciprocal lattice translations K_m which appear in 2.3.21 are necessarily zero and the irreducible representations of G(**k**) are again given by 2.3.22. Only for non symmorphic space groups, when **k** is on the surface of the Brillouin zone and coincides with a point of symmetry or lies on a line or plane of symmetry of the surface, may difficulties arise in finding out the irreducible representations of G(**k**). In these cases it is of practical advantage to introduce the concept of "ray" or "projective" representation of a point group defined as a matrix function of the group such that

$$\mathbf{D}(\alpha_i)\,\mathbf{D}(\alpha_j) = \omega_{ij}\,\mathbf{D}(\alpha_i\alpha_j) \qquad 2.3.23$$

which is a generalization of Eq. 2.3.21. α_i and α_j are elements of the group, $\mathbf{D}$ are matrices associated with them and ω_{ij} is a complex number which, for finite groups, can be chosen such that $|\omega_{ij}| = 1$. The numbers ω_{ij} form a factor system and only when all $\omega_{ij}=1$ the ray representation coincides with a regular vector representation. Obviously Eq. 2.3.21 defines a ray representation with a factor system $e^{iK_m \cdot \mathbf{v}(\alpha_j)}$ which is a function of elements α_i and α_j. If all ray representations of all crystallographic point groups are known, it is very simple to find all irre ducible representations of G(**k**) and G. The irreducible ray representations can be obtained since the reduction of ray representations is not more complicated that that for ordinary representations. Indeed, the irreducible ray representations of point groups are known and tabulated (see Ref.28 and 29) and it is not necessary to give further details her However, an important feature of ray representations should be pointed out. Unlike ordinary representations of point groups, among corresponding ray representations there may not exist any one-dimensional irreducible representation. This means that a given space group G(**k**) may have no monodimensional representation, although obviously the space group G always contains monodimensional irreducible representations among which the totally symmetric one. Indeed, the irreducible representations of G are induced by those of G(**k**) for all possible values of **k** in the Brillouin zone, and thus also for **k** = 0 for which a monodimensional, identity representation always occurs.

A complete description of the space group irreducible representations can be found in the book of Kovalev [30] in which the ray representation method is used. Complete tables for single and double valued representations of the groups of **k** were published by Bradley and Cracknel [27] and by Miller and Love [31] while only the character tables are given in the book of Zak et al. [32]. It is thus generally true that now the irreducible representations of any space group can be found with little algebraic effort from several published tables. However, for numerical computations it is much more convenient to have a simple method, to be applied as a routine in a computer program, which gives the matrices for all irreducible representations for any space group and for any val ue of **k**. This is accomplished in Ref. 33, in connection with group theo retical analysis of crystal vibrations, and in Refs. 34 and 35 for the irreducible representations themselves starting from one up to three generators of the space group.

In what follows a sample of computer generated irreducible representations of space groups is given, for a space group T_h^6 (Pa3) belonging to the cubic system, primitive lattice (P), for several values of the **k** vector. They correspond to symmetry points, shown in Fig. 2.3, on the surface of the Brillouin zone. In terms of basic translations in

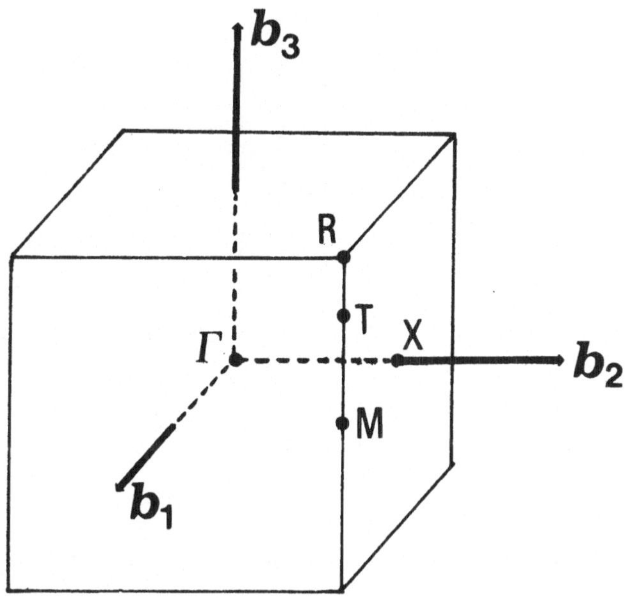

Fig. 2.3 Brillouin zone for a simple cubic lattice. Points M,T,X,R on the surface of the zone are shown. Γ is a point at the origin of the zone.

the reciprocal space the selected **k** vectors are : (0.5,0.5,0.) point M; (0.5,0.5,0.1) point T on line MR; (0.,0.5,0.) point X; (0.5,0.5,0.5) point R. The irreducible representations of the corresponding groups of **k** are given in Table 2.1. In this table space group elements, i.e. coset representatives of the space group, are also given. C_n (C2, C3, etc.) represents an anticlockwise rotation of 360°/n, S_n (S4,S6) the corresponding roto-reflection , MP are mirror planes and i (INV) is the in-

Table 2.1 Computer generated irreducible representations for a cubic space group, T_h^6 (Pa3), simple cubic system, primitive lattice. Selected **k** vectors correspond to points M, T, X, R of Fig. 2.3. See text for an explanation of the symbols used.

```
              CUBIC SYSTEM
          SIMPLE LATTICE (P)

 TRANSLATIONAL VECTORS (CARTESIAN REFERENCE FRAME) FOR
DIRECT LATTICE                                   RECIPROCAL LATTICE

T1 = ( A  , 0  , 0  )                            B1 = ( 1/A , 0   , 0   )
T2 = ( 0  , A  , 0  )                            B2 = ( 0   , 1/A , 0   )
T3 = ( 0  , 0  , A  )                            B3 = ( 0   , 0   , 1/A )
```

SIMPLE CUBIC LATTICE, SPACE GROUP TH(6) OR PA3, N.205

SYMBOL DEFINITION FOR ROTATIONAL ELEMENTS (CARTESIAN COORDINATES ARE USED TROUGHOUT)!
(FRACTIONAL TRANSLATIONS RELATIVE TO UNIT VECTORS OF DIRECT LATTICE)

EL.NO.	SYMBOL	EXT. SYMBOL	TRANSF.PROP.			SPACE GROUP COSET REPRESENTATIVE
N. 1	E	E	X	Y	Z	(E |0 ,0 ,0)
N. 2	C2X	C2(X)	X	-Y	-Z	(C2X |1/2,1/2,0)
N. 3	C2Y	C2(Y)	-X	Y	-Z	(C2Y |0 ,1/2,1/2)
N. 4	C2Z	C2(Z)	-X	-Y	Z	(C2Z |1/2,0 ,1/2)
N. 5	C3J	C3(X,Y,Z)	Z	X	Y	(C3J |0 ,0 ,0)
N. 6	C3K	C3(-X,-Y,Z)	-Z	X	-Y	(C3K |0 ,1/2,1/2)
N. 7	C3L	C3(-X,Y,-Z)	Z	-X	-Y	(C3L |1/2,1/2,0)
N. 8	C3W	C3(X,-Y,-Z)	-Z	-X	Y	(C3W |1/2,0 ,1/2)
N. 9	-C3J	-C3(X,Y,Z)	Y	Z	X	(-C3J|0 ,0 ,0)
N.10	-C3K	-C3(-X,-Y,Z)	Y	-Z	-X	(-C3K|1/2,1/2,0)
N.11	-C3L	-C3(-X,Y,-Z)	-Y	-Z	X	(-C3L|1/2,0 ,1/2)
N.12	-C3W	-C3(X,-Y,-Z)	-Y	Z	-X	(-C3W|0 ,1/2,1/2)
N.25	INV	I	-X	-Y	-Z	(INV |0 ,0 ,0)
N.26	MPX	MP(X)	-X	Y	Z	(MPX |1/2,1/2,0)
N.27	MPY	MP(Y)	X	-Y	Z	(MPY |0 ,1/2,1/2)
N.28	MPZ	MP(Z)	X	Y	-Z	(MPZ |1/2,0 ,1/2)
N.29	-S6J	-S6(X,Y,Z)	-Z	-X	-Y	(-S6J|0 ,0 ,0)
N.30	-S6K	-S6(-X,-Y,Z)	Z	-X	Y	(-S6K|0 ,1/2,1/2)
N.31	-S6L	-S6(-X,Y,-Z)	-Z	X	Y	(-S6L|1/2,1/2,0)
N.32	-S6W	-S6(X,-Y,-Z)	Z	X	-Y	(-S6W|1/2,0 ,1/2)
N.33	S6J	S6(X,Y,Z)	-Y	-Z	-X	(S6J |0 ,0 ,0)
N.34	S6K	S6(-X,-Y,Z)	-Y	Z	X	(S6K |1/2,1/2,0)
N.35	S6L	S6(-X,Y,-Z)	Y	Z	-X	(S6L |1/2,0 ,1/2)
N.36	S6W	S6(X,-Y,-Z)	Y	-Z	X	(S6W |0 ,1/2,1/2)

Table 2.1 (continued)

K = (.50000, .50000, .00000)
POINT GROUP OF K = D2H
IRREDUCIBLE REPRESENTATIONS OF G(K)

		E	C2X	C2Y	C2Z	INV	MPX	MPY	MPZ
N. 1	(1,1)	1	-I	-I	-1	0	0	0	0
	(1,2)	0	0	0	0	1	-I	-I	-1
	(2,1)	0	0	0	0	1	-I	I	1
	(2,2)	1	-I	I	1	0	0	0	0
N. 2	(1,1)	1	I	-I	1	0	0	0	0
	(1,2)	0	0	0	0	1	I	-I	1
	(2,1)	0	0	0	0	1	I	I	-1
	(2,2)	1	I	I	-1	0	0	0	0

CHARACTER TABLE FOR G(K)

	E	C2X	C2Y	C2Z	INV	MPX	MPY	MPZ
N. 1	2	-2I	0	0	0	0	0	0
N. 2	2	2I	0	0	0	0	0	0

TEST FOR EXTRA DEGENERACY DUE TO TIME INVERSION:

EXTRA DEGENERACY (STICKING OF DIFFERENT REPS) FOR REPRESENTATIONS : 1 2
COUPLING OF REPS IS I = 1 AND 2 .

K = (.50000, .50000, .10000)
POINT GROUP OF K = C2V
IRREDUCIBLE REPRESENTATIONS OF G(K)

		E	C2Z	MPX	MPY
N. 1	(1,1)	1	E	0	0
	(1,2)	0	0	-I	IE
	(2,1)	0	0	-I	-IE
	(2,2)	1	-E	0	0

CHARACTER TABLE FOR G(K)

	E	C2Z	MPX	MPY
N. 1	2	0	0	0

LIST OF SYMBOLS USED

E = (.951057, -.309017)

TEST FOR EXTRA DEGENERACY DUE TO TIME INVERSION:

EXTRA DEGENERACY (SELF STICKING) FOR REPRESENTATIONS: 1

K = (.00000, .50000, .00000)
POINT GROUP OF K = D2H
IRREDUCIBLE REPRESENTATIONS OF G(K)

		E	C2X	C2Y	C2Z	INV	MPX	MPY	MPZ
N. 1	(1,1)	1	I	0	0	0	0	-1	-1
	(1,2)	0	0	-I	-I	I	I	0	0
	(2,1)	0	0	-I	I	-I	I	0	0
	(2,2)	1	-I	0	0	0	0	1	-1
N. 2	(1,1)	1	1	0	0	0	0	1	1
	(1,2)	0	0	-I	-I	-I	-I	0	0
	(2,1)	0	0	-I	I	I	-I	0	0
	(2,2)	1	-1	0	0	0	0	-1	1

CHARACTER TABLE FOR G(K)

	E	C2X	C2Y	C2Z	INV	MPX	MPY	MPZ
N. 1	2	0	0	0	0	0	0	-2
N. 2	2	0	0	0	0	0	0	2

TEST FOR EXTRA DEGENERACY DUE TO TIME INVERSION:
NO EXTRA DEGENERACY

Table 2.1 (continued)

K = (.50000, .50000, .50000)
POINT GROUP OF K = TH
IRREDUCIBLE REPRESENTATIONS OF G(K)

		E	C2X	C2Y	C2Z	C3J	C3K	C3L	C3W	-C3J	-C3K	-C3L	-C3W	INV	MPX	MPY	MPZ	-S6J	-S6K	-S6L	-S6W	S6J	S6K	S6L	S6W
N. 1	(1,1)	1	0	I	0	E	IE	E	IE	E*	IE*	-E*	IE*	1	0	I	0	E	IE	E	IE	E*	IE*	-E*	IE*
	(1,2)	0	-1	0	-I	IE	-E	-IE	E	-E*	IE*	-E*	-IE*	0	-1	0	-I	IE	-E	-IE	E	-E*	IE*	-E*	-IE*
	(2,1)	0	1	0	-I	-E	IE	E	-IE	-IE*	E*	-IE*	-E*	0	1	0	-I	-E	IE	E	-IE	-IE*	E*	-IE*	-E*
	(2,2)	1	0	-I	0	IE	E	IE	E	-IE*	-E*	IE*	-E*	1	0	-I	0	IE	E	IE	E	-IE*	-E*	IE*	-E*
N. 2	(1,1)	1	0	-I	0	E*	-IE*	E*	-IE*	E	-IE	-E	-IE	1	0	-I	0	E*	-IE*	E*	-IE*	E	-IE	-E	-IE
	(1,2)	0	-1	0	I	-IE*	-E*	IE*	E*	-E	-IE	-E	IE	0	-1	0	I	-IE*	-E*	IE*	E*	-E	-IE	-E	IE
	(2,1)	0	1	0	I	-E*	-IE*	E*	IE*	IE	E	IE	-E	0	1	0	I	-E*	-IE*	E*	IE*	IE	E	IE	-E
	(2,2)	1	0	I	0	-IE*	E*	-IE*	E*	IE	-E	-IE	-E	1	0	I	0	-IE*	E*	-IE*	E*	IE	-E	-IE	-E
N. 3	(1,1)	1	0	I	0	F	F*	F	F*	F*	-F	-F*	-F	1	0	I	0	F	F*	F	F*	F*	-F	-F*	-F
	(1,2)	0	-1	0	-I	F*	-F	-F*	F	-F*	-F	-F*	F	0	-1	0	-I	F*	-F	-F*	F	-F*	-F	-F*	F
	(2,1)	0	1	0	-I	-F	F*	F	-F*	F	F*	F	-F*	0	1	0	-I	-F	F*	F	-F*	F	F*	F	-F*
	(2,2)	1	0	-I	0	F*	F	F*	F	F	-F*	-F	-F*	1	0	-I	0	F*	F	F*	F	F	-F*	-F	-F*
N. 4	(1,1)	1	0	I	0	E	IE	E	IE	E*	IE*	-E*	IE*	-1	0	-I	0	-E	-IE	-E	-IE	-E*	-IE*	E*	-IE*
	(1,2)	0	-1	0	-I	IE	-E	-IE	E	-E*	IE*	-E*	-IE*	0	1	0	I	-IE	E	IE	-E	E*	-IE*	E*	IE*
	(2,1)	0	1	0	-I	-E	IE	E	-IE	-IE*	E*	-IE*	-E*	0	-1	0	I	E	-IE	-E	IE	IE*	-E*	IE*	E*
	(2,2)	1	0	-I	0	IE	E	IE	E	-IE*	-E*	IE*	-E*	-1	0	I	0	-IE	-E	-IE	-E	IE*	E*	-IE*	E*
N. 5	(1,1)	1	0	-I	0	E*	-IE*	E*	-IE*	E	-IE	-E	-IE	-1	0	I	0	-E*	IE*	-E*	IE*	-E	IE	E	IE
	(1,2)	0	-1	0	I	-IE*	-E*	IE*	E*	-E	-IE	-E	IE	0	1	0	-I	IE*	E*	-IE*	-E*	E	IE	E	-IE
	(2,1)	0	1	0	I	-E*	-IE*	E*	IE*	IE	E	IE	-E	0	-1	0	-I	E*	IE*	-E*	-IE*	-IE	-E	-IE	E
	(2,2)	1	0	I	0	-IE*	E*	-IE*	E*	IE	-E	-IE	-E	-1	0	-I	0	IE*	-E*	IE*	-E*	-IE	E	IE	E
N. 6	(1,1)	1	0	I	0	F	F*	F	F*	F*	-F	-F*	-F	-1	0	-I	0	-F	-F*	-F	-F*	-F*	F	F*	F
	(1,2)	0	-1	0	-I	F*	-F	-F*	F	-F*	-F	-F*	F	0	1	0	I	-F*	F	F*	-F	F*	F	F*	-F
	(2,1)	0	1	0	-I	-F	F*	F	-F*	F	F*	F	-F*	0	-1	0	I	F	-F*	-F	F*	-F	-F*	-F	F*
	(2,2)	1	0	-I	0	F*	F	F*	F	F	-F*	-F	-F*	-1	0	I	0	-F*	-F	-F*	-F	-F	F*	F	F*

CHARACTER TABLE FOR G(K)

	E	C2X	C2Y	C2Z	C3J	C3K	C3L	C3W	-C3J	-C3K	-C3L	-C3W	INV	MPX	MPY	MPZ	-S6J	-S6K	-S6L	-S6W	S6J	S6K	S6L	S6W
N. 1	2	0	0	0	G	G	G	G	G*	-G*	-G*	-G*	2	0	0	0	G	G	G	G	G*	-G*	-G*	-G*
N. 2	2	0	0	0	G*	G*	G*	G*	G	-G	-G	-G	2	0	0	0	G*	3*	G*	G*	G	-G	-3	-G
N. 3	2	0	0	0	-1	-1	-1	-1	-1	1	1	1	2	0	0	0	-1	-1	-1	-1	-1	1	1	1
N. 4	2	0	0	0	G	G	G	G	G*	-G*	-G*	-G*	-2	0	0	0	-G	-G	-G	-G	-G*	G*	G*	G*
N. 5	2	0	0	0	G*	G*	G*	G*	G	-G	-G	-G	-2	0	0	0	-G*	-3*	-G*	-G*	-G	G	G	G
N. 6	2	0	0	0	-1	-1	-1	-1	-1	1	1	1	-2	0	0	0	1	1	1	1	1	-1	-1	-1

LIST OF SYMBOLS USED

E = (.683013, .183013)   F = (-.500000, .500000)   G = (.500000, .866026)

TEST FOR EXTRA DEGENERACY DUE TO TIME INVERSION:

EXTRA DEGENERACY (SELF STICKING) FOR REPRESENTATIONS: 3 6

EXTRA DEGENERACY (STICKING OF DIFFERENT REPS) FOR REPRESENTATIONS : 1 2 4 5
COUPLING OF REPS IS : - 1 AND 2 - 4 AND 5 -

version element. Direction of the rotation axes are given together with coordinates of point P' into which point P(x,y,z) is rotated by each symmetry element of G_0. Irreducible representations are printed in symbolic form : I stands for the imaginary unit $\sqrt{-1}$, the other symbols E, F, G being explicitly defined in Table 2.1 as complex numbers, with E*, F*, G*, corresponding to complex conjugate quantities. Row and column indices (i,j) of elements for each matrix of each representation are given. The corresponding character tables are also printed out, together with the Schönflies symbol for the appropriate point group $G_0(\mathbf{k})$ and the results of a test for extra degeneracy due to time inversion (see next paragraph). We note that, for all the four **k** vectors selected, no monodimensional representation exists and in one case, point T, all eigenvalues of the dynamical matrix are expected to be fourfold degenerate due to the fact that they belong to a two-dimensional representation which 'sticks with itself' because of time inversion. We note that fourfold degeneracy cannot occur with crystallographic point groups. This,however, is not the maximum degeneracy which can be found for space groups. For instance,several space groups of k of the cubic system show sixfold degenerate irreducible representations, where the degeneracy is now due to spatial symmetry as opposed to time inversion symmetry.

Finally an example of the application of the procedure leading to 2.3.12 is given in Table 2.2 in which a 6×6 irreducible representation of the full space group T_h^6 (Pa3) is shown, induced by an irreducible representation of dimensionality 2 of the group of **k** = X. This representation, as any other for the full space group, is of little practical value but it may be useful to elucidate the relation with the representations of G(k). Although wave functions within a crystal are classified according to the irreducible representations of the full space group G, these are determined by the irreducible representations of the groups of k and then labelling of eigenfunctions follows accordingly. The eigenvalues of a crystal Hamiltonian are regarded as being distributed among wave vectors **k** in the Brillouin zone and labels of the irreducible representations of G(**k**)'s are used to identify phonon dispersion curves. Projection techniques are correspondingly employed to provide symmetry-adapted bases, with projection operators given by

Table 2.2 Irreducible representation of the space group T_h^6 (Pa3) induced by an irreducible representation (N.1 of Table 2.1) of the space group of $\mathbf{k}_1=(0\frac{1}{2}0)$, with $\mathbf{k}_2=(\frac{1}{2}00)$ and $k_3=(00\frac{1}{2})$. Block structure is emphasized by dots replacing 2×2 blocks of zeros.

{E |000}

1	0	•	•	•	•
0	1	•	•	•	•
•	•	1	0	•	•
•	•	0	1	•	•
•	•	•	•	1	0
•	•	•	•	0	1

{C2X |½½0}

1	0	•	•	•	•
0	-1	•	•	•	•
•	•	0	-i	•	•
•	•	-i	0	•	•
•	•	•	•	0	-i
•	•	•	•	i	0

{C2Y |0½½}

0	-i	•	•	•	•
-i	0	•	•	•	•
•	•	0	-i	•	•
•	•	i	0	•	•
•	•	•	•	1	0
•	•	•	•	0	-1

{C2Z |½0½}

0	-i	•	•	•	•
i	0	•	•	•	•
•	•	1	0	•	•
•	•	0	-1	•	•
•	•	•	•	0	-i
•	•	•	•	-i	0

{C3J |000}

•	•	1	0	•	•
•	•	0	1	•	•
•	•	•	•	1	0
•	•	•	•	0	1
1	0	•	•	•	•
0	1	•	•	•	•

{C3K |0½½}

•	•	0	-i	•	•
•	•	-i	0	•	•
•	•	•	•	0	-i
•	•	•	•	i	0
1	0	•	•	•	•
0	-1	•	•	•	•

{C3L |½½0}

•	•	1	0	•	•
•	•	0	-1	•	•
•	•	•	•	0	-i
•	•	•	•	-i	0
0	-i	•	•	•	•
i	0	•	•	•	•

{C3W |½0½}

•	•	0	-i	•	•
•	•	i	0	•	•
•	•	•	•	1	0
•	•	•	•	0	-1
0	-i	•	•	•	•
-i	0	•	•	•	•

{-C3J|000}

•	•	•	•	1	0
•	•	•	•	0	1
1	0	•	•	•	•
0	1	•	•	•	•
•	•	1	0	•	•
•	•	0	1	•	•

{-C3K|½½0}

•	•	•	•	1	0
•	•	•	•	0	-1
0	-i	•	•	•	•
-i	0	•	•	•	•
•	•	0	-i	•	•
•	•	i	0	•	•

{-C3L|½0½}

•	•	•	•	0	-i
•	•	•	•	i	0
1	0	•	•	•	•
0	-1	•	•	•	•
•	•	0	-i	•	•
•	•	-i	0	•	•

{-C3W|0½½}

•	•	•	•	0	-i
•	•	•	•	-i	0
0	-i	•	•	•	•
i	0	•	•	•	•
•	•	1	0	•	•
•	•	0	-1	•	•

{INV |000}

0	i	•	•	•	•
-i	0	•	•	•	•
•	•	0	i	•	•
•	•	-i	0	•	•
•	•	•	•	0	i
•	•	•	•	-i	0

{MPX |½½0}

0	i	•	•	•	•
i	0	•	•	•	•
•	•	-1	0	•	•
•	•	0	1	•	•
•	•	•	•	-1	0
•	•	•	•	0	-1

{MPY |0½½}

-1	0	•	•	•	•
0	1	•	•	•	•
•	•	-1	0	•	•
•	•	0	-1	•	•
•	•	•	•	0	i
•	•	•	•	i	0

{MPZ |½0½}

-1	0	•	•	•	•
0	-1	•	•	•	•
•	•	0	i	•	•
•	•	i	0	•	•
•	•	•	•	-1	0
•	•	•	•	0	1

{-S6J|000}

•	•	0	i	•	•
•	•	-i	0	•	•
•	•	•	•	0	i
•	•	•	•	-i	0
0	i	•	•	•	•
-i	0	•	•	•	•

{-S6K|0½½}

•	•	-1	0	•	•
•	•	0	1	•	•
•	•	•	•	-1	0
•	•	•	•	0	-1
0	i	•	•	•	•
i	0	•	•	•	•

{-S6L|½½0}

•	•	0	i	•	•
•	•	i	0	•	•
•	•	•	•	-1	0
•	•	•	•	0	1
-1	0	•	•	•	•
0	-1	•	•	•	•

{-S6W|½0½}

•	•	-1	0	•	•
•	•	0	-1	•	•
•	•	•	•	0	i
•	•	•	•	i	0
-1	0	•	•	•	•
0	1	•	•	•	•

{S6J |000}

•	•	•	•	0	i
•	•	•	•	-i	0
0	i	•	•	•	•
-i	0	•	•	•	•
•	•	0	i	•	•
•	•	-i	0	•	•

{S6K |½½0}

•	•	•	•	0	i
•	•	•	•	i	0
-1	0	•	•	•	•
0	1	•	•	•	•
•	•	-1	0	•	•
•	•	0	-1	•	•

{S6L |½0½}

•	•	•	•	-1	0
•	•	•	•	0	-1
0	i	•	•	•	•
i	0	•	•	•	•
•	•	-1	0	•	•
•	•	0	1	•	•

{S6W |0½½}

•	•	•	•	-1	0
•	•	•	•	0	1
-1	0	•	•	•	•
0	-1	•	•	•	•
•	•	0	i	•	•
•	•	i	0	•	•

$$P^{\mathbf{k}p}_{\mu\nu} = \frac{l_p}{q} \sum_{\{\beta|\mathbf{b}\}} \Gamma^p_{\mathbf{k}}(\{\beta|\mathbf{b}\})^*_{\mu\nu} \; \hat{O}(\{\beta|\mathbf{b}\}) \qquad 2.3.24$$

with the sum extended to all coset representatives in G(**k**). $\Gamma^p_{\mathbf{k}}$ is an irreducible representation of dimensionality l_p. Eq. 2.3.24 projects out the part of any function belonging to the νth row of the **k**p irreducible representation. A simple application of this technique for translational subgroups immediately gives translationally symmetrized Cartesian coordinates anticipated in 1.2.20.

2.4 TIME REVERSAL

Up to now only spatial symmetry was considered under which the Hamiltonian is invariant but, as first pointed out by Wigner, there is an additional symmetry connected to the transformation $t \to -t$ involving the time. It implies reversing of all velocities and it is usually referred to as "time reversal" or "time inversion". When the time proceeds in the reverse direction classically the system runs back but the forces do not change since they involve second derivatives with respect to time. In quantum mechanics, if H is a real operator, that is if the spin is neglected, the time inversion operator simply changes ψ into ψ^*. In fact if we consider the time dependent Schrödinger equation

$$H\psi = i\hbar \frac{\partial \psi}{\partial t} \qquad 2.4.1$$

taking the complex conjugate and using the fact that H is real we have

$$H\psi^* = i\hbar \frac{\partial \psi^*}{\partial(-t)} \qquad 2.4.2$$

Thus ψ^* will evolve in the positive direction of t exactly as ψ would have evolved in the -t direction, with the density probability $|(\psi,\psi)|$ unaffected by this operator. Consequently $\psi(\mathbf{r},t)$ and $\psi^*(\mathbf{r},t)$ obey the same equation and for a stationary state

$$H\psi = E\psi \quad \text{and} \quad HK\psi = EK\psi \qquad 2.4.3$$

Therefore ψ and $\psi^*=K\psi$ are degenerate eigenfunctions with the same eigenvalue E. In 2.4.3 K is the operator of complex conjugation coincident with the operator of time reversal, for a scalar Hamiltonian in which spin is ignored, and applies to the study of symmetry of excitations in a crystal for a given wave vector **k**. It acts on a dynamical matrix of the kind 1.2.14 by changing **k** into -**k**. Thus if $\{\alpha|\mathbf{a}\}$ is an element of the space group which changes **k** into -k such that $\alpha\mathbf{k}\equiv-\mathbf{k}$, the symmetry of the dynamical matrix for a given vector **k** is given by the group

$$F(\mathbf{k}) = G(\mathbf{k}) + K\{\alpha|\mathbf{a}\}G(\mathbf{k}) \qquad 2.4.4$$

As usual, here $G(\mathbf{k})$ is the group of **k** while $\{\alpha|\mathbf{a}\}$ is an element of G such that $\alpha\mathbf{k}$ is in the star of **k**. It is important to point out that K is not a linear operator as, by definition, if c is a constant $Kc\psi = c^*K\psi$ while for linear opearators $cK\psi$ would have been produced. An operator with such property is said to be "antilinear". If it also leaves the transition probability between any two states ψ and φ invariant, i.e. if $|(\psi,\varphi)| = |(K\psi,K\varphi)|$, then it is called"antiunitary". The group $F(\mathbf{k})$ with the above properties is called"antiunitary group", with half its element unitary. If u is one of the unitary elements of $G(\mathbf{k})$ and a one of the antiunitary elements and if $\psi_1, \psi_2, \ldots, \psi_d$ form a basis for an irreducible representation of $G(\mathbf{k})$, then we have by definition

$$u\psi_\mu = \sum_\nu \psi_\nu \Gamma(u)_{\nu\mu} \qquad 2.4.5$$

On the other hand, let define d functions of the kind

$$\Phi_\mu = a\psi_\mu \qquad 2.4.6$$

By acting on these new functions with the unitary operators of $F(\mathbf{k})$ we have

$$u\Phi_\mu = ua\psi_\mu = a(a^{-1}ua)\psi_\mu = a\sum_\nu \psi_\nu \Gamma(a^{-1}ua)_{\nu\mu} = \sum_\nu \Phi_\nu \Gamma^*(a^{-1}ua)_{\nu\mu} \qquad 2.4.7$$

where use is made of the fact that a is antiunitary and $(a^{-1}ua)$ a member of G(**k**). In matrix form, 2.4.5 and 2.4.7 reduce to

$$u\ \widehat{\psi_1 \ldots \psi_d\ \Phi_1 \ldots \Phi_d} = \widehat{\psi_1 \ldots \psi_d\ \Phi_1 \ldots \Phi_d} \begin{pmatrix} \Gamma(u) & \mathbf{0} \\ \mathbf{0} & \Gamma^*(a^{-1}ua) \end{pmatrix} \qquad 2.4.8$$

By definition the d functions ψ are linearly independent as they form a basis for an irreducible representation of G(**k**). The same must necessarily hold for the d functions $a\psi$, the two representations being given by matrices $\Gamma(u)$ and $\Gamma^*(a^{-1}ua)$, respectively. It is, however, possible that the functions $a\psi$ are linearly dependent on the ψ's, that is

$$a\psi_\mu = \sum_\nu \psi_\nu \mathbf{U}_{\nu\mu} \qquad 2.4.9$$

from which it follows that

$$\Gamma^*(a^{-1}ua) = \mathbf{U}^{-1}\, \Gamma(u)\, \mathbf{U} \qquad 2.4.10$$

and the two representations are equivalent. In this case, case a, no additional degeneracy is introduced by time reversal symmetry as no additional functions are added by the antiunitary elements of F(**k**). However, even if the two sets ψ and $a\psi$ are independent, it is still possible that the two representations Γ and Γ^* are connected by relation 2.4.10. This is case b in which the functions ψ and $a\psi$ are linearly independent, the representations Γ and Γ^* equivalent and both sets are needed in the basis of the full group including K, and the degeneracy is doubled. Case c is similar to case b but now Γ and Γ^* are inequivalent representations and the degeneracy is again doubled. These three different cases correspond to a reality classification of the irreducible representation Δ of the full space group G induced by an irreducible representation Γ of G(**k**) : case a thus refers to an irreducible representation of the first kind in which Δ and Δ^* are equivalent to the same real irreducible representation; in case b, Δ is of second kind, Δ is equivalent to Δ^* but not to any group of real matrices; in case c, Δ is of the third kind and it is not equivalent to Δ^*.

The addition of time inversion symmetry may cause extra degeneracy produced from energy levels corresponding to a given wave vector **k** only in case b and c. Case b gives a doubled degeneracy produced by two different energy levels, both belonging to the same representation Γ, which become degenerate. This situation can be loosely described as self-sticking of eigenvalues belonging to the same irreducible representation. In case c, two different representations stick together and the degeneracy is again doubled. In order to be able to find out which of the three cases applies for a given irreducible representation of the space group of **k**, the following criterion was derived [36]

$$\sum_a \chi(a^2) \quad = \quad \begin{matrix} g \text{ in case a} \\ -g \text{ in case b} \\ 0 \text{ in case c} \end{matrix} \qquad 2.4.11$$

The sum is extended to the characters of the squares of all antiunitary coset representatives of the group F(**k**), that is if $\{\beta|\mathbf{b}\}$ belongs to G(**k** and if $\{\alpha|\mathbf{a}\}$is a given element of G which sends **k** into -**k**, the sum is extended to the characters of the squares of the elements $\{\alpha|\mathbf{a}\}\{\beta|\mathbf{b}\}$. The test 2.4.11 was applied to the irreducible representations previousl reported in Table 2.1 and the results were given in terms of sticking of representations. We note that when $\mathbf{k}\equiv-\mathbf{k}$, which happens for those points of symmetry for which **k** is zero or half a reciprocal lattice vector, the sum 2.4.11 reduces to a sum over all elements of G(**k**). This is the only case in which the analysis of reality of irreducible representations of G(**k**) has a physical meaning in assessing the degeneracy of phonon states

2.5 SYMMETRY OF THE DYNAMICAL MATRIX

The dynamical matrix was defined in Chapter 1 as the force constant matrix for translationally symmetrized, mass weighted Cartesian coordinates. Elements of the dynamical matrix are given by 1.2.14 as a function of atomic force constants. Because of the property 2.2.13 of reciprocal basis vectors it readily follows that

$$\mathbf{D}_{\alpha\beta}\left(\begin{smallmatrix}\mu i\\ \nu j\end{smallmatrix}\middle|\mathbf{k}\right) = \mathbf{D}_{\alpha\beta}\left(\begin{smallmatrix}\mu i\\ \nu j\end{smallmatrix}\middle|\mathbf{k}+\mathbf{K}_m\right) \qquad 2.5.1$$

Thus $\mathbb{D}$ has the periodicity of the reciprocal lattice. A corresponding property must also be valid for the eigenfrequencies and eigenvectors of $\mathbb{D}$, that is

$$\omega_{p\mathbf{k}} = \omega_{p(\mathbf{k}+K_m)} \qquad 2.5.2$$

and

$$e_\alpha(\mu i|p\mathbf{k}) = e_\alpha(\mu i|p(\mathbf{k}+K_m)) \qquad 2.5.3$$

as these quantities enter in the basic equation 1.2.16. On the other hand, because of the definition 1.2.14 and as anticipated in 1.2.18a it must be

$$\mathbb{D}_{\alpha\beta}(\begin{smallmatrix}\mu i\\ \nu j\end{smallmatrix}|\mathbf{k}) = \mathbb{D}^*_{\alpha\beta}(\begin{smallmatrix}\mu i\\ \nu j\end{smallmatrix}|-\mathbf{k}) \qquad 2.5.4$$

Then, if the point group associated with the space group G contains an operation which sends $\mathbf{k}$ into $-\mathbf{k}$, it is possible to combine this operation with the complex conjugation thus producing an operator which commutes with the dynamical matrix. $\mathbb{D}$ is also an Hermitian matrix as it is readily demonstrated starting from the basic definition 1.2.14 and 1.2.10

$$\mathbb{D}_{\beta\alpha}(\begin{smallmatrix}\nu j\\ \mu i\end{smallmatrix}|\mathbf{k}) = \frac{1}{\sqrt{m_i m_j}} \sum_m (\partial^2 V/\partial U^{0\nu j}_{\beta}\partial U^{m\mu i}_{\alpha})_0\, e^{i\mathbf{k}\cdot\mathbf{R}_m} =$$

$$= \frac{1}{\sqrt{m_i m_j}} \sum_m (\partial^2 V/\partial U^{0\mu i}_{\alpha}\partial U^{-m\nu j}_{\beta})_0\, e^{i\mathbf{k}\cdot\mathbf{R}_m} = \mathbb{D}^*_{\alpha\beta}(\begin{smallmatrix}\mu i\\ \nu j\end{smallmatrix}|\mathbf{k}) \qquad 2.5.5$$

Properties 1.2.5 and 1.2.6 of Cartesian force constants were used. It follows that $\mathbb{D}$ has real eigenvalues $\omega^2_{p\mathbf{k}}$. If we replace $\mathbf{k}$ by $-\mathbf{k}$ in 1.2.16, take the complex conjugate of the resulting equation and use the property 2.5.4 of $\mathbb{D}$, we obtain

$$\sum_\beta\sum_\nu\sum_j \mathbb{D}_{\alpha\beta}(\begin{smallmatrix}\mu i\\ \nu j\end{smallmatrix}|\mathbf{k})\; e^*_\beta(\nu j|p\text{-}\mathbf{k}) = \omega^2_{p-\mathbf{k}}\, e^*_\alpha(\mu i|p\text{-}\mathbf{k}) \qquad 2.5.6$$

As $\omega^2_{p\mathbf{k}}$ and $\omega^2_{p-\mathbf{k}}$ are eigenvalues of the same dynamical matrix, then necessarily

$$\omega^2_{p-\mathbf{k}} = \omega^2_{p\mathbf{k}} \qquad 2.5.7$$

and this gives a prescription for labelling frequencies at $-\mathbf{k}$ in terms of those at $\mathbf{k}$. As also eigenvectors $\mathbf{e}(p\mathbf{k})$ and $\mathbf{e}^*(p-\mathbf{k})$ satisfy the same eigenvalue equation, they may differ at most by an arbitrary factor of modulus 1. Hence, if one selects a phase factor equal to 1 it follows

$$e^*_\alpha(\mu i|p-\mathbf{k}) = e_\alpha(\mu i|p\mathbf{k}) \qquad 2.5.8a$$

or, from the definition 1.2.25

$$Q^*_{p\mathbf{k}} = Q_{p-\mathbf{k}} \qquad 2.5.8b$$

We will now analyze the transformation law for polarization vectors, with components $e_\alpha(\mu i|p\mathbf{k})$, under a crystal symmetry operation $\{\gamma|\mathbf{g}\}$ whio leaves the crystal invariant, moving every atom into an equivalent one. This operation is described by its action on a particular vector $\mathbf{r}^{m\mu i}$ written as $\mathbf{r}^{m\mu i} = \mathbf{r}^{0\mu i} + \mathbf{R}_m$, where $\mathbf{r}^{0\mu i}$ is a vector from the origin to atom i of molecule μ in the (0,0,0) unit cell and $\mathbf{R}_m$ is a lattice translation which identifies the mth unit cell. By acting on a vector $\mathbf{r}^{m\mu i}$ with an element $\{\gamma|\mathbf{g}\}$of G let

$$\mathbf{r}^{n\nu j} = \{\gamma|\mathbf{g}\}\mathbf{r}^{m\mu i} \qquad 2.5.9$$

where j is an atom equivalent to i and $\mathbf{R}_n$ is such that

$$\mathbf{R}_n = \{\gamma|\mathbf{g}\}\mathbf{r}^{0\mu i} - \mathbf{r}^{0\nu j} + \gamma\mathbf{R}_m \qquad 2.5.10$$

On the other hand, let $\hat{O}(\{\gamma|\mathbf{g}\})U^{n\nu j}_\alpha$ be a displacement obtained by acting with the symmetry operator $\hat{O}(\{\gamma|\mathbf{g}\})$ on a component of a displacement on atom j, molecule $n\nu$. The rotated displacement must be related to the corresponding displacement of atom i, molecule $m\mu$, of the unrotated crys tal according to the definition 2.5.9 of the element $\{\gamma|\mathbf{g}\}$ of G. Thus, we have

$$\hat{O}(\{\gamma|\mathbf{g}\})U_\alpha^{n\nu j} = \sum_\beta \mathbb{R}(\gamma)_{\alpha\beta}U_\beta^{m\mu i} \qquad 2.5.11$$

where $\mathbb{R}(\gamma)$ is a 3×3 orthogonal matrix associated with the point group rotation γ and 2.5.11 represents just a transformation law for vector fields. It will be immediately used to derive transformation properties of translationally symmetrized coordinates 1.2.20. We have

$$\hat{O}(\{\gamma|\mathbf{g}\})U_\alpha(\nu j|\mathbf{k}) = \frac{1}{\sqrt{L}}\sum_n e^{-i\mathbf{k}\cdot\mathbf{R}_n}\,\hat{O}(\{\gamma|\mathbf{g}\})U_\alpha^{n\nu j} =$$

$$= \frac{1}{\sqrt{L}}\sum_m e^{-i\mathbf{k}\cdot(\{\gamma|\mathbf{g}\}\mathbf{r}^{0\mu i}-\mathbf{r}^{0\nu j}+\gamma\mathbf{R}_m)}\sum_\beta \mathbb{R}(\gamma)_{\alpha\beta}U_\beta^{m\mu i} =$$

$$= e^{-i\mathbf{k}\cdot(\{\gamma|\mathbf{g}\}\mathbf{r}^{0\mu i}-\mathbf{r}^{0\nu j})}\sum_\beta \mathbb{R}(\gamma)_{\alpha\beta}\frac{1}{\sqrt{L}}\sum_m e^{-i\gamma^{-1}\mathbf{k}\cdot\mathbf{R}_m}U_\beta^{m\mu i} =$$

$$= e^{-i\mathbf{k}\cdot(\{\gamma|\mathbf{g}\}\mathbf{r}^{0\mu i}-\mathbf{r}^{0\nu j})}\sum_\beta \mathbb{R}(\gamma)_{\alpha\beta}U_\beta(\mu i|\gamma^{-1}\mathbf{k}) \qquad 2.5.12$$

In order to derive the final result 2.5.12 definition 2.5.10 was adopted for $\mathbf{R}_n$. If now we replace $\mathbf{k}$ by $\gamma\mathbf{k}$ and make use of the following identity

$$\gamma\mathbf{k}\cdot(\{\gamma|\mathbf{g}\}\mathbf{r}^{0\mu i}-\mathbf{r}^{0\nu j}) = -\mathbf{k}\cdot(\{\gamma|\mathbf{g}\}^{-1}\mathbf{r}^{0\nu j}-\mathbf{r}^{0\mu i}) \qquad 2.5.13$$

2.5.12 is readily transformed into

$$\hat{O}(\{\gamma|\mathbf{g}\})U_\alpha(\nu j|\gamma\mathbf{k}) = e^{i\mathbf{k}\cdot(\{\gamma|\mathbf{g}\}^{-1}\mathbf{r}^{0\nu j}-\mathbf{r}^{0\mu i})}\sum_\beta \mathbb{R}(\gamma)_{\alpha\beta}U_\beta(\mu i|\mathbf{k}) \qquad 2.5.14$$

Eq. 2.5.14 gives components of the rotated vector and can be rewritten in matrix form as

$$\mathbf{U}'(\gamma\mathbf{k}) = \mathbb{H}_\mathbf{k}(\{\gamma|\mathbf{g}\})\,\mathbf{U}(\mathbf{k}) \qquad 2.5.15$$

where $\mathbf{U}$ is a vector with 3NZ components $U_\alpha(\mu i|\mathbf{k})$ (α=1,2,3; i=1,2,..,N; μ=1,2,..,Z) and $\mathbf{U}'$ is the corresponding vector of rotated components. $\mathbb{H}_\mathbf{k}$ is a 3NZ×3NZ matrix in block form, each block having dimensionality 3×3 so that only one block per row block (or column block) is different from zero. For each couple j and i, the non zero block is uniquely identified according to 2.5.10 and is given by the transformation matrix $\mathbb{R}(\gamma)$ multiplied by $e^{i\mathbf{k}\cdot R_n}$ with $R_n = \{\gamma|\mathbf{g}\}^{-1}\mathbf{r}^{0\nu j} - \mathbf{r}^{0\mu i}$. Obviously if the Fourier transformed matrix has eigenvectors $\mathbf{e}(p\mathbf{k})$ with components given by $e_\alpha(\mu i|p\mathbf{k})$, the corresponding matrix for a crystal rotated through $\{\gamma|\mathbf{g}\}$ with displacements $\mathbf{U}'(\gamma\mathbf{k})$ has eigenvectors

$$\mathbf{e}'(p\gamma\mathbf{k}) = \mathbb{H}_{\mathbf{k}}(\{\gamma|g\})\mathbf{e}(p\mathbf{k}) \qquad 2.5.16$$

or, in component form, according to the explicit form of $\mathbb{H}_{\mathbf{k}}$ given by Eq. 2.5.14

$$e'_\alpha(\nu j|p\gamma\mathbf{k}) = e^{i\mathbf{k}\cdot(\{\gamma|\mathbf{g}\}^{-1}\mathbf{r}^{0\nu j}-\mathbf{r}^{0\mu i})}\sum_\beta \mathbb{R}(\gamma)_{\alpha\beta}\, e_\beta(\mu i|p\mathbf{k}) \qquad 2.5.17$$

As $\mathbb{R}(\gamma)$'s are orthogonal matrices, all $\mathbb{H}_{\mathbf{k}}$ are unitary and define a unitary transformation for the dynamical matrix

$$\mathbb{H}_{\mathbf{k}}(\{\gamma|\mathbf{g}\})\,\mathbb{D}(\mathbf{k})\,\mathbb{H}^\dagger_{\mathbf{k}}(\{\gamma|\mathbf{g}\}) = \mathbb{D}(\gamma\mathbf{k}) \qquad 2.5.18$$

with eigenvectors $\mathbf{e}(p\gamma\mathbf{k})$ and eigenvalues $\omega^2_{p\gamma\mathbf{k}} = \omega^2_{p\mathbf{k}}$. Using the definition 2.5.15 it must be

$$\mathbf{U}'(\delta\mathbf{k}) = \mathbb{H}_{\mathbf{k}}(\{\delta|\mathbf{d}\})\mathbf{U}(\mathbf{k}) \qquad 2.5.19$$

and for a vector $\bar{\mathbf{k}} = \delta\mathbf{k}$ and an element $\{\gamma|\mathbf{g}\}$

$$\mathbf{U}''(\gamma\bar{\mathbf{k}}) = \mathbb{H}_{\bar{\mathbf{k}}}(\{\gamma|\mathbf{g}\})\mathbf{U}'(\bar{\mathbf{k}}) = \mathbb{H}_{\bar{\mathbf{k}}}(\{\gamma|\mathbf{g}\})\,\mathbb{H}_{\mathbf{k}}(\{\delta|\mathbf{d}\})\mathbf{U}(\mathbf{k}) \qquad 2.5.20$$

that is

$$\mathbf{U}''(\gamma\delta\mathbf{k}) = \mathbb{H}_{\delta\mathbf{k}}(\{\gamma|g\})\,\mathbb{H}_{\mathbf{k}}(\{\delta|\mathbf{d}\})\mathbf{U}(\mathbf{k}) \qquad 2.5.21$$

However the basic definition 2.5.15 requires

$$\mathbf{U}''(\gamma\delta\mathbf{k}) = \mathbb{H}_{\mathbf{k}}(\{\gamma|g\}\{\delta|\mathbf{d}\})\mathbf{U}(\mathbf{k}) \qquad 2.5.22$$

which, after comparison with Eq. 2.5.21, leads to the following multiplication properties of the representation by $\mathbb{H}$ matrices of G

$$\mathbb{H}_{\mathbf{k}}(\{\gamma|\mathbf{g}\}\{\delta|\mathbf{d}\}) = \mathbb{H}_{\delta\mathbf{k}}(\{\gamma|\mathbf{g}\})\ \mathbb{H}_{\mathbf{k}}(\{\delta|\mathbf{d}\}) \qquad 2.5.23$$

Because of the factor $\gamma\mathbf{k}$ which occurs in the first term on the right-hand side of 2.5.23 the matrices $\mathbb{H}$ do not form a regular representation of G. Only if we restrict our attention to space groups $G(\mathbf{k})$ do we obtain a unitary representation of dimensionality 3NZ in terms of matrices $\mathbb{H}_{\mathbf{k}}(\{\beta|\mathbf{b}\})$, for all $\{\beta|\mathbf{b}\}$ such that $\beta\mathbf{k}\equiv\mathbf{k}$, which commute with the dynamical matrix $\mathbb{D}(\mathbf{k})$ because of 2.5.18. This commutation property of matrices $\mathbb{H}_{\mathbf{k}}$ for the elements of $G(\mathbf{k})$ can be exploited in the sense that if $\mathbb{E}(\mathbf{k})$ are "polarization matrices" containing the eigenvectors of $\mathbb{D}(\mathbf{k})$, as given in 1.2.18, then also the relation

$$\mathbb{D}(\mathbf{k})\ \mathbb{H}_{\mathbf{k}}(\{\beta|\mathbf{b}\})\ \mathbb{E}(\mathbf{k}) = \mathbb{H}_{\mathbf{k}}(\{\beta|\mathbf{b}\})\ \mathbb{E}(\mathbf{k})\mathbf{\Omega}^2(\mathbf{k}) \qquad 2.5.24$$

must hold. This means the the columns of $\mathbb{H}_{\mathbf{k}}(\{\beta|\mathbf{b}\})\mathbb{E}(\mathbf{k})$ are also eigenvectors of $\mathbb{D}(\mathbf{k})$ associated to eigenvalues $\omega^2_{p\mathbf{k}}$, and can be written as a linear combination of the columns of $\mathbb{E}(\mathbf{k})$ corresponding to the same eigenvalues. This situation can be represented by

$$\mathbb{H}_{\mathbf{k}}(\{\beta|\mathbf{b}\})\ \mathbb{E}(\mathbf{k}) = \mathbb{E}(\mathbf{k})\ \mathbb{B}_{\mathbf{k}}(\{\beta|\mathbf{b}\}) \qquad 2.5.25$$

Barring accidental degeneracy, the set of vectors in $\mathbb{E}(\mathbf{k})$ is irreducible and then the eigenvector matrix completely reduces the representation of $G(\mathbf{k})$ based on matrices $\mathbb{H}_{\mathbf{k}}$. This result can be conveniently rewritten in the form

$$\mathbb{H}_{\mathbf{k}}(\{\beta|\mathbf{b}\})\ \mathbf{e}(p_s\mathbf{k}) = \sum_t \mathbf{e}(p_t\mathbf{k})\ \mathbb{B}_{p\mathbf{k}}(\{\beta|\mathbf{b}\})_{ts} \qquad 2.5.26$$

where s and t are degeneracy indices. The sum extends to all l_p degenerate eigenvectors corresponding to the same eigenfrequency $\omega_{p\mathbf{k}}$. Eq. 2.5.26 could also be written starting with a proper linear combination of eigenvectors in such a way that each block $\mathbb{B}_{p\mathbf{k}}$ furnishes an irreducible representation of $G(\mathbf{k})$, with eigenvectors belonging to the sth row of the representation.

In concluding this Section, we mention the fact that the transformation properties of the eigenvectors can also be analyzed for a dynamical matrix written in molecular coordinates defined in Section 1.5. A simplification would occur for internal displacements which transform one into the other. Translations given by 1.5.5a transform precisely as Cartesian components so that 2.5.14 applies, while rotations 1.5.5b transform like components of a pseudo vector, and in this case $\mathbb{R}(\gamma)$ must be replaced by $-\mathbb{R}(\gamma)$ for improper rotational elements of G. A more detailed account of the symmetry properties of the dynamical matrix is given in Ref. 37 in terms of space group multiplier representations while the symmetry properties of external coordinates are considered specifically in Ref. 38.

2.6 SYMMETRY PROPERTIES OF VIBRATIONAL STATES

In order to determine selection rules for processes involving phonons, it is necessary to study the transformation properties of vibrational eigenfunctions, which are in turn better analyzed by adopting a representation based on occupation numbers. As shown in Section 1.2, this means writing the Hamiltonian in terms of phonon creation, $a^{\dagger}_{p\mathbf{k}}$, and phonon annihilation, $a_{p\mathbf{k}}$, operators with vibrational eigenfunctions given as a product

$$\Psi = \prod_{p\mathbf{k}} (a^{\dagger}_{p\mathbf{k}})^{n_{p\mathbf{k}}} \Psi^0 \qquad 2.6.1$$

where Ψ^0 is the vacuum state $|0,0, .. , 0\rangle$, each vibrational state being characterized by $n_{p\mathbf{k}}$ phonons in the modes identified by p and $\mathbf{k}$. Since the vacuum state is invariant under all symmetry operations of the crystal, the transformation properties of eigenfunctions are completely determined by those of the creation and annihilation operators.

Using the basic definition 1.2.39 and 1.2.40 for such operators and adopting the properties 2.5.7 and 2.5.8 of normal coordinates and eigenvectors, it is easy to see that

$$\Omega_{pk} = \left(\frac{\hbar}{2\omega_{p_{\mathbf{k}}}}\right)^{\frac{1}{2}} (a_{pk} + a^{\dagger}_{p-k}) \qquad 2.6.2$$

Using 1.2.26 the Cartesian displacements are given in terms of creation and annihilation operators by

$$W^{m\mu i}_{\alpha} = \sum_p \sum_{\mathbf{k}}' \left(\frac{\hbar}{2L\omega_{pk}}\right)^{\frac{1}{2}} e_{\alpha}(\mu i|p\mathbf{k}) \; (a_{p\mathbf{k}} + a^{\dagger}_{p-\mathbf{k}}) \; e^{i\mathbf{k}\cdot\mathbf{R}_m} +$$

$$+ \sum_p \sum_{\mathbf{k}}' \left(\frac{\hbar}{2L\omega_{p-\mathbf{k}}}\right)^{\frac{1}{2}} e_{\alpha}(\mu i|p\text{-k})\,(a_{p-\mathbf{k}} + a^{\dagger}_{p\mathbf{k}}) \; e^{-i\mathbf{k}\cdot\mathbf{R}_m} \qquad 2.6.3$$

where the sum over all values of k was split into two separate sums for positive and negative values of the wave vector. $\sum'$ stands for a sum restricted to positive values of **k**. Collecting terms and using 2.5.7 , the right-hand side of the last equation can be rewritten as

$$\sum_p \sum_{\mathbf{k}}' \left(\frac{\hbar}{2L\omega_{p\mathbf{k}}}\right)^{\frac{1}{2}} \{ [e_{\alpha}(\mu i|p\mathbf{k}) \; a_{p\mathbf{k}} \; e^{i\mathbf{k}\cdot\mathbf{R}_m} + e_{\alpha}(\mu i|p\text{-}\mathbf{k}) \; a_{p-\mathbf{k}} \; e^{-i\mathbf{k}\cdot\mathbf{R}_m}] +$$

$$+ [e_{\alpha}(\mu i|p\text{-}\mathbf{k}) \; a^{\dagger}_{p\mathbf{k}} \; e^{-i\mathbf{k}\cdot\mathbf{R}_m} + e_{\alpha}(\mu i|p\mathbf{k}) \; a^{\dagger}_{p-\mathbf{k}} e^{i\mathbf{k}\cdot\mathbf{R}_m}] \} \qquad 2.6.4$$

The terms in the first square bracket can be collected under a single summation over all positive and negative values of **k** and the same is done for the second square bracket although in this case 2.5.8a is first used for the eigenvectors. Thus, finally, we have Cartesian, mass weighted displacements expressed as

$$W^{m\mu i}_{\alpha} = \frac{1}{\sqrt{L}} \sum_p \sum_{\mathbf{k}} \left(\frac{\hbar}{2\omega_{pk}}\right)^{\frac{1}{2}} [e_{\alpha}(\mu i|p\mathbf{k}) a_{p\mathbf{k}} \; e^{i\mathbf{k}\cdot\mathbf{R}_m} + e^{*}_{\alpha}(\mu i|p\mathbf{k}) a^{\dagger}_{p\mathbf{k}} \; e^{-i\mathbf{k}\cdot\mathbf{R}_m}] \qquad 2.6.5$$

which gives a field vector in terms of creation and annihilation operators.

From 2.5.11 we already know the effect of a crystal symmetry element on Cartesian displacements and this defines the transformation law of a vector operator **U** under a symmetry operator $\hat{O}(\{\gamma|\mathbf{g}\})$ as

$$\hat{O}(\{\gamma|g\}^{-1}) \; U^{n\nu j}_{\alpha} \hat{O}(\{\gamma|g\}) = \sum_{\beta} \mathbb{R}(\gamma)_{\alpha\beta} U^{m\mu i}_{\beta} \qquad 2.6.6$$

Multiplying 2.6.6 on the left by $\hat{O}(\{\gamma|g\})$ and on the right by $\hat{O}(\{\gamma|g\}^{-1})$

and using the orthogonality of the 3×3 matrices $\mathbb{R}(\gamma)$, one obtains

$$\hat{O}(\{\gamma|\mathbf{g}\})\ U_\alpha^{m\mu i}\ \hat{O}(\{\gamma|\mathbf{g}\}^{-1}) = \sum_\beta \mathbb{R}(\gamma)_{\beta\alpha}\, U_\beta^{n\nu j} \qquad 2.6.7$$

This defines the rotational transform of $U_\alpha^{m\mu i}$ if $\{\gamma|\mathbf{g}\}$is an element of G which transforms a vector $\mathbf{r}^{m\mu i}$ into an equivalent one $\mathbf{r}^{n\nu j} = \{\gamma|g\}\mathbf{r}^{m\mu i}$ Using 2.6.5, Eq. 2.6.7 may also be written as

$$\sum_p\sum_\mathbf{k}\frac{1}{\sqrt{\omega_{p\mathbf{k}}}}e_\alpha(\mu i|p\mathbf{k})\ e^{i\mathbf{k}\cdot\mathbf{R}_m}\ \hat{O}(\{\gamma|\mathbf{g}\})\ a_{p\mathbf{k}}\ \hat{O}(\{\gamma|\mathbf{g}\}^{-1}) + \text{complex conjugate} =$$

$$=\sum_p\sum_\mathbf{k}\frac{1}{\sqrt{\omega_{p\mathbf{k}}}}e^{i\mathbf{k}\cdot(\{\gamma|g\}\mathbf{r}^{0\mu i}-\mathbf{r}^{0\nu j}+\gamma\mathbf{R}_m)}\sum_\beta\mathbb{R}(\gamma)_{\beta\alpha}e_\beta(\nu j|p\mathbf{k})a_{p\mathbf{k}} + \text{complex conjugate} \qquad 2.6.8$$

Here the expression 2.5.10 for $\mathbf{R}_n$ was used. Since symmetry operators transform annihilation operators among themselves, as well as creation operators into one another, we may omit the complex conjugate parts from 2.6.8. The resulting equation will be used to obtain the transformation properties of $a_{p\mathbf{k}}$ from which those of $a^\dagger_{p\mathbf{k}}$ are derived by taking the Hermitian adjoint counterpart. If both sides of 2.6.8 are multiplied by $e^{-i\mathbf{k}'\cdot\mathbf{R}_m}$ and summed over m, using the result 2.2.25 and replacing $\mathbf{k}'$ by $\mathbf{k}$, it follows that for each value of $\mathbf{k}$ it must be

$$\sum_p\frac{1}{\sqrt{\omega_{p\mathbf{k}}}}\ e_\alpha(\mu i|p\mathbf{k})\ \hat{O}(\{\gamma|\mathbf{g}\})\ a_{p\mathbf{k}}\ \hat{O}(\{\gamma|\mathbf{g}\}^{-1}) =$$

$$= \sum_p\frac{1}{\sqrt{\omega_{p\mathbf{k}}}}\ e^{i\gamma\mathbf{k}\cdot(\{\gamma|\mathbf{g}\}\mathbf{r}^{0\mu i}-\mathbf{r}^{0\nu j})}\ \sum_\beta\mathbb{R}(\gamma)_{\beta\alpha}\ e_\beta(\nu j|p\gamma\mathbf{k})\ a_{p\gamma\mathbf{k}} \qquad 2.6.9$$

In the right-hand side of this equation it is possible to recognize elements of the matrix $\mathbb{H}^\dagger_\mathbf{k}(\{\gamma|\mathbf{g}\})$ defined in 2.5.15. Therefore, in matrix notation, if $\mathbb{E}(\mathbf{k})$ is the matrix of the eigenvectors and $\mathbf{\Omega}^2(\mathrm{k})$ the diagonal matrix of the eigenvalues as in 1.2.18, Eq. 2.6.9 can be rewritten as

$$\mathbb{E}(\mathbf{k})\ \mathbf{\Omega}^{-\frac{1}{2}}(\mathbf{k})\ \hat{O}(\{\gamma|\mathbf{g}\})\ \mathbf{a}_\mathbf{k}\ \hat{O}(\{\gamma|\mathbf{g}\}^{-1}) = \mathbb{H}^\dagger_\mathbf{k}(\{\gamma|\mathbf{g}\})\ \mathbb{E}(\gamma\mathbf{k})\ \mathbf{\Omega}^{-\frac{1}{2}}(\mathbf{k})\ \mathbf{a}_{\gamma\mathbf{k}} \qquad 2.6.10$$

where $\mathbf{a}_\mathbf{k}$ is the column vector of the annihilation operators and, as derived from 2.5.18, the result $\omega_{p\mathbf{k}} = \omega_{p\gamma\mathbf{k}}$ was used. The latter equality 2.6.10 can be transformed into

$$\hat{O}(\{\gamma|g\})\, \mathbf{a}_{\mathbf{k}}\, \hat{O}(\{\gamma|g\}^{-1}) = \mathbf{\Omega}^{\frac{1}{2}}(\mathbf{k})\, [\mathbb{H}_{\mathbf{k}}(\{\gamma|g\})\mathbb{E}(\mathbf{k})]^{\dagger}\mathbb{E}(\gamma\mathbf{k})\mathbf{\Omega}^{-\frac{1}{2}}(\mathbf{k})\mathbf{a}_{\gamma k} \qquad 2.6.11$$

which readily becomes

$$\hat{O}(\{\beta|\mathbf{b}\})\, \mathbf{a}_{\mathbf{k}}\, \hat{O}(\{\beta|\mathbf{b}\}^{-1}) = \mathbb{B}^{\dagger}_{\mathbf{k}}(\{\beta|\mathbf{b}\})\, \mathbf{a}_{\mathbf{k}} \qquad 2.6.12$$

for all $\{\beta|\mathbf{b}\} \in G(\mathbf{k})$. This result was obtained using 2.5.25 and the fact that eigenvectors corresponding to naturally degenerate eigenfrequencies are rotated into a linear combination of the others. This means that $\mathbb{B}_{\mathbf{k}}$ is a block diagonal matrix which commutes with the diagonal matrix $\mathbf{\Omega}(\mathbf{k})$ of similar block structure. For creation operators we obtain from 2.6.12

$$\hat{O}(\{\beta|\mathbf{b}\})\, \mathbf{a}^{\dagger}_{\mathbf{k}}\, \hat{O}(\{\beta|\mathbf{b}\}^{-1}) = \mathbf{a}^{\dagger}_{\mathbf{k}}\, \mathbb{B}_{\mathbf{k}}(\{\beta|\mathbf{b}\}) \qquad 2.6.13$$

for all $\{\beta|\mathbf{b}\} \in G(\mathbf{k})$. Hence, the transformation properties of annihilation and creation operators are established for all elements of $G(\mathbf{k})$ in terms of a completely reduced representation defined by 2.5.25 for the eigenvectors of the dynamical matrix.

Going back to 2.6.11 we note that, as it is seen from 2.5.18, the matrix $\mathbb{D}(\gamma\mathbf{k})$ has eigenvectors $\mathbb{H}_{\mathbf{k}}(\{\gamma|\mathbf{g}\})\mathbb{E}(\mathbf{k})$. Therefore the product $[\mathbb{H}_{\mathbf{k}}(\{\gamma|g\})\mathbb{E}(\mathbf{k})]^{\dagger}\mathbb{E}(\gamma\mathbf{k})$ is a matrix in block form with the same structure as $\mathbf{\Omega}(\mathbf{k})$ and commutes with the matrix of the eigenvalues. When this result is introduced in 2.6.11 we have

$$\hat{O}(\{\gamma|\mathbf{g}\})\, \mathbf{a}_{\mathbf{k}}\, \hat{O}(\{\gamma|\mathbf{g}\}^{-1}) = \mathbb{E}^{\dagger}(\mathbf{k})\, \mathbb{H}^{\dagger}_{\mathbf{k}}(\{\gamma|\mathbf{g}\})\, \mathbb{E}(\gamma\mathbf{k})\, \mathbf{a}_{\gamma k} \qquad 2.6.14$$

and

$$\hat{O}(\{\gamma|\mathbf{g}\})\, \mathbf{a}^{\dagger}_{\mathbf{k}}\, \hat{O}(\{\gamma|\mathbf{g}\}^{-1}) = \mathbf{a}^{\dagger}_{\gamma k}\, \mathbb{E}^{\dagger}(\gamma\mathbf{k})\mathbb{H}_{\mathbf{k}}(\{\gamma|\mathbf{g}\})\, \mathbb{E}(\mathbf{k}) \qquad 2.6.15$$

We see that creation and annihilation operators associated with different points of the star of $\mathbf{k}$ are transformed into one another for a generic element $\{\gamma|\mathbf{g}\}$ of G. Once the eigenvectors of the dynamical matrix are known for a given vector $\mathbf{k}_1$, say, 2.5.16 is used for the definition of different eigenvectors for other points of the star

$$\mathbb{E}(\alpha_1^i \mathbf{k}_1) = \mathbb{H}_{\mathbf{k}_1}(\{\alpha_1^i|\mathbf{a}_1^i\})\ \mathbb{E}(\mathbf{k}_1) \qquad 2.6.16$$

This can be done as eigenvectors are not uniquely defined and thus for different values of $\mathbf{k}$ matrices $\mathbb{E}$ can be arranged using the list of coset representatives of G. With this choice of eigenvectors it will be shown that operators $a^{\dagger}_{p\mathbf{k}}$ form a basis for a representation, of the space group G, induced from a representation of $G(\mathbf{k}_1)$. The latter is formed by matrices $\mathbb{B}_{\mathbf{k}_1}$ which obey 2.5.25. Let $\mathbb{L}(\{\gamma|\mathbf{g}\})$ be a matrix for such a representation of G with a transformation law

$$\hat{O}(\{\gamma|\mathbf{g}\})\ \mathbf{a}^{\dagger}_{p\mathbf{k}_j}\ \hat{O}(\{\gamma|\mathbf{g}\}^{-1}) = \sum_l \sum_q a^{\dagger}_{q\mathbf{k}_l}\ \mathbb{L}(\{\gamma|\mathbf{g}\})_{lq,jp} \qquad 2.6.17$$

According to the general property 2.6.15, if γ is such that $\gamma\mathbf{k}_j = \mathbf{k}_i$ then only the blocks identified by indices i and j are different from zero in the matrix $\mathbb{L}$ representing $\{\gamma|\mathbf{g}\}$in 2.6.17. Using arguments given in Section 2.3 let

$$(\alpha_1^i)^{-1}\ \gamma\ \alpha_1^j = \alpha_f^1 \qquad \text{for } \gamma\mathbf{k}_j = \mathbf{k}_i \qquad 2.6.18$$

where α_f^1 is an element of the group $G_0(\mathbf{k}_1)$. Then

$$\begin{aligned}\mathbb{L}(\{\gamma|\mathbf{g}\})_{ij} &= \mathbb{L}(\{\alpha_1^i|\mathbf{a}_1^i\}\{\alpha_f^1|\mathbf{a}_f^1\}\{\alpha_1^j|\mathbf{a}_1^j\}^{-1}) = \\ &= \sum_p \sum_q \mathbb{L}(\{\alpha_1^i|\mathbf{a}_1^i\})_{ip}\ \mathbb{L}(\{\alpha_f^1|\mathbf{a}_f^1\})_{pq}\ \mathbb{L}(\{\alpha_1^j|\mathbf{a}_1^j\}^{-1})_{qj} = \\ &= \mathbb{L}(\{\alpha_1^i|\mathbf{a}_1^i\})_{i1}\ \mathbb{L}(\{\alpha_f^1|\mathbf{a}_f^1\})_{11}\ \mathbb{L}(\{\alpha_1^j|\mathbf{a}_1^j\}^{-1})_{1j}\end{aligned} \qquad 2.6.19$$

Here use was made of the property that, for a given element, only one block of $\mathbb{L}$ can be different from zero as discussed above. On the other hand, from 2.6.15 we have

$$\hat{O}(\{\alpha_1^i|\mathbf{a}_1^i\})\mathbf{a}_{\mathbf{k}_1}^{\dagger}\hat{O}(\{\alpha_1^i|\mathbf{a}_1^i\}^{-1}) = \mathbf{a}_{\mathbf{k}_i}^{\dagger}\,\mathbb{E}^{\dagger}(\mathbf{k}_i)\,\mathbb{H}_{\mathbf{k}_1}(\{\alpha_1^i|\mathbf{a}_1^i\})\,\mathbb{E}(\mathbf{k}_1) =$$

$$= \mathbf{a}_{\mathbf{k}_i}^{\dagger}\,[\mathbb{H}_{\mathbf{k}_1}(\{\alpha_1^i|\mathbf{a}_1^i\})\mathbb{E}(\mathbf{k}_1)]^{\dagger}\mathbb{H}_{\mathbf{k}_1}(\{\alpha_1^i|\mathbf{a}_1^i\})\mathbb{E}(\mathbf{k}_1) =$$

$$= \mathbf{a}_{\mathbf{k}_i}^{\dagger} \qquad 2.6.20$$

where the particular choice 2.6.16 of eigenvectors was made. It follows that $\mathbb{L}(\{\alpha_1|\mathbf{a}_1\})_{i1} = \mathbb{E}$ for the representation 2.6.17. Substituting this result in 2.6.19 we finally obtain, according to 2.6.13,

$$\mathbb{L}(\{\gamma|\mathbf{g}\}) = \mathbb{L}(\{\alpha_f^1|\mathbf{a}_f^1\})_{11} = \mathbb{B}_{\mathbf{k}_1}(\{\alpha_f^1|\mathbf{a}_f^1\}) \qquad 2.6.21$$

and thus the statement on the representation 2.6.17 is proved. As $\mathbb{B}_{\mathbf{k}}$ are matrices of a completely reduced representation, creation operators $a_{p\mathbf{k}}^{\dagger}$ classify into irreducible sets, each set being associated with the same frequency $\omega_{p\mathbf{k}}$.

7. SELECTION RULES

The representation theory is very useful for determining whether a quantum transition from one state to another is allowed or forbidden in the first approximation of the perturbation theory. General rules called selection rules can be derived to determine, for instance, whether a matrix element, which measure the transition probability among two states in consequence of a perturbation H_1 and is given by $(\psi, H_1\psi')$, vanishes simply for reasons of symmetry. These matrix elements have the general form

$$(\psi_{i'}, \hat{O}_{i''}\psi_i) \qquad 2.7.1$$

where $\psi_{i'}$ and ψ_i are basic functions for the representations Γ' and Γ, respectively, of the group and $\hat{O}_{i''}$ is an appropriate operator. By definition, for each element α of the group the transformation properties are given by

$$\hat{O}(\alpha)\psi_i = \sum_j \psi_j \Gamma(\alpha)_{ji} \qquad 2.7.2$$

and similar relations hold for $\psi_{i'}$. $\hat{O}_{i''}$ is a general tensor operator which tranforms as

$$\hat{O}(\alpha)\ \hat{O}_{i''}\ \hat{O}(\alpha^{-1}) = \sum_{j''} \hat{O}_{j''}\Gamma''(\alpha)_{j''i''} \qquad 2.7.3$$

in terms of a representation Γ'' of the same group. Group theoretical methods can be used to see whether matrix elements 2.7.1 vanish and how many independent elements exists among all those possible when i, i' and i" assume all values allowed by the dimensionalities of the corresponding representations. If a symmetry operator, selected among the elements of the common group, is applied to 2.7.1, the following relation, based on the general requirement that a transition probability cannot be affected by a symmetry operation, is found :

$$(\psi_{i'},\hat{O}_{i''}\psi_i) = \sum_{j'}\sum_{j''}\sum_j \Gamma'^*(\alpha)_{j'i'}\Gamma''(\alpha)_{j''i''}\Gamma(\alpha)_{ji}\ (\psi_{j'},\hat{O}_{j''}\psi_j) \qquad 2.7.4$$

As a result, a representation is produced which is in general reducible into irreducible representations of the group. Using a general theorem which states the orthogonality of basis functions, belonging to different irreducible representations, matrix elements 2.7.1 are zero if the identity representation does not occur in the reduction of the representation 2.7.4. Or, alternatively, if Γ' does not occur in the reduction of the representation formed by the direct product, $\Gamma''\times\Gamma$, of the other two representations.

In general, if ψ_i belongs to a representation Γ of dimensionality d and $\psi_{i''}$ belongs to a representation Γ'' of dimensionality d", the representation built on the basis $\psi_i\psi_{i''}$ of dimensionality d×d" is

$$\hat{O}(\alpha)\psi_i\psi_{i''} = \sum_j\sum_{j''}\psi_j\psi_{j''}\Gamma(\alpha)_{ji}\Gamma''(\alpha)_{j''i''} \qquad 2.7.5$$

This is called the "direct product" or "Kronecker product" representation whose matrice are

$$\mathbb{P}(\alpha)_{jj'',ii''} = \mathbf{\Gamma}(\alpha)_{ji}\ \mathbf{\Gamma}''(\alpha)_{j''i''} \qquad 2.7.6$$

This representation is in general reducible. If the two sets of functions are distinct, the character of this reducible representation is given by

$$\chi_{\Gamma\times\Gamma''}(\alpha) = \chi(\alpha)\ \chi''(\alpha) \qquad 2.7.7$$

If the direct product representation is formed by multiplying two identical sets, there are only d×(d+1)/2 independent functions which form the basis of a so-called "symmetrized" Kronecker product whose character $\chi_{(2)}(\alpha)$ is given by

$$\chi_{(2)}(\alpha) = \frac{1}{2}\ \{|\chi(\alpha)|^2 + \chi(\alpha^2)\} \qquad 2.7.8$$

These definitions will now be first applied to a crystal in which the common symmetry group for ψ, ψ' and $\hat{0}''$ is the translational group. In this case the representations are labelled by vectors of the stars of **k**, **k'** and **k"** and the character of the direct product representation is given by

$$\chi(\{E|\mathbf{R}_n\}) = e^{i(-\mathbf{k}'_{i'}+\mathbf{k}''_{i''}+\mathbf{k}_i)\cdot\mathbf{R}_n} \qquad 2.7.9$$

This representation can be reduced by means of a general expression which gives the number of times, N_j, a given irreducible representation Γ_j appears in the reduction of a reducible representation with characters $\chi(\alpha)$. This general result of group theory can be written as

$$N_j = \frac{1}{g}\sum_\alpha \chi_j^*(\alpha)\ \chi(\alpha) \qquad 2.7.10$$

where g is the order of the group, $\chi_j(\alpha)$ the character of an element α in the jth irreducible representation and the sum is over all elements of the group. Applying 2.7.10 to the characters 2.7.9 one obtains the number of times the identity representation appears in the reduction of the direct product of three representations of the translational group. This number is given by

$$\frac{1}{L}\sum_n e^{i(-\mathbf{k}'_{i'}+\mathbf{k}''_{i''}+\mathbf{k}_i)\cdot R_n} \qquad 2.7.11a$$

which, according to the general result 2.2.23, is zero unless

$$-\mathbf{k}'_{i'} + \mathbf{k}''_{i''} + \mathbf{k}_i = \mathbf{K}_m \qquad 2.7.11b$$

In this case, the number is equal to 1. $\mathbf{K}_m$ is a reciprocal lattice translation and 2.7.11b represents a **k**-selection rule which must be satisfied by any matrix element. This rule holds even if the common group is actually a space group, that is, even if it contains rotational symmetry, in which case a more complicated expression needs to be considered when analyzing matrix elements 2.7.4. We first recall from 2.3.12 that a representation of the full space group G is defined by transformations of the kind

$$\hat{O}(\{\alpha|a\})\psi^{\nu}_{\mathbf{k}_i} = \sum_{\mu}\psi^{\mu}_{\alpha\mathbf{k}_i}\mathbf{\Gamma}_{\mathbf{k}}(\{\alpha_f|\mathbf{a}_f\})_{\mu\nu} \qquad 2.7.12$$

where $\mathbf{k}_i$ and $\alpha\mathbf{k}_i$ are vectors of the same star, the latter being obtained by acting on $\mathbf{k}_i$ with the rotational part of the element $\{\alpha|\mathbf{a}\}$. $\psi^{\nu}_{\mathbf{k}_i}$ is a function belonging to a representation Γ of the full space group, with ν and $\mathbf{k}_i$ being used to identify functions of the basis. $\Gamma_{\mathbf{k}}$ is a p-dimensional representation of the space group of k, G(**k**), where **k** is a given vector of the star. A basis of $\Gamma_{\mathbf{k}}$ is formed by the p functions listed as $\psi^1_{\mathbf{k}}, \psi^2_{\mathbf{k}}, \ldots, \psi^p_{\mathbf{k}}$. Hence, if the star of **k** has d points, there is a total of p×d functions $\psi^{\nu}_{\mathbf{k}_i}$. The p×d -dimensional representation Γ of G is induced by the p-dimensional representation $\Gamma_{\mathbf{k}}$ of the space group of **k** and $\{\alpha_f|\mathbf{a}_f\}$ is the appropriate element of G(**k**) determined as done in 2.3.10. and 2.3.11. The same procedure can be adopted for two other stars of **k'** and **k"**, with specific vectors $\mathbf{k}'_{i'}$ and $\mathbf{k}''_{i''}$ and irreducible representations of the space group Γ' and Γ'' induced, respectively, by irreducible representations $\Gamma'_{\mathbf{k}'}$ of G(**k'**) and $\Gamma''_{\mathbf{k}''}$ of G(**k"**). $\psi'^{\nu'}_{\mathbf{k}'_{i'}}$ thus belongs to Γ' and the operator $\hat{O}''_{\mathbf{k}''_{i''}}$ belongs to Γ'' so that matrix elements 2.7.4 become

$$(\psi'^{\nu'}_{\mathbf{k}_{i'}}, \hat{O}''^{\nu''}_{\mathbf{k}''_{i''}} \psi^{\nu}_{\mathbf{k}_i}) = \sum_{\mu'} \sum_{\mu''} \sum_{\mu} \boldsymbol{\Gamma}'^{*}_{\mathbf{k}'}(\{\alpha'_f|\mathbf{a}'_f\})_{\mu'\nu'} \boldsymbol{\Gamma}''_{\mathbf{k}''}(\{\alpha''_f|\mathbf{a}''_f\})_{\mu''\nu''} \boldsymbol{\Gamma}_{\mathbf{k}}(\{\alpha_f|\mathbf{a}_f\})_{\mu\nu} \times$$

$$\times (\psi'^{\mu'}_{\alpha\mathbf{k}'_{i'}}, \hat{O}''^{\mu''}_{\alpha\mathbf{k}''_{i''}} \psi^{\mu}_{\alpha\mathbf{k}_i}) \qquad 2.7.13$$

upon action with the space group element $\{\alpha|\mathbf{a}\}$. Expression 2.7.10 is applied to the determination of the number of occurrences of the identity representation in the reduction of the direct product representation formed through 2.7.13. To do this, let first write the character $\chi(\{\alpha|\mathbf{a}\})$ for a full space group representation. According to 2.3.12 it is given by

$$\chi(\{\alpha|a\}) = \sum_i \chi_{\mathbf{k}_1}(\{\alpha_i|\mathbf{v}(\alpha_i)\}^{-1} \{\alpha|\mathbf{a}\} \{\alpha_i|\mathbf{v}(\alpha_i)\}) \delta_{\alpha\mathbf{k}_i,\mathbf{k}_i} \qquad 2.7.14$$

The sum is over all vectors of the star of $\mathbf{k}$, α_i is the usual rotation which sends $\mathbf{k}_1$ into $\mathbf{k}_i$ and the delta makes explicit the fact that only diagonal blocks of the full space group representation contribute to the character. A short notation is now introduced for convenience. Let a_i represent a space group coset representative with fractional translation of minimum length

$$a_i = \{\alpha_i|\mathbf{v}(\alpha_i)\} \qquad 2.7.15$$

According to this simplified notation we write

$$\{\alpha|\mathbf{a}\} = \{\alpha|\mathbf{v}(\alpha)+\mathbf{R}_n\} = \{E|\mathbf{R}_n\}\{\alpha|\mathbf{v}(\alpha)\} = \{E|\mathbf{R}_n\}a \qquad 2.7.16a$$

and

$$\{\alpha_i|\mathbf{v}(\alpha_i)\}^{-1}\{\alpha|\mathbf{a}\}\{\alpha_i|\mathbf{v}(\alpha_i)\} = \{E|\alpha_i^{-1}\mathbf{R}_n\}(a_i^{-1} a a_i) \qquad 2.7.16b$$

If we define

$$\delta_{\alpha i,i} = \delta_{\alpha\mathbf{k}_i,\mathbf{k}_i} \qquad 2.7.17$$

which is equal to 1 if α belongs to $G_0(\mathbf{k}_i)$, the character 2.7.14 can be rewritten in a compact form as

$$\chi(\{\alpha|a\}) = \sum_i \chi_{\mathbf{k}_1}(a_i^{-1} aa_i)\, e^{i\alpha_i \mathbf{k}_1 \cdot \mathbf{R}_n}\, \delta_{\alpha i,i} \qquad 2.7.18$$

Using 2.7.7 and 2.7.10 and the characters 2.7.18 for any element and for any irreducible representation of the space group, it is simple to write an expression which gives the number of times the identity representation occurs in the reduction of the representation 2.7.13 when **k**, k' and **k**" are vectors of three different stars. This number is

$$N = \frac{1}{gL}\sum_a \sum_{i'} \sum_{i''} \sum_i \chi'^*_{\mathbf{k}'_1}(a'^{-1}_{i'} aa'_{i'})\chi''_{\mathbf{k}''_1}(a''^{-1}_{i''} aa''_{i''})\chi_{\mathbf{k}_1}(a_i^{-1} aa_i)\delta_{\alpha i',i'}\delta_{\alpha i'',i''}\delta_{\alpha i,i} \times$$

$$\times \sum_n e^{i(-\alpha'_{i'}\mathbf{k}'_1 + \alpha''_{i''}\mathbf{k}''_1 + \alpha \mathbf{k}_1)\cdot \mathbf{R}_n} \qquad 2.7.19$$

where gL is the order of the space group, as a product of the number of pure translations times the number of coset representatives. The sums in 2.7.19 involve all coset representatives and all different vectors in each star of k. However, not all triads of wave vectors are allowed since the **k**-selection rule appropriate to 2.7.13 must be obeyed. Let us suppose that for some $a_0 = a_i$, $a'_0 = a'_{i'}$ and $a''_0 = a''_{i''}$ one has

$$-\alpha'_0\mathbf{k}_1 + \alpha''_0\mathbf{k}''_1 + \alpha_0\mathbf{k}_1 = \mathbf{K}_m \qquad 2.7.20$$

For this particular triad of vectors the sum over n in 2.7.19 gives L and a contribution to N comes only from coset representatives whose rotations are, because of the three deltas, members of the intersection group

$$G_i = G_0(\alpha'_0\mathbf{k}'_1) \cap G_0(\alpha''_0\mathbf{k}''_1) \cap G_0(\alpha_0 k_1) \qquad 2.7.21$$

The corresponding contribution to N is

$$\frac{1}{g}\sum_a \chi'^*_{\mathbf{k}'_1}(a'^{-1}_0 aa'_0)\chi''_{\mathbf{k}''_1}(a''^{-1}_0 aa''_0)\chi_{\mathbf{k}_1}(a_0^{-1} aa_0) \qquad 2.7.22$$

Other triads which obey **k**-selection rule can be obtained acting with a rotation of G_0, not belonging to G_i, on the three vectors 2.7.20. However it can be readily verified that each of these triads gives a contribution identical to that already seen in 2.7.22. If h is the order of the intersection group, there will be g/h different triads allowed in 2.7.19 each derived from 2.7.20 using rotations of G_0. Therefore the total value of N is just g/h times the contribution 2.7.22 and one has

$$N = \frac{1}{h} \sum_a \chi'^*_{\mathbf{k}'_1} (a'^{-1}_0 a a'_0)\ \chi''_{\mathbf{k}''_1} (a''^{-1}_0 a a''_0)\ \chi_{\mathbf{k}_1} (a_0^{-1} a a_0) \qquad 2.7.23$$

where the sum runs over the h elemements of the intersection group 2.7.21 and the vectors $\mathbf{k}_1$, $\mathbf{k}'_1$ and $\mathbf{k}''_1$ obey 2.7.20. This result is valid if the three functions in 2.7.13 belong to three different representations of the space group. We also assumed that all possible triads which obey **k**-selection rules are derived by acting with the element of G_0 on a particular triad but this may not always be the case. If a triad exists which is not obtained by symmetry from a given one, it is necessary to repeat the procedure with this new triad and add to 2.7.23 the number obtained in this way. It may, however, be shown that the additional sum can only lead to a factor multiplying the result 2.7.23 and thus does not change the selection rule.

If matrix elements 2.7.13 are symmetric for the interchange of two indices, i and i" say, that is if Γ and Γ" identify the same irreducible representation of the full space group, the symmetrized Kronecker product $\Gamma_{(2)}$, rather than Γ×Γ", occurs. For any element $\{\alpha|\mathbf{a}\}$ of the full space group the character for a representation based on a symmetrized Kronecker product can be obtained using 2.7.8 and 2.7.14. A simplified notation 2.7.15 for coset representatives is adopted, recalling that α_i is a selected rotation such that $\alpha_i \mathbf{k}_1 = \mathbf{k}_i$. The result is

$$\chi_{(2)}(\{\alpha|\mathbf{a}\}) = \frac{1}{2}\Big[\sum_i \chi_{\mathbf{k}_1} (a_i^{-1} \{\alpha|\mathbf{a}\} a_i)\delta_{\alpha i,i} \sum_{i''} \chi_{\mathbf{k}_1} (a_{i''}^{-1} \{\alpha|\mathbf{a}\} a_{i''})\delta_{\alpha i'',i''} +$$

$$+ \sum_i \chi_{\mathbf{k}_1} (a_i^{-1} \{\alpha|\mathbf{a}\}^2 a_i)\delta_{\alpha^2 i,i}\Big] \qquad 2.7.24$$

This can be used to compute the number of times the identity representation occurs in the reduction of the representation $\Gamma'^{*}\times\Gamma_{(2)}$ appropriate to 2.7.13 in this case. Using 2.7.16a and the corresponding quantity

$$(a_i^{-1}\{\alpha|a\}^2 a_i) = \{E|\alpha_i^{-1}(R_n+\alpha R_n)\}(a_i^{-1}a^2 a_i) \qquad 2.7.25$$

we have

$$N = \frac{1}{2gL}\Big[\sum_a\sum_{i'}\sum_{i''}\sum_i \chi'^{*}_{k_1'}(a'^{-1}_{i'}aa'_{i'})\chi_{k_1}(a_i^{-1}aa_i)\chi_{k_1}(a_{i''}^{-1}aa_{i''})\delta_{\alpha i',i'}\delta_{\alpha i,i}\delta_{\alpha i'',i''}\times$$

$$\times\sum_n e^{i(-\alpha'_{i'}k'_1+\alpha_i k_1+\alpha''_i k_1)\cdot R_n} + \qquad 2.7.26$$

$$+\sum_a\sum_{i'}\sum_i \chi'^{*}_{k_1'}(a'^{-1}_{i'}aa'_{i'})\chi_{k_1}(a_i^{-1}a^2a_i)\delta_{\alpha i',i'}\delta_{\alpha^2 i,i}\sum_n e^{i(-\alpha'_{i'}k'_1+\alpha_i k_1+\alpha^{-1}\alpha_i k_1)\cdot R_n}\Big]$$

The first of the two terms in the last equation has been already encountered in 2.7.19 and, apart from the factor 1/2, will give a contribution which was already computed in 2.7.23. The second term is different from zero only if

$$-\alpha'_{i'}k'_1 + \alpha_i k_1 + \alpha^{-1}\alpha_i k_1 = K_m \qquad 2.7.27$$

On the other hand, because of the $\delta_{\alpha i',i'}$, the sum over a is restricted to coset representatives which belong to $G(k'_{i'})$. The other restriction imposed by the second delta, $\delta_{\alpha^2 i,i}$, is implied by the k-selection rule 2.7.27 and need not be further considered. Conditions on indices i and i' imposed by 2.7.27 can be stated in a more convenient form if one defines

$$\alpha_i k_1 = \beta\alpha_0 k_1$$

$$\alpha^{-1}\alpha_i k_1 = \alpha^{-1}\beta\alpha_0 k_1 = \beta\alpha''_0 k_1 \qquad 2.7.28$$

$$\alpha'_{i'}k'_1 = \beta\alpha'_0 k'_1$$

where β belongs to G_0 and is a rotation associated with an element b of the space group. Since α is an element of $G_0(k'_{i'})$ it must be $\alpha k'_{i'} = k'_{i'}$

or, using the definition $k'_{i'} = \alpha'_{i'} k'_1$ and the third of 2.7.28,

$$\alpha\beta\alpha'_0 k_1 = \beta\alpha'_0 k'_1 \qquad 2.7.29$$

that is

$$(\beta^{-1}\alpha\beta) \in G_0(\alpha'_0 k'_1) \qquad 2.7.30$$

Besides, from the second of 2.7.28 it follows that

$$\alpha_0^{-1}(\beta^{-1}\alpha\beta)\alpha''_0 k_1 = k_1 \qquad 2.7.31$$

This means that

$$\alpha_0^{-1}(\beta^{-1}\alpha\beta)\alpha''_0 \in G_0(k_1) \qquad 2.7.32$$

or

$$(\beta^{-1}\alpha\beta) \in \alpha_0 G_0(k_1)\alpha''^{-1}_0 \qquad 2.7.33$$

Thus rather than summing over the allowed values of i and i', in the second term of 2.7.26, it is possible to sum over elements b which define allowed triads of the kind 2.7.27 starting from a given one written as in 2.7.20. Introducing in the second term of 2.7.26 the definition $a_i = ba_0$ and $a'_{i'} = ba'_0$, the sum over all pure translations can be carried out at once and one is left with

$$\frac{1}{2g}\sum_a \sum_b \chi'^{*}_{k'_1}(a'^{-1}_0 b^{-1} aba'_0)\ \chi_{k_1}(a_0^{-1} b^{-1} aba_0) \qquad 2.7.34$$

However from 2.7.30 and 2.7.33 it is evident that the list of elements $(\beta^{-1}\alpha\beta)$ does not depend on β. Hence, if g/h possible different triads are generated starting with 2.7.20, the sum over b in 2.7.34 gives just g/h times the contribution obtained from a given element b. Therefore the final result for N, as given in 2.7.26, is

$$N = \frac{1}{2h}\sum_a \chi'^*_{\mathbf{k}'_1}(a'^{-1}_0 aa'_0)\chi_{\mathbf{k}_1}(a''^{-1}_0 aa''_0)\chi_{\mathbf{k}_1}(a^{-1}_0 aa_0) \quad +$$

$$+ \frac{1}{2h}\sum_{\bar{a}} \chi'^*_{\mathbf{k}'_1}(a'^{-1}_0 \bar{a}a'_0)\chi_{\mathbf{k}_1}(a^{-1}_0 \bar{a}^2 a_0) \qquad 2.7.35$$

In 2.7.35 the sum over a concerns coset representatives defined by the intersection group $G_0(\alpha'_0\mathbf{k}'_1)\cap G_0(\alpha''_0\mathbf{k}_1)\cap G_0(\alpha_0\mathbf{k}_1)$. Coset representatives $\bar{a}$, appearing in the second term, must satisfy both 2.7.30 and 2.7.33 an do not, in general, form a group.

Results 2.7.23 and 2.7.35 will now be applied to the interaction of radiation with lattice vibrations in order to give a practical example of the application of selection rules. We will consider vibrational selection rules for infrared absorption or Raman scattering of a crystal. The discussion is restricted to the description of the interaction between a crystal and a radiation field in the electric dipole approximation. When the perturbation operator is expanded in this approximation, terms linear in the field vector of the radiation give rise to ordinary emission or absorption. Quadratic terms are connected to two photon effects, the most important of which for the present purposes is Raman scattering. Matrix elements involved in an optical process are of the form

$$\langle f|\hat{O}|i\rangle \qquad 2.7.36$$

where $\langle f|$ is the final state and $|i\rangle$ the initial state which will be taken as the ground state. For infrared dipole absorption $\hat{O} = \nabla$ = the gradient operator; for Raman scattering $\hat{O}$ is the polarizability tensor. In the assumption of infinite wavelength, these operators transform like irreducible representations of $G(\mathbf{0})$. If only one phonon is excited, one has one-phonon states which are described by a wave function

$$\psi = a^{\dagger}_{p\mathbf{k}_i}\psi_0 \qquad 2.7.37$$

Since ψ_0 is total symmetric, ψ transforms according to an irreducible representation of the full space group, as given by 2.6.17. The operator

$\hat{O}$ transforms like an irreducible representation for $\mathbf{k} = 0$. Hence, using the general selection rule, only one-phonon states with $\mathbf{k} = 0$ are infrared or Raman active.

Two-phonon states are described by a vibrational function

$$\psi = a^{\dagger}_{p\mathbf{k}_j} a^{\dagger}_{p''\mathbf{k}''_{j''}} \psi_0 \qquad 2.7.38$$

where the nomenclature is appropriate for two different stars k and k", with the two creation operators belonging to Γ and Γ'' of G. If the frequencies $\omega_{p\mathbf{k}_j}$ and $\omega_{p''\mathbf{k}''_{j''}}$ are different, ψ transforms like the direct product $\Gamma\times\Gamma''$ and will be reduced with selection rules given in 2.7.23. This is certainly the case if $\mathbf{k}_j$ and $\mathbf{k}''_{j''}$ belong to two different stars but, even if this is not the case, it is still possible that the two creation operators belong to different irreducible sets. In these cases the two-phonon states correspond to combination vibrations. Instead, if the two creation operators belong to the same irreducible set, the vibrational function transforms like the symmetrized Kronecker product, $\Gamma_{(2)}$, and the corresponding state is classified as overtone vibration.

For two-phonon combination processes a wave function 2.7.38 can be introduced, as the final state, in 2.7.36 and the number of occurences of the identity representation is given by 2.7.23. By assumption, the operator $\hat{O}$ transforms according to the irreducible representation Γ' induced by $\mathbf{k} = 0$. Therefore the obvious choice α'_0 = identity is made and the same convenient choice can be made for α''_0 (or for α_0) in order to reduce the **k**-selection rule 2.7.11b to

$$\mathbf{k}''_1 + \alpha_0\mathbf{k}_1 = \mathbf{K}_m \qquad 2.7.39$$

The number of occurrences of the identity representation is thus

$$N = \frac{1}{h}\sum \chi'^{*}_0(a)\ \chi''_{\mathbf{k}''_1}(a)\chi_{\mathbf{k}_1}(a_0^{-1}aa_0) \qquad 2.7.40$$

where the sum is over the space group coset representatives defined by the intersection group $G_0(\mathbf{k}''_1)\cap G_0(\alpha_0\mathbf{k}_1)$ of order h. In 2.7.40 $\chi'_0(a)$ can be taken as $\chi'(\alpha)$ which is the character for the corresponding element in an irreducible representation of the point group G_0 associated with the space group G.

If $\mathbf{k}_1$ and $\mathbf{k}''_1$ are vectors of different stars, it is possible to select $\mathbf{k}_1$ such that also α_0 coincides with the identity. Then necessarily $\mathbf{k}''_1 = -\mathbf{k}_1 + K_m$, and one star is the inverse of the other. As the two star are supposed to be different, G_0 cannot contain the inversion element. If $\mathbf{k}_1$ and $\mathbf{k}''_1$ are vectors of the same star, one possibility is that $\mathbf{k}_1 \equiv \mathbf{k}''_1$ and α_0=identity: this can happen only if $\mathbf{k}_1 = \mathbf{0}$ or if it has components such that $2\mathbf{k}_1 = K_m$, corresponding to particular points on the surface of the Brillouin zone. The second possibility is that the star has an even number of points: for any $\mathbf{k}$ there is a corresponding $-\mathbf{k}$ with α_0 defined, like the inversion, as the element which moves each $\mathbf{k}_i$ into $-\mathbf{k}_i$.

For overtone vibrations $\mathbf{k}_1$ and $\mathbf{k}''_1$ must be necessarily points of the same star hence, as it was remarked above, two cases are possible: $\mathbf{k}_1 \equiv -\mathbf{k}_1$ and $\mathbf{k}_1 \not\equiv -\mathbf{k}_1$. In the first case a_0=identity, the intersection group coincides with $G_0(\mathbf{k}_1)$ and elements $(\beta^{-1}\alpha\beta)$, see 2.7.30 and 2.7.33, are also members of this group. It follows that

$$N = \frac{1}{2h} \sum_{\alpha \in G_0(\mathbf{k}_1)} \chi_0'^{*}(a)\{[\chi_{\mathbf{k}_1}(a)]^2 + \chi_{\mathbf{k}_1}(a^2)\} \qquad 2.7.41$$

In the second case a_0 is different from the identity and 2.7.35 reduces to

$$N = \frac{1}{2h}\sum_{\alpha \in G_0(\mathbf{k}_1)} \chi_0'^{*}(a)\chi_{\mathbf{k}_1}(a)\chi_{\mathbf{k}_1}(a_0^{-1}aa_0) + \frac{1}{2h}\sum_{\alpha \in \alpha_0 G_0(\mathbf{k}_1)} \chi_0'^{*}(a)\chi_{\mathbf{k}_1}(a_0^{-1}a^2a_0)$$
$$2.7.42$$

Time reversal symmetry may also have effect on selection rules, which were derived here using the invariance of matrix elements 2.7.1 under unitary transformations of the group. Time reversal is, however, an anti unitary operator which cannot leave the integral unchanged and thus, in general, cannot produce additional selection rules. If, due to time reversal, the wave function of the final state is related to that of the initial state, and if these two sets of functions span the same space, then additional selection rules may result. For a more detailed discussion of this aspect, the reader is referred to a detailed mathematical treatment given in Refs. 25 and 37.

CHAPTER 3

INTERMOLECULAR POTENTIALS

3.1 THE CRYSTAL POTENTIAL

A basic problem in the dynamics of molecular crystals is the nature of the crystal potential. The occurrence in the crystal of well-defined chemical entities, the molecules themselves bound together by weak intermolecular forces, can be correctly represented only by a crystal potential in which the intra- and the intermolecular contributions are separated in order to play different roles in determining the crystal properties.

We have already anticipated in Chapter 1 that a convenient form of the crystal potential is of the type

$$V = V_M + V_I \tag{3.1.1}$$

where V_M and V_I represent the intra- and the intermolecular contributions, respectively.

The total intramolecular potential V_M is the sum of all the intramolecular contributions $V(m\mu)$ of the molecules in the crystal and has thus the form

$$V_M = \sum_{m\mu} V(m\mu) \tag{3.1.2}$$

where $V(m\mu)$ is the intramolecular potential of molecule μ in the unit cell m. In principle the potential $V(m\mu)$ is different from the intramolecular potential of a truly isolated molecule in the gas phase at very low pressure. $V(m\mu)$ represents in fact the intramolecular potential of the molecule with the effective geometry and electronic structure that it possesses at the site μ in the crystal, under the influence of the surrounding molecules. In many cases the structural changes are negligibly small and one can reasonably assume that the same is true for the internal potential. In most of the cases, however, minor variations in the bond lengths and angles are observed between the gas and the crystal and these are indicative of definite variations in the electronic structure and thus in the force field. Finally in a limited number of cases the gas phase structure is very different from the structure that is stable in the crystal, especially for systems with low barriers to the rotation around a single bond. These considerations show that great care must be exerted in transferring the intramolecular potential from the gas to the crystalline phase and that this can be an acceptable procedure only if one is satisfied with a qualitative fit of the internal spectrum. For precise calculations the intramolecular potential must be derived using the crystal frequencies with all the difficulties that this procedure implies.

The intermolecular potential V_I includes in principle all the pairwise interactions as well as three- and multi-body interactions between different molecules in the crystal. Owing to the fact that three- and multi-body interactions are, except for the specific case of very simple crystals such as those of the rare gases, extremely weak, it is normal practice to neglect them in dynamical calculations considering only the pairwise interactions. We shall later discuss three-body forces that in some molecular crystals display small but not negligible effects. Apart from these cases, the use of pairwise interactions only represents an excellent approximation of the intermolecular potential. For this reason we can express the total intermolecular potential in the form

$$V_I = \frac{1}{2}\sum_{m\mu}\sum_{n\nu}{}' V\binom{m\mu}{n\nu} \qquad 3.1.3$$

where $V\binom{m\mu}{n\nu}$ represents the interaction potential between molecule $m\mu$ and molecule $n\nu$ and where the prime indicates that the sum over $n\nu$ omits the term $m\mu$.

By substitution of 3.1.1 in the definition 1.6.14 of the harmonic force constants we obtain

$$F_{\ell\ell'}\binom{m\mu}{n\nu} = K_{\ell\ell'}\delta_{mn}\delta_{\mu\nu} + (\partial^2 V_I/\partial R_\ell^{m\mu}\partial R_{\ell'}^{n\nu})_0 \qquad 3.1.4$$

where $K_{\ell\ell'}$ are intramolecular force constants defined as

$$K_{\ell\ell'} = (\partial^2 V_M / \partial R_\ell^{m\mu} \partial R_{\ell'}^{m\mu})_0 \qquad 3.1.5$$

which are equal to zero for $\ell, \ell' = 1,2,\ldots,6$,i.e. if at least one of the molecular coordinate is external. In the particular case $m\mu = n\nu$, we obtain from 3.1.3 and 3.1.4

$$F_{\ell\ell'}\binom{m\mu}{m\mu} = K_{\ell\ell'} + \sum_{n\nu}{}' [\partial^2 V\binom{m\mu}{n\nu} / \partial R_\ell^{m\mu} \partial R_{\ell'}^{m\mu}]_0 \qquad 3.1.6$$

This relation shows clearly the role that the intermolecular potential plays in the block of the dynamical matrix involving the internal coordinates. It acts actually as a perturbation on the intramolecular force constants giving rise to shifts and splittings of the internal frequencies. Furthermore it couples internal and external modes since, according to 3.1.4, non-zero terms will appear in the blocks of the dynamical matrix involving one internal and one external coordinate.

The intermolecular potential is instead completely responsible of the external frequencies since in the corresponding block of the dynamical matrix all the $K_{\ell\ell'}$ are zero.

By substitution of 3.1.1 in the crystal equilibrium conditions 1.6.15 and using 3.1.4 and 3.1.6 we obtain

$$[\partial V(m\mu) / \partial R_\ell^{m\mu}]_0 + \sum_{n\nu}{}' [\partial V\binom{m\mu}{n\nu} / \partial R_\ell^{m\mu}]_0 = 0 \qquad 3.1.7$$

For an isolated molecule in the gas phase the first derivatives of $V\binom{m\mu}{n\nu}$ with respect to internal coordinates are all exactly zero since they define the equilibrium conditions of the free molecule. In the crystal, however, this is not necessarily true since the equilibrium conditions 3.1.7 involve the total crystal potential and not separately the intra- and the intermolecular part. In the crystal the molecule may well be under strain as long as the linear terms of the intramolecular potential are exactly cancelled by the linear terms of the intermolecular potential as required by 3.1.7. The implications of the equilibrium conditions 3.1.7 are discussed in Reference 70.

3.2 THE INTRAMOLECULAR POTENTIAL

The intramolecular potential $V(m\mu)$ is known in analytical form only for diatomic and for a few triatomic molecules. For polyatomic molecules the intramolecular potential is normally expanded in power series of the internal coordinates $S_\ell^{m\mu}$ in the same way as done for a free molecule. We then have

$$V(m\mu) = \sum_\ell K_\ell S_\ell^{m\mu} + \frac{1}{2} \sum_{\ell\ell'} K_{\ell\ell'} S_\ell^{m\mu} S_{\ell'}^{m\mu} + \frac{1}{3!} \sum_{\ell\ell'\ell''} K_{\ell\ell'\ell''} S_\ell^{m\mu} S_{\ell'}^{m\mu} S_{\ell''}^{m\mu} + \frac{1}{4!} \sum_{\ell\ell'\ell''\ell'''} K_{\ell\ell'\ell''\ell'''} S_\ell^{m\mu} S_{\ell'}^{m\mu} S_{\ell''}^{m\mu} S_{\ell'''}^{m\mu} + \ldots \qquad 3.2.1$$

where

$$K_\ell = [\partial V(m\mu)/\partial S_\ell^{m\mu}]_0$$

$$K_{\ell\ell'} = [\partial^2 V(m\mu)/\partial S_\ell^{m\mu} \partial S_{\ell'}^{m\mu}]_0 \qquad 3.2.2$$

$$K_{\ell\ell'\ell''} = [\partial^3 V(m\mu)/\partial S_\ell^{m\mu} \partial S_{\ell'}^{m\mu} \partial S_{\ell''}^{m\mu}]_0$$

etc.

As discussed in the previous Section, the linear terms K_ℓ are not necessarily zero since the molecule can be in a strained situation under the influence of the surrounding molecules in the crystal. This is not a serious problem for the solution of the equations of motion of the crystal since in the total crystal potential they are exactly cancelled by the linear terms of the intermolecular potential. For this reason the linear terms in the expansion 3.2.1 are always neglected in actual calculations. In the harmonic approximation we neglect also cubic and higher terms and write the intramolecular potential in the well-known form

$$V(m\mu) = \frac{1}{2} \sum_{\ell\ell'} K_{\ell\ell'} S_\ell^{m\mu} S_{\ell'}^{m\mu} \qquad 3.2.3$$

The calculation of the harmonic force constants $K_{\ell\ell'}$ is a standard problem of vibrational spectroscopy[11,13] and will not be discussed here. In the following discussion we shall therefore assume that the intramolecular potential is always known.

3.3 THE INTERMOLECULAR POTENTIAL

Analogous to the intramolecular potential one could use for the intermolecular potential a "brute force" approach and develop V_I in a power series of the molecular coordinates. The only difference with the previous case is that the intermolecular potential is a function of both the internal and the external coordinates of the molecules and that the derivatives with respect to two molecular coordinates on two different molecules are non-zero in this case.

This approach actually used by some authors[7,44,45] for the calculation of crystal frequencies does not involve any assumption on the nature of the intermolecular potential and has the advantage of extreme simplicity. It has, however, two main disadvantages that strongly limit its interest. The first is that the dynamical properties of the crystal remain completely uncorrelated with other crystal properties such as the structure and the energy. The second is that the number of intermolecular force constants that one has to deal with is extremely high and drastic assumptions have to be made to reduce this number to a practical one. This amounts either to take into account only interactions between next neighboring molecules or to impose relations between force constants of the same type but involving molecules at different distances in the crystal.

A more convenient approach is to make full use of the results of the theory of intermolecular forces which furnishes suitable analytical expressions for the intermolecular potential. This permits not only to use the same potential for many physical properties of the crystal at the same time and thus to correlate static and dynamical behaviors of molecular solids but also to reduce the number of adjustable parameters. In most recent studies on the dynamics of molecular crystals it is thus normal practice to adjust analytical intermolecular potentials to reproduce correctly not only all the vibrational frequencies of the crystal but also at least the crystal structure and energy. If the data are available, an even larger range of crystal properties can be considered including sound velocity, crystal compressibility coefficients, the Grünaisen parameters, elastic moduli, etc.

The theory of intermolecular forces permits a number of physical mechanisms that contribute to the total intermolecular interaction to be distinguished. A detailed discussion of the general theory of intermolecular forces is beyond the scope of this book. Here we are essentially concerned with the dynamical properties of molecular crystals and thus the discussion will be limited to those aspects of the theory that are peculiar of the crystalline state and to a critical survey of models of the intermolecular potentials that are suited and have been actually used for the calculation of static and dynamic properties of molecular crystals.

A convenient classification of intermolecular interactions between neutral molecules is obtained from the standard perturbation theory[40-43]. Accordingly we classify each contribution to the total intermolecular interaction in terms of its specific nature and of its range of activity. This classification is summarized in Table 3.1

Table 3.1 Classification of intermolecular forces

Interaction	Range	Dependence on r
Repulsive	short	e^{-Br} or r^{-m} $(m > 9)$
Dispersion	medium	r^{-6}
Polarization	medium-short	r^{-6} r^{-7} r^{-8} etc
Electrostatic	long to short	r^{-3} r^{-4} r^{-5} r^{-6} etc

The first-order perturbation theory furnishes an interaction term that corresponds to the classical electrostatic interaction between the char distributions of the molecules. From the second-order perturbation theory, we obtain instead the repulsive as well as the dispersion and polarization interactions. A detailed discussion on the perturbation trea ment can be found in Refs. 40 - 43. We notice here that the division int short, medium and long-range forces is quite arbitrary. Most of the authors use, for instance, only the short- and long-range classification. Here by short-range we mean that the interaction occurs only between adjacent molecules, by medium-range that it may involve also the second next-neighbors and by long-range that the interaction extends far away in the crystal.

According to this schematic classification of the interactions, the total molecule-molecule interaction potential between molecule A ($A = m\mu$) and molecule B ($B = n\nu$) is given by

$$V^{AB} = V^{AB}_{rep} + V^{AB}_{disp} + V^{AB}_{pol} + V^{AB}_{el} \qquad 3.3.1$$

where the four terms represent the repulsive, the dispersion, the polarization (or induction), and the electrostatic interactions, respectively. We discuss now briefly analytical forms of these interaction potentials that are normally utilized in the calculation of the dynamical properties of molecular crystals.

a) Repulsive and dispersion interactions.

The repulsive and dispersion interactions are the most difficult ones to be represented in analytical form. In principle these interactions can be calculated from a knowledge of the electronic distribution in the molecules. Detailed quantum-mechanical calculations have actually been made since more than thirty years but only recently, thanks to the development of large electronic computers, accurate results were obtained with ab-initio methods. These calculations are extremely difficult since they involve the use of very extended orbital bases and the inclusion of a very large number of configuration interactions. Furthermore, in order to apply the ab-initio potentials to the calculation of phonon frequencies, it is necessary to evaluate the interaction energy for a variety of intermolecular distances and relative molecular orientations and to fit then functional expressions to these energies. The ab-initio calculations are certainly very promising but for the moment are limited to very small molecular systems and are far from being easily extended to large molecules. For this reason semiempirical interaction potentials are normally utilized in lattice dynamics calculations. The theory furnishes actually concrete indications of both the functional form and the r dependence of phenomenological potentials which reasonably account for the repulsive and dispersion interactions. These are always treated together and thus we shall use a comprehensive symbol for both

$$V_{vdw}^{AB} = V_{rep}^{AB} + V_{disp}^{AB} \qquad 3.3.2$$

collecting them under the general name of Van der Waals interactions.

Most of the molecule-molecule potentials of this type originate from the simpler case of the interaction between two neutral atoms. We shall therefore briefly discuss this case first.

Several analytical functions have been proposed for the description of the Van der Waals interaction between atoms. A well-known expression is the so-called Lennard-Jones potential[40-43,46-50]

$$V_{vdw}^{ij} = 4\varepsilon_{ij}[(\rho_{ij}/r_{ij})^{n} - (\rho_{ij}/r_{ij})^{6}] \qquad 3.3.3$$

where ε_{ij} is the depth of the potential well, r_{ij} the distance between atom i and atom j and ρ_{ij} the value of r_{ij} for which $V_{vdw} = 0$.

The attractive part of the potential, which is proportional to r_{ij}^{-6}, represents the induced dipole-induced dipole interaction due to the fluctuating electrons in the atoms and describes the dispersion interaction. For mathematical convenience the repulsive contribution is simulated by an inverse power term. The exponent n is normally taken equal to 12 but values ranging from 9 to 13 have been used in the literature in several cases [51].

The best known analytical potential for the interaction between two atoms is, however, the modified Buckingham or 6-exp potential[40-43,46-50]

$$V_{vdw}^{ij} = \frac{1}{\mu_{ij} - 1}\,\varepsilon_{ij}\{\exp[-\,6\mu_{ij}(r_{ij}\rho_{ij}^{-1} - 1)] - \mu_{ij}\rho_{ij}^{6}r_{ij}^{-6}\} \qquad 3.3.4$$

often written in the simpler form

$$V_{vdw} = A_{ij}\exp(-B_{ij}r_{ij}) - C_{ij}r_{ij}^{-6} \qquad 3.3.5$$

In 3.3.4 ε_{ij} is the equilibrium depth of the potential, ρ_{ij} the equilibrium distance and μ_{ij} the steepness of the potential. The coefficients A_{ij}, B_{ij} and C_{ij} are related to ε_{ij}, ρ_{ij} and μ_{ij} by the expressions

$$A_{ij} = [\varepsilon_{ij}/(\mu_{ij}-1)]\exp(6\mu_{ij})$$

$$B_{ij} = 6\mu_{ij}/\rho_{ij} \qquad 3.3.6$$

$$C_{ij} = \varepsilon_{ij}\mu_{ij}\rho_{ij}^{6}/(\mu_{ij} - 1)$$

In principle, the coefficients A_{ij}, B_{ij} and C_{ij} can be calculated from the electronic structure of the interacting atoms. Calculations of this type have been recently made by Gordon and his group using the free electron gas model [118]. In most of the actual calculations of phonon frequencies of rare gas crystals these coefficients have been, however, taken as empirical parameters and their numerical values were obtained by refining potentials of the type 3.3.3 or 3.3.6 for the best fit of experimental data.

Several approaches have been utilized to describe the van der Waals interaction between molecules. Small diatomic or even triatomic molecules are often treated as single atoms using molecule-molecule potentials of the type described above for atoms. The molecular shape is taken into account by introducing appropriate angular dependences of the coefficients. A typical molecule-molecule potential, limited essentially to small molecular systems, has been proposed by Kihara[42,52,94] and has been widely utilized by his coworkers. It is obtained by assigning to the molecules a given simplified shape, assumed as rigid and impenetrable (hard core), and by treating the interaction in terms of an atom-atom potential in which the atom-atom distance is replaced by the shortest distance between the two hard cores. Spherical- rod- or spindle-like shapes have been utilized. A difficulty inherent in the model is just the analytical representation of the shortest distance between two hard cores arbitrarily oriented. For instance, the shortest distance between the two spindle-like hard cores of Fig.3.1 is a complicated function of

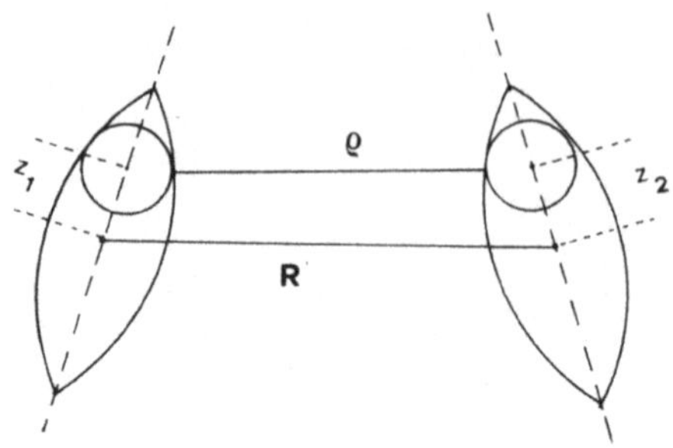

Fig.3.1 Interaction between spindle-like hard cores

distance R between the centers of mass, of distances z_1 and z_2 shown in figure, of length l of the spindles and of polar angles Θ and ϕ giving the orientation of the spindle axis with respect to the crystal-fixed axes.

Another molecule-molecule potential has been recently proposed by Berne and Pechukas [53]. The model assimilates the molecules to spheroids with a charge distribution given by spheroidal Gaussians. The repulsive part of the potential is then assumed to be proportional to the intermolecular overlap volume. Along the same line, Luty has recently [93] proposed another type of molecule-molecule potential in which the electronic charge density function of a molecule is represented by a superposition of functions centered on the nuclei. Other more or less sophisticated molecule-molecule potentials have been proposed for specific cases but none of them is of general validity and can be easily extended to complex polyatomic molecules.

Intermolecular potentials of general validity must actually fulfill two basic requirements. They must be simple and easily transferable from one molecule to another and they must possess analytical forms in which the number and the nature of the constituent atoms is clearly evidentiated. By far the most popular intermolecular potentials are based on the assumption that the molecule-molecule interaction can be partitioned into a sum of atom-atom contributions involving all the interactions between each atom of one molecule and all atoms of the other molecule. The molecule-molecule potential is thus expressed in the form [54,63]

$$V^{AB}_{vdw} = \sum_i \sum_j V\binom{Ai}{Bj} \qquad \begin{array}{l} i \text{ belonging to A} \\ j \text{ belonging to B} \end{array} \qquad 3.3.7$$

where $V\binom{Ai}{Bj}$ is an atom-atom potential of the type described before.

The validity of this assumption has been discussed by many authors. For instance, Longuet-Higgins and Salem [46] have shown that when the molecules have well localized orbitals and when the electronic correlation dies rapidly with the distance, the repulsive and dispersion interactions are locally additive. When, however, the molecules have delocalized orbitals which extend over the whole molecular framework, the local additivity is partially lost. Despite the drastic assumption involved in this model, there are few doubts that the atom-atom potentials represent the simplest and yet more powerful type of interaction potential that can be used for molecular systems. They have been widely utilized, either in conjunction with various types of electrostatic potentials or, more often, alone as representative of the total intermolecular potential to calculate a large number of crystal properties including crystal structures and energies, harmonic and anharmonic phonon frequencies,phonon lifetimes and phonon-phonon interactions .

A convenient although oversimplifying feature of the atom-atom potentials is that the parameters A_{ij},B_{ij} and C_{ij} are taken as characteristic of each pair of atoms, regardless of the molecular environment of the atoms. On this basis, many authors have proposed sets of potential parameters that, to a great extent, are transferable from one molecule to another, within classes of similar compounds. The most widely known sets of potential parameters are those proposed by Williams for aliphatic [47] and aromatic [48] hydrocarbons and by Kitaigorodsky [49,50] for several molecular systems. Tables of potential parameters are given by these authors.

In several practical applications of these phenomenological potentials, the atom-atom distance is taken as the distance between the atomic centers. In other cases, especially in the original papers of Williams [47,48] the interaction centers are slightly shifted in the bonds in order to account for the deplacement of the electron density in the bond. As discussed before, the potential parameters are obtained by refinement procedures to the best fit of crystal properties. It has been recently suggested [55] that in the refinement of the parameters, it is more convenient to use expression 3.3.4 instead of 3.3.5.

The main drawback of the atom-atom potential is its central nature that produces a completely isotropic interaction around the atoms. This is physically unrealistic for atoms linked by chemical bonds and more so for atoms with lone pairs or involved in delocalized π electron systems. Some attempts to develop anisotropic atom-atom potentials have been made very recently for simple crystals like solid nitrogen containing diatomic molecules. For instance, Filippini et al. [56] have proposed an anisotropic potential in which the anisotropy is introduced by multiplying the parameter A of the repulsive part by an appropriate cosine function. This treatment can be generalized to more complex systems by multiplying the potential parameters by appropriate combinations of Legendre polynomials that reproduce the desired shape of potential curves around the atoms. Quantum-mechanical calculations [141] have shown that the expansion in Legendre polynomials must be extended at least to the polynomial of order 4 to obtain a satisfactory description of the angular dependence.

Another type of anisotropic repulsive potential has been proposed by Raich et al.[57] as an extension of the Kihara potential. In this model the shape of a diatomic molecule is defined by three spherical cores, two located on the atoms and one in the center of the bond. This repulsive three-center potential has the form

$$V^{AB}_{rep} = \sum_i \sum_j \exp\{ - \alpha [r_{ij} - R^c_i - R^c_j]\} \qquad \begin{matrix} i \text{ belonging to A} \\ j \text{ belonging to B} \end{matrix} \qquad 3.3.8$$

where r_{ij} is the distance between two cores on different molecules and R^c_i the radius of the ith core. Also this type of multicore potential, which simulates the anisotropy by adding another interaction center in the bond, could be generalized to polyatomic molecules using for each atom X involved in a bond a minimum of two cores and possibly even three if there are lone pairs on the atom.

More sophisticated analytical functions can also be used for the attractive part of the potential. The second-order perturbation theory actually shows that, in addition to the leading term in r_{ij}^{-6}, which accounts for the second-order dipole-dipole interactions, terms in r_{ij}^{-8} and r_{ij}^{-10} should also be considered in 3.3.5. These terms originate from second-order dipole-quadrupole and quadrupole-quadrupole interactions, respectively. Their inclusion in the attractive part of the potential overemphasizes the inconvenient feature of the Buckingham potential of going to $-\infty$ when $r_{ij} \to 0$. In order to avoid this overestimate of the attraction at very short distances, a damping function can be introduced into the potential. A more general form of the atom-atom potential is then

$$V_{vdw}^{ij} = A_{ij}\exp(-B_{ij}r_{ij}) - f(r_{ij})\left[C_6^{ij}r_{ij}^{-6} + C_8^{ij}r_{ij}^{-8} + C_{10}^{ij}r_{ij}^{-10}\right] \qquad 3.3.9$$

where C_6^{ij}, C_8^{ij} and C_{10}^{ij} are adjustable parameters and $f(r_{ij})$ is a damping function. A convenient form for $f(r_{ij})$ is

$$\begin{aligned} f(r_{ij}) &= 1 && \text{for } r_{ij} > r_{ij}^0 \\ f(r_{ij}) &= \exp\left[-\alpha(r_{ij}^0 r_{ij}^{-1} - 1)^m\right] && \text{for } r_{ij} < r_{ij}^0 \end{aligned} \qquad 3.3.10$$

where α and m are integers and r_{ij}^0 is the value of r_{ij} at the potential minimum. Values of m equal to 2 and 3 are normally used. The parameter α is chosen to weight the damping efficiency.

The potential 3.3.9 is clearly more flexible than the simpler potential 3.3.5. It introduces, however, two additional parameters and these are not always well determined for complex systems. For this reason, expression 3.3.5 is normally preferred.

b) Electrostatic interactions

The electrostatic interaction between the charge distributions on the molecules in the crystal is correctly described by classical electrostatic theory. The simplest approach is to describe the charge distribution of a molecule by localizing charges or fractions of charges (often called "effective charges") on the atoms or in the bonds [51,58] and to use the Coulomb law. If the charges are located on the atoms, the electrostatic potential is partitioned, as the atom-atom potential discussed before, into atomic contributions

$$V_{el}^{AB} = \sum_i \sum_j q_i q_j / r_{ij} \qquad \begin{matrix} i \text{ belonging to A} \\ j \text{ belonging to B} \end{matrix} \qquad 3.3.11$$

where q_i and q_j are the effective charges on the atoms and r_{ij} is the same atom-atom distance used in the atom-atom potential. If the charges are located in the bonds, the expression of the interaction potential is the same, but r_{ij} is now a different distance.

Instead of charges one can locate point dipoles on the atoms or in the bonds. In this case, the electrostatic potential between two molecules becomes [59,60]

$$V_{el}^{AB} = \sum_i \sum_j (\mu_i \mu_j \cos\alpha - 3\mu_i \mu_j \cos\beta \cos\gamma)/r_{ij}^3 \qquad 3.3.12$$

where β and γ are the angles between the directions of the dipoles μ_i and μ_j and the distance r_{ij}, respectively, and α is the torsion angle between the planes defined by the two dipoles and by r_{ij}.

The use of an electrostatic potential of the type 3.3.11 or 3.3.12 is undoubtedly simple but involves two basic difficulties. The first is that the charge or the dipole partitioning on the atoms or in the bonds is somewhat arbitrary and in any case it is difficult to reproduce in this way the charge distribution of the molecule. The second difficulty that arises in the use of these potentials is that, especially the effective charge potential 3.3.11, they converge very slowly and thus crystal sums have to be extended to very large distances.

Another convenient approach consists in representing the charge distribution of the molecules by means of an expansion in terms of point multipoles located at a suitable origin in the molecule. This has the advantage that the molecular multipoles are often available experimentally or can be calculated by quantum-mechanical techniques.

We shall describe the multipole expansion potential in some detail for several reasons. The first is that the electrostatic theory furnishes a precise and convenient formulation of the interaction potential. The second is that exactly the same formalism can be used if one whishes to locate multipoles on the atoms or in the bonds. The third is that for small or even medium size molecules this represents the most convenient description of the electrostatic interaction with the great advantage that the specific contribution of each multipole-multipole term is clearly identified.

The most compact formulation of the electrostatic interaction is due to Jansen [43,61-64, 86]. In terms of Jansen's formalism, the electrostatic interaction between two molecules is given by

$$V_{el}^{AB} = \sum_{0}^{\infty}{}_{m}\sum_{0}^{\infty}{}_{n}(-1)^{m}\left[(2m-1)!!(2n-1)!!\right]^{-1}V_{mn}^{AB} \tag{3.3.13}$$

where m is the order of the multipole on molecule A, n is the order of the mutipole on molecule B, V_{mn}^{AB} is the multipole-multipole potential, and

$$(2m-1)!! = (2m-1)(2m-3)(2m-5)\ldots(1) \tag{3.3.14}$$

The multipole-multipole potential V_{mn}^{AB} is given by [63,86]

$$V_{mn}^{AB} = M_{m}^{A}\cdot T_{m+n}^{AB}\cdot M_{n}^{B} \tag{3.3.15}$$

where M_{m}^{A} and T_{m+n}^{AB} are tensors of rank m and m+n respectively and where the dots mean inner products of tensors.

The tensor T_{s}^{AB} is defined as

$$T_{s}^{AB} = \nabla^{s}R_{AB}^{-1} \tag{3.3.16}$$

where R_{AB} is the intermolecular distance and ∇ the operator $\partial/\partial\mathbf{R}_{AB}$, $\mathbf{R}_{AB}$ being the vectorial distance from A to B.

The zero-order multipole moment M_0 represents the net charge on the molecule and is different from zero only for ions. The first-order moment M_1 is the dipole moment of the molecule. Higher order moments are called quadrupole, octupole, hexadecapole etc. The nomenclature for the interaction terms is summarized in table 3.2 where the first few terms of the expansion are collected. Terms higher than the quadrupole-quadru-

Table 3.2 Multipole-Multipole interaction terms

m	n	M_m^A	M_n^B	T_{m+n}^{AB}	Name	
0	0	M_0	M_0	T_0	charge-charge	R^{-1}
0	1	M_0	M_1	T_1	charge-dipole	R^{-2}
1	0	M_1	M_0	T_1		
....						
1	1	M_1	M_1	T_2	dipole-dipole	R^{-3}
1	2	M_1	M_2	T_3	dipole-quadrupole	R^{-4}
2	1	M_2	M_1	T_3		
1	3	M_1	M_3	T_4	dipole-octupole	R^{-5}
3	1	M_3	M_1	T_4		
2	2	M_2	M_2	T_4	quadrupole-quadrupole	R^{-5}

pole are normally neglected in the expansion and will not be discussed here.

The normal molecular crystals of van der Waals type consist of neutral molecules so that charge-multipole terms are not involved in the electrostatic interaction. If the molecules possess a static dipole moment the leading term of the electrostatic potential is the dipole-dipole interaction. For molecules with a center of symmetry, even the dipole-dipole interaction is missing and the first term in the expansion is the quadrupole-quadrupole interaction. For a pair of neutral molecules, the electrostatic interaction potential is then,according to 3.3.13

$$V_{el}^{AB} = -M_1^A \cdot T_2^{AB} \cdot M_1^B + \frac{1}{3}(M_2^A \cdot T_3^{AB} \cdot M_1^B - M_1^A \cdot T_3^{AB} \cdot M_2^B) + \frac{1}{9} M_2^A \cdot T_4^{AB} \cdot M_2^B +$$

$$- \frac{1}{15}(M_1^A \cdot T_4^{AB} \cdot M_3^B + M_3^A \cdot T_4^{AB} \cdot M_1^B) + \ldots \qquad 3.3.17$$

The multipole moments M_m can be defined also in a compact form as

$$M_m = (-1)^m (m!)^{-1} \sum_i e_i r_i^{2m+1} T_m^i \qquad 3.3.18$$

where the sum is taken over all charges e_i in the molecule. Here r_i represents the distance of the charge e_i from a convenient origin within the molecule. For obvious reasons, the origin is generally taken at the molecular center of mass. In agrement with 3.3.13 the tensors T_m^i are defined as

$$T_m^i = \nabla^m r_i^{-1} \qquad 3.3.19$$

For the sake of clarity, we write in full the Cartesian components of the first few multipole terms in the crystal-fixed reference system. If we denote by X_α^i and X_α^0 the αth Cartesian coordinate of charge e_i and of origin O, respectively, the Cartesian components of r_i are given by

$$\Delta X_\alpha^i = X_\alpha^i - X_\alpha^0 \qquad 3.3.20$$

Therefore, from Eqs. 3.3.18 and 3.3.19, we obtain

$$\begin{aligned}
&\text{charge} && M_0 = \sum_i e_i \\
&\text{dipole} && M_\alpha = \sum_i e_i \Delta X_\alpha^i \\
&\text{quadrupole} && M_{\alpha\beta} = \frac{1}{2} \sum_i e_i (3\Delta X_\alpha^i \Delta X_\beta^i - r_i^2 \delta_{\alpha\beta}) \\
&\text{octopole} && M_{\alpha\beta\gamma} = \frac{1}{2} \sum_i e_i [5\Delta X_\alpha^i \Delta X_\beta^i \Delta X_\gamma^i - r_i^2 (\Delta X_\alpha^i \delta_{\beta\gamma} + \Delta X_\beta^i \delta_{\alpha\gamma} + \Delta X_\gamma^i \delta_{\alpha\beta})]
\end{aligned} \qquad 3.3.21$$

where the order of the multipole terms is specified directly by the number of indices.

In the same way, the Cartesian components of the T_s^{AB} tensors defined by 3.3.16 are given by

$$T_{\alpha\beta\gamma\ldots\rho}^{AB} = \left(\frac{\partial}{\partial R_\alpha} \frac{\partial}{\partial R_\beta} \frac{\partial}{\partial R_\gamma} \cdots \frac{\partial}{\partial R_\rho} \right) \frac{1}{R_{AB}} \qquad 3.3.22$$

where

$$R_{AB} = [\sum_\alpha (R_\alpha^{AB})^2]^{\frac{1}{2}} \qquad 3.3.23$$

with

$$R_\alpha^{AB} = X_\alpha^B - X_\alpha^A \qquad 3.3.24$$

The explicit form of the Cartesian components of the first few tensors is then

$$T_\alpha = -R_\alpha R^{-3}$$

$$T_{\alpha\beta} = (3!!R_\alpha R_\beta - 1!!\delta_{\alpha\beta}R^2)R^{-5}$$

$$T_{\alpha\beta\gamma} = (-5!!R_\alpha R_\beta R_\gamma + 3!!\sum_3 R_\alpha \delta_{\beta\gamma} R^2)R^{-7} \qquad 3.3.25$$

$$T_{\alpha\beta\gamma\delta} = (7!!R_\alpha R_\beta R_\gamma R_\delta - 5!!\sum_6 R_\alpha R_\beta \delta_{\gamma\delta} R^2 + 3!!\sum_3 \delta_{\alpha\beta}\delta_{\gamma\delta}R^4)R^{-9}$$

$$T_{\alpha\beta\gamma\delta\varepsilon} = (-9!!R_\alpha R_\beta R_\gamma R_\delta R_\varepsilon + 7!!\sum_{10} R_\alpha R_\beta R_\gamma \delta_{\delta\varepsilon}R^2 - 5!!\sum_{15} R_\alpha \delta_{\beta\gamma}\delta_{\delta\varepsilon}R^4)R^{-11}$$

$$T_{\alpha\beta\gamma\delta\varepsilon\theta} = (11!!R_\alpha \ldots R_\theta - 9!!\sum_{15} R_\alpha \ldots R_\delta \delta_{\varepsilon\theta}R^2 + 7!!\sum_{45} R_\alpha R_\beta \delta_{\gamma\delta}\delta_{\varepsilon\theta}R^4 +$$

$$- 5!!\sum_{15} \delta_{\alpha\beta}\delta_{\gamma\delta}\delta_{\varepsilon\theta}R^6)R^{-13}$$

The numbers on each sum indicate the number of terms in the sum obtained by permuting all indices in the typical term that is explicitly written. Again, the rank of each tensor is indicated directly by the number of indices. For simplicity the superscript AB has been omitted. These tensor elements have the relevant properties

$$T_{\alpha\beta}^{AB} = T_{\alpha\beta}^{BA} \qquad T_{\alpha\beta\gamma\delta}^{AB} = T_{\alpha\beta\gamma\delta}^{BA} \quad \text{etc.}$$

$$3.3.26$$

$$T_{\alpha\beta\gamma}^{AB} = -T_{\alpha\beta\gamma}^{BA} \qquad T_{\alpha\beta\gamma\delta\varepsilon}^{AB} = -T_{\alpha\beta\gamma\delta\varepsilon}^{BA} \quad \text{etc.}$$

By performing all the inner tensor products we can now express the electrostatic potential 3.3.17 in terms of the Cartesian components of the multipoles and of the T tensors. We have

$$V_{el}^{AB} = -\sum_{\alpha\beta} M_\alpha^A T_{\alpha\beta}^{AB} M_\beta^B + \frac{1}{3}\sum_{\alpha\beta\gamma}(M_{\alpha\beta}^A T_{\alpha\beta\gamma}^{AB} M_\gamma^B - M_\gamma^A T_{\alpha\beta\gamma}^{AB} M_{\alpha\beta}^B) + \qquad 3.3.27$$

$$+ \frac{1}{9} \sum_{\alpha\beta\gamma\delta} M^A_{\alpha\beta} T^{AB}_{\alpha\beta\gamma\delta} M^B_{\gamma\delta} - \frac{1}{15} \sum_{\alpha\beta\gamma\delta} (M^A_{\alpha} T^{AB}_{\alpha\beta\gamma\delta} M^B_{\beta\gamma\delta} + M^A_{\beta\gamma\delta} T^{AB}_{\alpha\beta\gamma\delta} M^B_{\alpha}) + \ldots$$

The crystal-fixed multipoles M are not, however, convenient quantities to use. Instead it is preferable to use molecular multipoles, defined in the molecule-fixed reference system, which are molecular quantities experimentally measurable.

For the molecular multipoles we shall employ, according to the standard nomenclature, the symbols μ, Θ and Ω to indicate the dipole, the quadrupole and the octupole moments, respectively. The crystal-fixed multipoles are easily expressed in terms of the molecular multipoles, using the relation 1.5.8 connecting the two bases. We have

$$M^A_{\alpha} = \sum_{\rho} \Gamma^A_{\alpha\rho} \mu_{\rho}$$

$$M^A_{\alpha\beta} = \sum_{\rho\sigma} \Gamma^A_{\alpha\rho} \Gamma^A_{\beta\sigma} \Theta_{\rho\sigma} \qquad 3.3.28$$

$$M^A_{\alpha\beta\gamma} = \sum_{\rho\sigma\tau} \Gamma^A_{\alpha\rho} \Gamma^A_{\beta\sigma} \Gamma^A_{\gamma\tau} \Omega_{\rho\sigma\tau}$$

where $\Gamma^A_{\alpha\rho}$ represents the direction cosine between the crystal-fixed axis α and the molecule-fixed axis ρ of molecule A. The molecular multipoles have no molecular notation since they are the same for all molecules in the crystal.They are defined in the molecular reference system by relations analogous to 3.3.21 with the only difference that the Δx^i_{α} are replaced by the corresponding coordinates x^i_{ρ} (see Section 1.5).

Using 3.3.28 the electrostatic potential can be rewritten in the form

$$V^{AB}_{el} = - \sum_{\alpha\beta} \sum_{\rho\sigma} \Gamma^A_{\alpha\rho} \Gamma^B_{\beta\sigma} T^{AB}_{\alpha\beta} \mu_{\rho} \mu_{\sigma} +$$

$$+ \frac{1}{3} \sum_{\alpha\beta\gamma} \sum_{\rho\sigma\tau} (\Gamma^A_{\alpha\rho} \Gamma^A_{\beta\sigma} \Gamma^B_{\gamma\tau} - \Gamma^B_{\alpha\rho} \Gamma^B_{\beta\sigma} \Gamma^A_{\gamma\tau}) T^{AB}_{\alpha\beta\gamma} \mu_{\tau} \Theta_{\rho\sigma} +$$

$$+ \frac{1}{9} \sum_{\alpha\beta\gamma\delta} \sum_{\rho\sigma\tau\xi} \Gamma^A_{\alpha\rho} \Gamma^A_{\beta\sigma} \Gamma^B_{\gamma\tau} \Gamma^B_{\delta\xi} T^{AB}_{\alpha\beta\gamma\delta} \Theta_{\rho\sigma} \Theta_{\tau\xi} + \qquad 3.3.29$$

$$- \frac{1}{15} \sum_{\alpha\beta\gamma\delta} \sum_{\rho\sigma\tau\xi} (\Gamma^A_{\alpha\rho} \Gamma^B_{\beta\sigma} \Gamma^B_{\gamma\tau} \Gamma^B_{\delta\xi} + \Gamma^B_{\alpha\rho} \Gamma^A_{\beta\sigma} \Gamma^A_{\gamma\tau} \Gamma^A_{\delta\xi}) T^{AB}_{\alpha\beta\gamma\delta} \mu_{\rho} \Omega_{\sigma\tau\xi} + \ldots$$

This form of the electrostatic potential is very convenient for lattice dynamics calculations. We notice in fact that the T tensor elements depend only on the intermolecular distance which varies only during a translational motion of the molecule. On the other hand the direction cosines depend only on the relative position of the molecular with respect to the crystal-fixed reference system and change only during the rotational motions of the molecules. Finally, the molecular moments are a function of the charge distribution within the molecule and vary only with the internal vibrations.

The functional dependence of the T tensor components on the intermolecular distances and hence on the translational coordinates is specified by the relations 3.3.23 to 3.3.25. The functional dependence of the direction cosines on the rotation angles has been discussed in Section 1.5 and is given by Equation 1.5.16.

The functional dependence of the multipole moments on the internal coordinates is not known in analytical form. In molecular spectroscopic problems, it is customary to expand the dipole moment in a Taylor series of the molecular normal coordinates

$$\mu = \mu^0 + \sum_i (\partial\mu/\partial q_i) q_i + \ldots \qquad 3.3.30$$

where μ^0 represents the static dipole moment of the molecule and the linear coefficients $(\partial\mu/\partial q_i)$ are related to the infrared absorption intensity and can thus be measured experimentally.

The same expansion can be used for the quadrupole and the octupole moments

$$\Theta = \Theta^0 + \sum_i (\partial\Theta/\partial q_i) q_i + \ldots \qquad 3.3.31$$

$$\Omega = \Omega^0 + \sum_i (\partial\Omega/\partial q_i) q_i + \ldots \qquad 3.3.32$$

but in this case the linear coefficients are not easily accessible experimentally and can thus be utilized only as additional parameters. Unless otherwise specified, in the following discussion we shall always refer to the static moments of the molecules. Only when dealing with factor group splittings of internal modes, is it actually necessary to consider the linear coefficients in the expansion of the molecular moments. In most cases it is, however, sufficient to consider only the variation of the dipole with the normal coordinates of the molecule. Only in very few cases, if the coefficient $(\partial\mu/\partial q_i)$ is zero, is it necessary to take into account the linear terms in the quadrupole expansion.

We consider now briefly the symmetry properties of the multipole components. We notice that in general a multipole moment of order m has 2m + 1 components. The multipole tensors are in fact symmetric with respect to interchange of indices and obey trace-like relations of the type [62-64]

$$\sum_\rho M_{\rho\rho\sigma\tau\ldots} = 0 \qquad 3.3.33$$

Molecular symmetry is very effective in reducing the number of independent non-zero components and thus in simplifying expression 3.3.29 of the electrostatic potential.

In general the static dipole moment has 3 independent components. If the molecule has a symmetry axis the dipole will be oriented along that axis that can be always taken coincident with the z axis of the molecule-fixed reference system. If the molecule has a center of symmetry, the dipole moment is equal to zero.

The quadrupole moment has 5 independent components. Total symmetry with respect to interchange of the indices in fact reduce the number of components from 12 to 6. The trace relation

$$\Theta_{xx} + \Theta_{yy} + \Theta_{zz} = 0 \qquad 3.3.34$$

introduces a constraint among the diagonal elements and thus reduces the number of independent components to 5.

The octupole has 7 independent components. Again, total symmetry with respect to interchange of indices lowers the number of components from 27 to 10. The trace-like relations

$$\Omega_{\rho xx} + \Omega_{\rho yy} + \Omega_{\rho zz} = 0 \qquad (\rho = x,y,z) \qquad 3.3.35$$

further reduce the number of independent components to 7.

As for the dipole moment, the number of independent components of the quadrupole and of the octupole moments is additionally reduced by the molecular symmetry. The non-zero components of the dipole, quadrupole and octupole moments for various point groups of interest for molecular systems are compiled in Table 3.3.

Table 3.3 Independent non-zero components of the permanent dipole,quadrupole and octupole moments

group	dipole	quadrupole	octopole
C_1	x y z	xx yy xy xz yz	xxx xxz xzz xyz yyy yzz zzz
C_i	0	xx yy xy xz yz	0
C_s	x y	xx yy xy	xxx yyy xzz yzz
C_2	z	xx yy xy	zzz zxx zxy
C_{2h}	0	xx yy xy	0
C_{2v}	z	xx yy	zzz zxx
D_2	0	xx yy	xyz
D_{2h}	0	xx yy	0
C_3	z	zz	zzz xxx yyy
C_{3v}	z	zz	zzz xxx
C_{3h}	0	zz	xxx yyy
D_3, D_{3h}	0	zz	xxx
C_n, C_{nv}	z	zz	zzz
D_{4h}, D_{6h}	0	zz	0
T, T_d	0	0	xxy

The electrostatic potential has been used by many authors in lattice dynamics calculations, either in conjunction with the atom-atom terms or even alone as representative of the total intermolecular potential. In several cases, when dealing with the splitting of internal vibrations of the molecules in the crystal, only the transition dipole-transition dipole interaction term

$$V_{el} = - \mu' \cdot T \cdot \mu' \qquad 3.3.36$$

has been employed with $\mu' = (\partial\mu/\partial q)$.

Mixed potentials including the quadrupole-quadrupole interaction and atom-atom potentials of the Lennard-Jones, Buckingham or multicore type have been widely utilized for diatomic molecular crystals. Mixed potentials of the same type have been also applied to the calculation of the lattice vibrations of polyatomic molecules.

c) Polarization interactions

A further contribution to the intermolecular potential arises from the polarization forces. Following the treatment of Buckingham, the polarization interactions can be also treated classically in terms of molecular polarizabilities. Using the same formalism utilized for the electrostatic terms, the polarization potential has the form [65,86]

$$\begin{aligned} V^{AB}_{pol} = \sum{}'_{C} [\, & M_1^A \cdot T_2^{AB} \cdot \alpha^B \cdot T_2^{BC} \cdot M_1^C - \tfrac{1}{3}(M_2^A \cdot T_3^{AB} \cdot \alpha^B \cdot T_2^{BC} \cdot M_1^C - M_1^A \cdot T_2^{AB} \cdot \alpha^B \cdot T_3^{BC} \cdot M_2^C) + \\ & - \tfrac{1}{9}(M_2^A \cdot T_3^{AB} \cdot \alpha^B \cdot T_3^{BC} \cdot M_2^C) + \\ & + \tfrac{1}{15}(M_1^A \cdot T_2^{AB} \cdot \alpha^B \cdot T_4^{BC} \cdot M_3^C + M_3^A \cdot T_4^{AB} \cdot \alpha^B \cdot T_2^{BC} \cdot M_1^C) + \dots \,] \end{aligned} \qquad 3.3.37$$

where all symbols were defined before, except α which represents the molecular polarizability in the crystal reference system. Performing all the inner tensor products and using, as before, molecular quantities, we obtain

$$\begin{aligned} V^{AB}_{pol} = \sum{}'_{C} \sum_{\alpha\beta\gamma\delta} \sum_{\rho\sigma\tau\xi} [\, & \Gamma^A_{\alpha\rho} \Gamma^B_{\beta\sigma} \Gamma^B_{\gamma\tau} \Gamma^C_{\delta\xi} T^{AB}_{\alpha\beta} T^{BC}_{\gamma\delta} \mu_\rho \mu_\delta \alpha_{\sigma\tau} + \\ & - \tfrac{1}{3} \sum_{\varepsilon} \sum_{\chi} (\Gamma^A_{\alpha\rho} \Gamma^B_{\beta\sigma} \Gamma^B_{\gamma\tau} \Gamma^C_{\delta\xi} \Gamma^C_{\varepsilon\chi} T^{AB}_{\alpha\beta} T^{BC}_{\gamma\delta\varepsilon} \mu_\rho \alpha_{\sigma\tau} \Theta_{\xi\chi} + \\ & \quad - \Gamma^C_{\alpha\rho} \Gamma^B_{\beta\sigma} \Gamma^B_{\gamma\tau} \Gamma^A_{\delta\xi} \Gamma^A_{\varepsilon\chi} T^{CB}_{\alpha\beta} T^{BA}_{\gamma\delta\varepsilon} \mu_\rho \alpha_{\sigma\tau} \Theta_{\xi\chi}) + \\ & - \tfrac{1}{9} \sum_{\varepsilon\theta} \sum_{\chi\psi} \Gamma^A_{\alpha\rho} \Gamma^A_{\beta\sigma} \Gamma^B_{\gamma\tau} \Gamma^B_{\delta\xi} \Gamma^C_{\varepsilon\chi} \Gamma^C_{\theta\psi} T^{AB}_{\alpha\beta\gamma} T^{BC}_{\delta\varepsilon\theta} \Theta_{\rho\sigma} \alpha_{\tau\xi} \Theta_{\chi\psi} + \dots \,] \end{aligned} \qquad 3.3.38$$

where the $\alpha_{\rho\sigma}$ are the components of the polarizability tensor in the molecular reference system connected to the components of α by the relations

$$\alpha_{\beta\gamma} = \sum_{\sigma\tau} \Gamma_{\beta\sigma} \Gamma_{\gamma\tau} \alpha_{\sigma\tau} \qquad 3.3.39$$

In these expressions the symbol $\sum_{C}'$ indicates that in the sum $C \neq B$. For symplicity only the interaction of induced dipoles is considered in this form of the polarization potential. It is worth pointing out that it includes both two- and three-body interactions. Normal two-body interactions correspond to the terms of 3.3.37 in which A = C. In this case the multipoles on molecule A induce a dipole on molecule B and this interacts with the multipoles on A.Three-body interactions are described instead by the terms in which $A \neq B \neq C$. In a typical three-body interaction, the multipoles on molecule A interact with the dipole induced on molecule B by all other molecules C in the crystal. Owing to the fact that the polarization interactions are relatively weak, it is sufficient to limit the expansion 3.3.37 to the first few terms corresponding to interactions with induced dipoles.

d) Hydrogen bonds

The four types of contributions to the intermolecular potential discussed above represent the normal interactions between polyatomic molecules. In some crystals, however, strong directional interactions[66] such those of the hydrogen bond or charge transfer type may be present. These interactions are not easily described by the potential terms discusse above and often require a separate treatment[66].

Typically,hydrogen bonds are localized only between neighboring molecules with the correct relative orientation in the crystal and involve an X——H group on one molecule and an Y atom on the other. The X——H distance is shorter than the normal van der Waals distances between non-bonded atoms in crystals.

For weak hydrogen bonds, one can reasonably describe the interaction by means of " ad hoc" X...Y and H...Y atom-atom potentials limited to the shortest hydrogen-bonded contacts and differing in the numerical values of the parameters, from the potentials utilized for the same type of interaction at larger distances.

For medium and strong hydrogen bonds, specific potential functions are often used. A potential function of the type [67]

$$V(r) = D_0\{ 1 - \exp[-n(r - r_0)^2/2r]\} \qquad 3.3.40$$

has been successfully utilized by Lippincott and Schrader for the calculation of many properties of hydrogen-bonded systems. Here, D_0 is the bond dissociation energy, r_0 the equilibrium distance and n a parameter related to the ionization potentials of the atoms involved in the bond and given by

$$n = K_0 r_0/D_0 \qquad 3.3.41$$

where K_0 is the stretching force constant of the bond.

The Lippincott potential has been further modified[68-69] to take into account the fact that repulsive and dispersion terms are still present in the hydrogen bond potential and that both the X—Y and the H—Y distances are involved in the displacement of the proton. A modified Lippincott potential has the form [68,69]

$$V(r,R) = - D_0'\exp[- \frac{n'(R - r - r_0')^2}{2(R - r)}] + A'\exp(-B'R) - C'/R^6 \qquad 3.3.42$$

where R is the X—Y distance and the parameters D_0', n', A', B' and C' are adjusted empirically.

The Lippincott potential, in its standard form or in modified versions, has been rather popular in the past. In recent treatments[51], however it has been replaced by an atom-atom (12 - 6 or 9 - 6 Lennard-Jones) plus an electrostatic potential or by a Morse function. Such potentials are easier to use in lattice dynamics calculations and correctly reproduce the anharmonic behavior of the hydrogen bond. In addition Morse potentials have also been used for charge transfer interactions.

3.4 INTERMOLECULAR FORCE CONSTANTS

For the solution of the equations of motion of a molecular crystal we need the elements 1.6.33 of the dynamical matrix. These contain the second derivatives $F_{ts}\binom{m\mu}{n\nu}$ of the intermolecular potential with respect to the normal coordinates q_t of the isolated molecules. In this Section we discuss the general method of construction of these derivatives using the model potentials considered in the previous section. We limit our exposition to the atom-atom and to the electrostatic potentials since it is easy to extend this treatment to more specific cases.

We recall from Section 3.1 that the intermolecular potential, assuming only pairwise interactions, has the form

$$V_I = \tfrac{1}{2}\sum_A \sum_B{}' V^{AB} \qquad 3.4.1$$

where the prime indicates that the sum over B omits the term A. From this basic relation we have

$$(\partial^2 V_I/\partial q_t^A \partial q_s^B)_0 = (\partial^2 V^{AB}/\partial q_t^A \partial q_s^B)_0$$

$$3.4.2$$

$$(\partial^2 V_I/\partial q_t^A \partial q_s^A)_0 = \sum_B{}'(\partial^2 V^{AB}/\partial q_t^A \partial q_s^A)_0$$

where, for convenience, we have written separately typical diagonal and off-diagonal elements.

We consider first the case of the atom-atom potential. From 3.3.7 we obtain

$$(\partial^2 V^{AB}/\partial q_t^A \partial q_s^B)_0 = \sum_{ij}[\partial^2 V\binom{Ai}{Bj}/\partial q_t^A \partial q_s^B]_0 \qquad 3.4.3$$

where each atom-atom potential $V\binom{Ai}{Bj}$ is a function of only one interatomic distance. In order to obtain the derivatives 3.4.3, it is only necessary to establish a relationship between the atom-atom distance and the normal coordinates. This is easily performed by using as intermediates the crystal-fixed Cartesian coordinates of the atoms. The atom-atom distance is expressed, in terms of the Cartesian coordinates of the atoms by the relation

$$r_{ij} = [\sum_\alpha (X^{Bj}_\alpha - X^{Ai}_\alpha)^2]^{\frac{1}{2}}_0 \qquad 3.4.4$$

and hence

$$[\partial^2 V(^{Ai}_{Bj})/\partial q^A_t \partial q^B_s]_0 = -\sum_{\alpha\beta} P^{AiBj}_{\alpha\beta} X^{Ai}_{\alpha t} X^{Bj}_{\beta s} \qquad 3.4.5$$

$$[\partial^2 V(^{Ai}_{Bj})/\partial q^A_t \partial q^A_s]_0 = \sum_{\alpha\beta} P^{AiBj}_{\alpha\beta} X^{Ai}_{\alpha t} X^{Ai}_{\beta s} - \sum_\alpha N^{AiBj}_\alpha X^{Ai}_{\alpha ts} \qquad 3.4.6$$

where

$$X^{Ai}_{\alpha t} = (\partial X^{Ai}_\alpha / \partial q^A_t)_0$$

$$X^{Ai}_{\alpha ts} = (\partial^2 X^{Ai}_\alpha / \partial q^A_t \partial q^A_s)_0 \qquad 3.4.7$$

and

$$P^{AiBj}_{\alpha\beta} = [\partial^2 V(^{Ai}_{Bj})/\partial r^2_{ij}]_0 r^{ij}_\alpha r^{ij}_\beta r^{-2}_{ij} +$$

$$+ [\partial V(^{Ai}_{Bj})/\partial r_{ij}]_0 (\delta_{\alpha\beta} - r^{ij}_\alpha r^{ij}_\beta r^{-2}_{ij})/r_{ij} \qquad 3.4.8$$

$$N^{AiBj}_\alpha = [\partial V(^{Ai}_{Bj})/\partial r_{ij}]_0 r^{ij}_\alpha / r_{ij}$$

The coefficients 3.4.7 of the transformation from Cartesian to normal coordinates for translations, rotations and internal normal coordinates are given in Ref. 63 .

We now consider the derivatives of the electrostatic potential expressed by the multipole expansion 3.3.13. Each multipole-multipole interaction potential V^{AB}_{mn} has, according to 3.3.29, the general form

$$V^{AB}_{mn} = \sum_{\alpha\beta\gamma\ldots} \sum_{\alpha'\beta'\gamma'\ldots} \sum_{\rho\sigma\tau\ldots} \sum_{\rho'\sigma'\tau'\ldots} T^{AB}_{\alpha\beta\gamma\ldots\alpha'\beta'\gamma'\ldots} \times$$

$$\times \Gamma^A_{\alpha\rho} \ldots \Gamma^B_{\alpha'\rho'} \ldots M^A_{\rho\sigma} \ldots M^B_{\rho'\sigma'} \qquad 3.4.9$$

where the first two sums extend over m crystal-fixed indices α, β, γ ...

for molecule A and over n similar indices for molecule B and the second two sums over m molecule-fixed indices $\rho,\sigma,\tau,\ldots$ for molecule A and n similar indices for molecule B. The dots after the direction cosine $\Gamma^A_{\alpha\rho}$ indicate m such factors and after $\Gamma^B_{\alpha'\rho'}$ n such factors in total.

As discussed in Section 3.3 the electrostatic potential is factored in this way into a product of terms, each depending on only one of the three types of molecular normal coordinates used. In particular the T tensor elements only depend on the intermolecular distance, the direction cosines only on the molecular orientation and the multipole moments only on the molecular structure.

From the known relationships between the terms occurring in 3.4.9 and the vibrational coordinates, the derivatives are easily obtained. We show here the basic procedure.

For translations we have, from 1.5.2, 1.5.18, 1.6.9a and from the definition 3.3.22 of the T tensor elements

$$(\partial T^{AB}_{\alpha\beta\gamma\ldots\alpha'\beta'\gamma'\ldots}/\partial t^A_\nu)_0 = -M^{-\frac{1}{2}}\sum_\omega T^{AB}_{\alpha\beta\gamma\ldots\alpha'\beta'\gamma'\ldots\omega}\Lambda^A_{\omega\nu}$$
$$(\partial T^{AB}_{\alpha\beta\gamma\ldots\alpha'\beta'\gamma'\ldots}/\partial t^B_\nu)_0 = M^{-\frac{1}{2}}\sum_\omega T^{AB}_{\alpha\beta\gamma\ldots\alpha'\beta'\gamma'\ldots\omega}\Lambda^B_{\omega\nu} \qquad 3.4.10$$

and therefore

$$(\partial^2 T^{AB}_{\alpha\beta\gamma\ldots\alpha'\beta'\gamma'\ldots}/\partial t^A_\nu\partial t^A_\mu)_0 = M^{-1}\sum_{\omega\psi} T^{AB}_{\alpha\beta\gamma\ldots\alpha'\beta'\gamma'\ldots\omega\psi}\Lambda^A_{\omega\nu}\Lambda^A_{\psi\mu}$$
$$(\partial^2 T^{AB}_{\alpha\beta\gamma\ldots\alpha'\beta'\gamma'\ldots}/\partial t^A_\nu\partial t^B_\mu)_0 = -M^{-1}\sum_{\omega\psi} T^{AB}_{\alpha\beta\gamma\ldots\alpha'\beta'\gamma'\ldots\omega\psi}\Lambda^A_{\omega\nu}\Lambda^B_{\psi\mu} \qquad 3.4.11$$

We notice that the elements of rank m+n in the potential become elements of a tensor of rank m+n+1 in the first derivatives and of rank m+n+2 in the second derivatives with respect to translations. Using these expressions, the derivatives of the multipole potential with respect to translational normal coordinates are easily obtained by substitution of 3.4.11 in place of the corresponding T tensor elements in Eq. 3.4.9.

For rotations we have, from 1.5.16 and 1.6.9b

$$(\partial\Gamma^A_{\alpha\rho}/\partial r^A_\nu)_0 = I_\nu^{-\frac{1}{2}}\sum_\pi \Lambda^A_{\alpha\pi}\delta_{\pi\rho\nu} \qquad 3.4.12$$

$$(\partial^2\Gamma^A_{\alpha\rho}/\partial r^A_\nu \partial r^A_\mu)_0 = -\frac{1}{2}(I_\nu I_\mu)^{-\frac{1}{2}}\sum_{\pi\xi}\Lambda^A_{\alpha\xi}(\delta_{\xi\pi\nu}\delta_{\rho\pi\mu} + \delta_{\rho\pi\nu}\delta_{\xi\pi\mu}) \qquad 3.4.13$$

Substituting these relations in 3.4.9 in place of the corresponding direction cosines, the derivatives of the multipole potential with respect to rotational normal coordinates are easily obtained.

The functional dependence of the multipoles on the internal coordinates was discussed in Section 3.3. The derivatives of the multipole potential with respect to the internal normal coordinates are obtained by substitution of the static multipoles with their first derivatives in 3.4.9. The first-order derivatives are obtained from 3.3.30 to 3.3.32.

In order to clarify the procedure outlined above we work out here the explicit expressions of the derivatives for the specific case of the dipole-dipole interaction potential. According to Eq. 3.3.29, the dipole-dipole potential is

$$V^{AB}_{\mu\mu} = -\sum_{\alpha\beta}\sum_{\rho\sigma}\Gamma^A_{\alpha\rho}\Gamma^B_{\beta\sigma}T^{AB}_{\alpha\beta}\mu_\rho\mu_\sigma \qquad 3.4.14$$

and thus, using the previous equations we obtain

$$(\partial^2 V^{AB}_{\mu\mu}/\partial t^A_\nu \partial t^A_\lambda)_0 = -M^{-1}\sum_{\alpha\beta\gamma\delta}\sum_{\rho\sigma}(\Lambda^A_{\alpha\rho}\Lambda^A_{\gamma\nu}\Lambda^A_{\delta\lambda}\Lambda^B_{\beta\sigma}T^{AB}_{\alpha\beta\gamma\delta}\mu_\rho\mu_\sigma)$$

$$(\partial^2 V^{AB}_{\mu\mu}/\partial t^A_\nu \partial t^B_\lambda)_0 = M^{-1}\sum_{\alpha\beta\gamma\delta}\sum_{\rho\sigma}(\Lambda^A_{\alpha\rho}\Lambda^A_{\gamma\nu}\Lambda^B_{\delta\lambda}\Lambda^B_{\beta\sigma}T^{AB}_{\alpha\beta\gamma\delta}\mu_\rho\mu_\sigma)$$

$$(\partial^2 V^{AB}_{\mu\mu}/\partial r^A_\nu \partial r^A_\eta)_0 = \frac{1}{2}(I_\nu I_\eta)^{-\frac{1}{2}}\sum_{\alpha\beta}\sum_{\rho\sigma\pi\xi}T^{AB}_{\alpha\beta}(\delta_{\xi\pi\nu}\delta_{\rho\pi\eta} + \delta_{\rho\pi\nu}\delta_{\xi\pi\eta})\Lambda^A_{\alpha\xi}\Lambda^B_{\beta\sigma}\mu_\rho\mu_\sigma$$

$$(\partial^2 V^{AB}_{\mu\mu}/\partial r^A_\nu \partial r^B_\eta)_0 = -(I_\nu I_\eta)^{-\frac{1}{2}}\sum_{\alpha\beta}\sum_{\rho\sigma\pi\xi}T^{AB}_{\alpha\beta}\delta_{\xi\rho\nu}\delta_{\pi\sigma\eta}\Lambda^A_{\alpha\xi}\Lambda^B_{\beta\pi}\mu_\rho\mu_\sigma \qquad 3.4.15$$

$$(\partial^2 V^{AB}_{\mu\mu}/\partial t^A_\nu \partial r^A_\eta)_0 = (MI_\eta)^{-\frac{1}{2}}\sum_{\alpha\beta\gamma}\sum_{\rho\sigma\pi} T^{AB}_{\alpha\beta\gamma}\Lambda^A_{\alpha\pi}\Lambda^A_{\gamma\nu}\Lambda^B_{\beta\sigma}\delta_{\pi\rho\eta}\mu_\rho\mu_\sigma$$

$$(\partial^2 V^{AB}_{\mu\mu}/\partial t^A_\nu \partial r^B_\eta)_0 = (MI_\eta)^{-\frac{1}{2}}\sum_{\alpha\beta\gamma}\sum_{\rho\sigma\pi} T^{AB}_{\alpha\beta\gamma}\Lambda^A_{\alpha\rho}\Lambda^A_{\gamma\nu}\Lambda^B_{\beta\pi}\delta_{\pi\sigma\eta}\mu_\rho\mu_\sigma$$

The corresponding expressions for derivatives of higher terms of multipole-multipole potential can be found in Ref. 63.

Very recently a diagrammatic method leading to compact expressions for the multipole-multipole interaction potential and for the corresponding derivatives has been developed by Burgos and Bonadeo in terms of permutation polynomials.[71].

3.5 LATTICE SUMS AND EWALD'S METHOD

Lattice dynamics calculations invariably involve the evaluation of some sort of lattice sums which should be in principle computed over an infinite crystal or at least in such a way that the periodic boundary conditions are fulfilled. In several cases, which have already been dis cussed, the intermolecular potentials depend on very high powers of the intermolecular separation (e.g. atom-atom or high-order multipole-multi pole potentials). Thus, the problem simplifies since, in practice, it i sufficient to determine the lattice sums by direct summation in the rea space, within a reasonable interaction radius. For instance, in the cas of the atom-atom potential, it has been shown [72] that a good convergenc is ensured by summation within a range of 6 - 10 A. The evaluation of such lattice sums is not particularly time-consuming and can be performed in a straightforward manner. When, however, the intermolecular interaction has a long-range character, the evaluation of the lattice sum is not a simple problem.This is particularly the case with dipole-dipol and charge-charge interactions. The dipole-dipole potential depends on $1/r^3$ and,owing to the fact that the volume on which the direct summatio is performed varies with r^3, it is clear that an absolute convergence of the direct summation cannot be obtained. To be more precise, the summation of these types of interactions is only conditionally convergent, in the sense that the result depends on the order of the indices in the summation or on the external shape of the sample. For instance, the inadequacy of the direct summation over a spherical sample for dipole-dipole interactions is easily realized on the basis of the following considerations. With strong dipole-dipole interactions, the $\mathbf{k} = 0$ vibrational modes may show in the infrared and Raman spectra a significant splitting into longitudinal and transverse components. This effect arises specifically from the long-range part of the interaction and cannot be obtained by direct summation over a sphere where long-range coupling is explicitly neglected. It is therefore necessary in this case to consider more general methods of evaluation of the lattice sums which on the one side ensure a fast convergence and on the other side permit features to be reproduced such as the LO-TO splitting or the angular frequency-dispersion which specifically originate from long-range interactions.

The problem of the evaluation of general lattice sums, for any dependence of the intermolecular distance has been discussed by many authors [3,4,73,74]. Here, it is sufficient to briefly discuss this problem with particular reference to the summation of the $\nabla^2(1/R)$ tensor which enters the expressions of the force constants for both dipole-dipole and charge-charge interactions.

Two summation methods are commonly used: the planewise [75,76] and the Ewald-Kornfeld [77,78] methods. These methods give exactly the same result if equivalent boundaries are selected at the crystal.[79]. The planewise summation method, as the name clearly indicates, essentially groups together the lattice points within each crystalline plane perpendicular to some selected z crystal axis and sums all contributions, plane by plane. The main advantage of the method is that only the summation over two indices (in the xy plane) is actually involved while the summation over the z direction can be carried on analytically. It is evident that the method is particularly suitable for some types of crystals.

The second method, by far more utilized in lattice dynamics, was originally developed by Ewald for a crystalline array of point charges and later extended by Kornfeld to dipolar and quadrupolar arrays. The basic Ewald idea is that in an array of point charges z_λ, a Gaussian charge distribution is added and subtracted to each point charge, the various terms being then arranged in rapidly and slowly converging series. By exploiting the symmetry properties of the lattice, the slowly converging series is Fourier transformed into the reciprocal space where its convergence is increased rapidly. The convergence depends on a parameter P which defines the width of the Gaussian charge distribution $\rho(r)$

$$\rho(r) = z_\lambda (P/\pi)^{3/2} e^{-Pr^2} \qquad 3.5.1$$

The lattice sum

$$S^{\mu\nu}_{\alpha\beta}(\mathbf{k}) = \sum_b \nabla^2_{\alpha\beta}\left(\frac{1}{R^{\nu}_b - R^{\mu}_0}\right) e^{i\mathbf{k}\cdot(\mathbf{R}^{\nu}_b - \mathbf{R}^{\mu}_0)} \quad 3.5.2$$

is then given by

$$S^{\mu\nu}_{\alpha\beta}(\mathbf{k}) = G^{\mu\nu}_{\alpha\beta}(\mathbf{k}) + R^{\mu\nu}_{\alpha\beta}(\mathbf{k}) + \frac{4\pi}{V}\left\{ \frac{k_\alpha k_\beta}{|k|^2} + \frac{k_\alpha k_\beta}{|k|^2}\left[\exp(-k^2/4P) - 1\right]\right\} \quad 3.5.3$$

where $G^{\mu\nu}_{\alpha\beta}(\mathbf{k})$ and $R^{\mu\nu}_{\alpha\beta}(\mathbf{k})$ are the contributions in the direct and reciprocal space, respectively. These are given by

$$G^{\mu\nu}_{\alpha\beta}(\mathbf{k}) = P^{3/2} \sum_b \left\{ \frac{R^{\nu\mu}_{\alpha} R^{\nu\mu}_{\beta}}{(R^{\nu}_b - R^{\mu}_0)^2}\left[\frac{3}{x^3}\operatorname{erfc}(x) + 2\pi^{-\frac{1}{2}}\left(\frac{3}{x^2} + 2\right)e^{-x^2} + \right.\right.$$
$$\left.\left. - \delta_{\alpha\beta}\left(\frac{1}{x^3}\operatorname{erfc}(x) + 2\pi^{-\frac{1}{2}}\frac{1}{x^2}e^{-x^2}\right)\right]\right\} e^{-i\mathbf{k}\cdot(\mathbf{R}^{\nu}_b - \mathbf{R}^{\mu}_0)} \quad 3.5.4$$
$$+ \frac{4}{3}\pi^{-\frac{1}{2}} P^{-3/2} \delta_{\alpha\beta}\delta_{\mu\nu}$$

and

$$R^{\mu\nu}_{\alpha\beta}(\mathbf{k}) = \frac{4\pi}{V}\sum_\tau \frac{(\tau_\alpha + k_\alpha)(\tau_\beta + k_\beta)}{|\tau + k|^2} e^{-|\tau + k|^2/4P} e^{-i\boldsymbol{\tau}\cdot(\mathbf{R}^{\nu}_b - \mathbf{R}^{\mu}_0)} \quad 3.5.5$$

where $R^{\nu\mu}_{\alpha}$ is the αth component of $\mathbf{R}^{\nu}_b - \mathbf{R}^{\mu}_0$, x is a reduced intermolecular distance given by $x = (R^{\nu}_b - R^{\mu}_0)P^{\frac{1}{2}}$, V is the volume of the unit cell $\boldsymbol{\tau}$ is a vector in the reciprocal space and erfc(x) is the complementary error function

$$\operatorname{erfc}(x) = 2\pi^{-\frac{1}{2}}\int_x^\infty e^{-s^2} ds \quad 3.5.6$$

We observe that the last term in 3.5.3 is singular at $\mathbf{k} = 0$. Its value in the limit of $\mathbf{k} \to 0$ depends on the direction in which $\mathbf{k}$ approaches zero and this introduces a shape dependence of the lattice sums and hence of the calculated frequencies whenever the dipole-dipole coupling is active. One of the main advantages of the Ewald method is that of considering separately two rapidly converging series and a singular term

The simultaneous rapid convergence of the two series is obtained by a suitable choice of the convergence parameter P. It has been found that a value of $P^{\frac{1}{2}}$ of the order of the inverse nearest neighbor distance is appropriate but, in general, the results of the summations are not particularly sensitive to the value of P.

The singular term in Eq.3.5.3, which arises from the long-range macroscopic field, ensures the possibility of calculating LO and TO frequencies. The singular term is only effective in the long wavelength limit and gives no contribution for larger values of **k**, as model calculations have also proved [73,80].

The origin of the LO-TO splittings can be understood qualitatively by symmetry arguments. It is sufficient to consider that wave vectors of interest in optical experiments are not exactly zero although they are very small. This gives rise to a breaking of the symmetry of the various phonon branches of the dispersion curves which is not predicted by the factor group selection rules [80,81].

As already noticed, the Ewald method can be extended to intermolecular potentials with any dependence on the intermolecular distance. In practice, however, the method has not been used in lattice dynamical calculations of molecular crystals, except for dipole-dipole and charge-charge interactions.

3.6 CALCULATIONS OF PHONON FREQUENCIES

Experimental information on the vibrational modes of a molecular crystal is obtained by a variety of techniques. By far the most direct and powerful methods of investigation of the vibrational properties of solids are infrared, Raman and neutron scattering spectroscopy.

The infrared and Raman spectra of a molecular crystal can be measured on powdered samples or on single crystals. With modern spectrophotometers and interpherometers these measurements offer no special difficulties apart from those arising from the growth of single crystals and from the preparation of suitable samples with the correct orientation of the optical axes. The measurements can be easily made at any desired temperature and, using special cells, even under very high pressures. Infrared and Raman data are,however, limited to optically active phonons, i.e. to crystal vibrations in which all unit cells vibrate in phase since only these crystal modes can interact with the electromagnetic radiation. Thus in a phonon dispersion curve only the point at the origin Γ of the Brillouin zone is accessible by optical spectroscopic methods.

Neutron scattering experiments are order of magnitude more difficult and time-consuming. With modern neutron spectrometers and high-flux neutron beams, the situation has drastically been improved as compared with few years ago, but the measurement of a complete phonon spectrum of a molecular crystal remains a major experimental problem and is still far from being a routine type of work.

The great advantage of neutron scattering measurements is that they are not subject to limitations arising from selection rules so that all types of phonons, with any value of **k**, can be determined experimentally. For these measurements, single crystals of size much greater than that required for optical spectroscopy are necessary and this adds a further complication to the technique. Less detailed but still useful information can be obtained from incoherent with respect to coherent neutron scattering measurements. In incoherent measurements one obtains a spectrum in the frequency space in the form of a density of phonon states. From coherent measurements one obtains instead the complete set of phonon dispersion curves in the **k** space.

The infrared, Raman and neutron scattering data, when properly interpreted, represent the major source of information on the vibrational properties of a molecular crystal. These data can be used to check the validity of models of intermolecular potentials or the latter can, in turn, be applied to the interpretation of the data. In most cases only the phonon frequencies are utilized but, as discussed in Chapter 6, infrared and Raman intensities offer a not yet fully investigated source of important information on the dynamical behavior of the molecular solids.

A large amount of work on the theoretical calculation of crystal vibrations has been accumulated in the last two decades, using intermolecular potentials of the type discussed in Section 3.3. Although some calculations of the complete phonon spectrum of molecular crystals have been made, most of the work is limited to the calculation of external vibrations in the rigid-body approximation. The main reason for this limited interest in the internal vibrations of the molecules in the crystal is that these vibrations are dominated by the internal force field, the intermolecular potential acting only as a small perturbation on it. On the contrary, the external vibrations are completely controlled by the intermolecular potential and thus represent a more stringent and powerful test than the shifts and the splittings of internal modes.

Excellent books[73,82,83] and reviews [84-87] on the dynamics of molecular crystals are available. The latest review by Colombo and Mathieu[87] covers the most recent production. Additional references can be found in the book by Poulet and Mathieu [82] and in the review by Schrader [85]. For this reason in this section we will give a discussion of few selected examples covering a variety of intermolecular potentials, rather than an up-to-date and complete review of all recent works. In addition we shall briefly consider some recent papers that were not included in the reviews mentioned above. The discussion will be limited to harmonic calculations since anharmonic ones are discussed in Chapter 4.

Solid nitrogen is one of the most studied molecular crystals [88] since it represents the ideal system for a test of intermolecular potentials, owing to the fact that the molecule is very simple and has only one internal vibration at 2330 cm^{-1}, well separated from the lattice vibrations. It therefore represents an excellent system for a rigid body treatment.

N_2 exists in three crystalline modifications [88] known as α, β and γ crystal phases. The phase diagram is well known and is shown in Fig.3.2.

The low-temperature phase α is stable at normal pressure below 35.4 K and possesses a cubic structure (space group Pa3, T_h^6)with four molecules per unit cell. The second low-temperature phase γ exists only at high pressures above 3.5 kbar and is tetragonal (space group $P4_2/mnm$, D_{4h}^{14}) with two molecules per unit cell. The high-temperature phase β is hexagonal(space group $P6_3/mmc$) and shows a considerable amount of orientational disorder characteristic of plastic crystals. Owing to the fact that standard lattice dynamics is not a convenient tool for treating plastic phases, which are more effectively studied by different theoretical methods, we shall not discuss here the dynamics of this crystalline form of N_2.

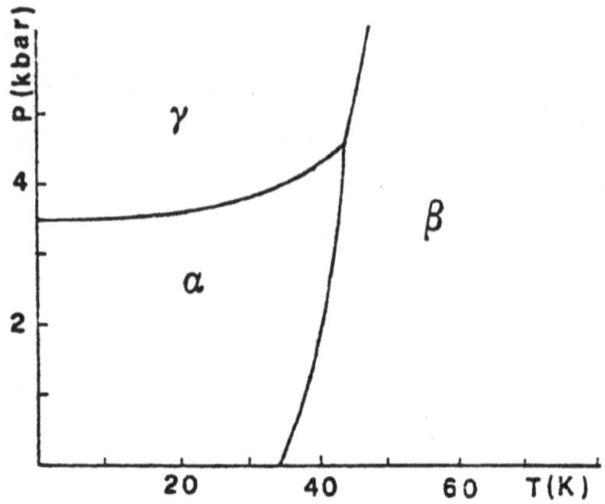

Fig.3.2 Phase diagram of solid Nitrogen

The lattice vibrations of the α phase have been extensively investigated by infrared, Raman and neutron scattering spectroscopy. The γ phase has been less studied but infrared and Raman spectra have recently become available.

The temperature dependence of the frequency and intensity of the lattice vibrations [88,90,95] shows a considerable anharmonic contribution, especially in the case of librational modes. Despite this inconvenient feature of the dynamics of crystalline N_2, a great number of harmonic calculations has been made, using several kinds of intermolecular potentials. Since it would be impossible to discuss all pubshed results, we shall report here only the most recent developments.

Table 3.4 collects the optically active lattice frequencies of the α phase measured at 15 K by Kjems and Dolling [89], together with the calculated values with different intermolecular potentials. A detailed discussion of the relative merits and drawbacks of these potentials is included in almost all recent papers on the subject. We summarize the essential points. Standard atom-atom potentials, either of the Lennard-Jones [90] or of the Buckingham [91] type, give rise to a reasonable fit of the optically active lattice modes, except for the T_g^+ libration which is always calculated at a frequency lower than that observed. Furthermore these potentials fail in predicting the stability of the γ phase at high pressure and lead to calculated Grünaisen parameters which are much higher than those observed. Addition of a quadrupole-quadrupole electrostatic interaction to the atom-atom potential [92] improves the agreement with the T_g^+ mode whereas the $\alpha \to \gamma$ transition still remains unexplained. The instability of the γ phase is related to the fact that all these potentials give rise to negative eigenvalues for one of the acoustic branches in the (110) direction. Filippini et al. [56] have recently shown that this difficulty can be eliminated by use of an aniso-

Table 3.4 Observed and Calculated Optically Active Lattice Frequencies of Crystalline α-N_2 in cm^{-1}

Symm.	Experimental	I	II	III	IV	V	VI
E_g	32.3 ± 0.8	33.6	33.0	32.4	34.6	32.0	37.5
T_g^-	36.3 ± 0.8	37.8	37.7	38.7	38.5	36.6	47.7
T_g^+	59.7 ± 1.6	45.8	46.4	55.0	47.6	44.6	75.2
T_u^-	48.4 ± 0.8	45.5	48.0	45.2	46.6	45.0	47.7
T_u^+	69.4 ± 1.6	67.1	70.0	66.0	69.7	68.6	69.5
A_u	46.8 ± 1.6	42.6	44.7	44.6	45.0	41.6	45.9
E_u	54.0 ± 2.4	50.4	53.0	48.7	52.3	46.3	54.0

I - Lennard-Jones potential.Ref.90

II - Buckingham potential. Ref.91

III- Lennard-Jones plus quadrupole-quadrupole interaction.Ref.92

IV - Buckingham potential with anisotropic repulsive part.Ref.56

V - Shell model potential (pseudoatom approximation).Ref.93

VI - Raich and Gillis potential.Ref.57

tropic repulsive atom-atom potential of the type

$$V_{rep} = A'\left[1 + P(\cos^2\theta_1 + \cos^2\theta_2)\right]\exp(-B'r) \qquad 3.6.1$$

where θ_1 and θ_2 are the angles between the line joining the interacting non-bonded atoms and the molecular axes, P is a scaling factor and A', B' are reparametrized empirical constants.

An interesting application of the shell model to the calculation of the dynamical properties of solid N_2 has been suggested by Luty [93]. Using the so-called pseudoatom approximation, Luty has obtained a screened nuclear potential that reproduces many dynamical properties of solid N_2.

More complex phenomenological potentials have been proposed by Raich and Gillis and by Kobashi and Kihara. The Raich-Gillis [57] potential includes, in addition to a quadrupole-quadrupole electrostatic term a repulsive potential of the type 3.3.8 as well as an attractive dispersion term

$$V_{disp} = - \varepsilon \left[\frac{C_6}{\delta + r^6} \quad \frac{C_8}{\delta + r^8} \quad \frac{C_{10}}{\delta + r^{10}} \right] \qquad 3.6.2$$

where ε is a parameter and $r = R/R_{min}$. The Kobashi-Kihara [94] potential is obtained by combining a Kihara spindle-core potential (see Section 3.3) and a quadrupole-quadrupole electrostatic interaction. Both potentials correctly account for the $\alpha \rightarrow \gamma$ transition and for the dynamical properties of solid N_2.

The observed and calculated lattice frequencies of the γ phase are collected in Table 3.5. The comparison between experimental and calculated frequencies is less significant in this case since the number of observed lattice frequencies is limited.

Table 3.5 Observed and Calculated Optically Active Lattice Frequencies of Crystalline γ-N_2 in cm^{-1}

Symm.	Experimental		I	II	III	IV	V	VI
	5 kbar	4 kbar	---4 kbar---		5 kbar	-------4 kbar-------		
E_g	58.4	55.0	53.1	53.0	56.7	50.5	56.9	50.5
B_{1g}	103.6	98.1	87.6	68.7	93.1	88.6	97.2	74.8
A_{2g}	inactive		100.2	84.0	112.5	100.2	111.2	105.1
E_u	---	65.0	63.4	64.3	64.0	61.8	65.6	58.3
B_{1u}	inactive		105.9	103.3	104.1	102.7	109.9	103.1

The potentials are numbered as in Table 3.4, except potential I which is taken from Ref. 95.

Crystalline ammonia offers another convenient example of a molecular crystal for which extensive dynamical calculations have been made.

NH_3 crystallizes in the simple cubic system (space group $P2_13$, T^4), with four molecules per unit cell on sites of C_3 symmetry. A schematic view along the cube diagonal of the spatial arrangement of the molecules in the crystal is shown in Fig.3.3. Each molecule has six neighbors, three above and three below the plane formed by the hydrogen atoms so that the trigonal molecular symmetry is preserved in the crystal. In this way each molecule is involved in six hydrogen bonds. These bonds

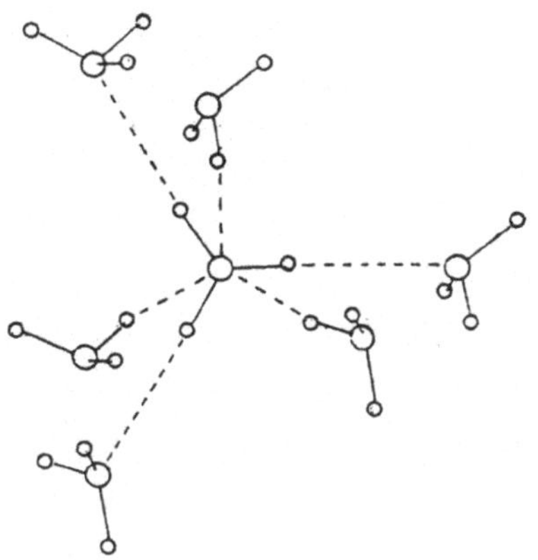

Fig.3.3 Schematic drawing of the spatial arrangement of the ammonia molecules in the crystal. View along the cube diagonal.

are very weak, the closest N...N distance being at the lower limit and the crystal energy at the upper limit of the conventional values taken as typical of van der Waals crystals.

The infrared and Raman spectra of crystalline ammonia and of ammonia-d_3 have been measured and interpreted by Binbrek and Anderson [96]. A reassignment of some lattice modes has recently been proposed [17] and later confirmed by intensity calculations [97].

The lattice vibrations of both crystalline NH_3 and ND_3 have been calculated over the complete Brillouin zone, using a mixed potential which includes an atom-atom part representative of the repulsive and dispersion interactions and an electrostatic part. The atom-atom potential is of the conventional type (see Eq.3.3.5) for each of the three types of atom-atom contacts, N...N, N...H and H...H occurring in the crystal. In addition, in order to account for the hydrogen bonds existing between neighboring molecules, another N...H atom-atom potential has been introduced to distinguish between hydrogen-bonded N...H interactions and interactions of the same type but between molecules which are far apart in the crystal. In this way the atom-atom potential includes 12 independent parameters. The refined values of these parameters are compiled in Table 3.6.

The electrostatic potential is expressed in terms of multipole-multipole interactions (see 3.3.29) and includes dipole-dipole, dipole-quadrupole, quadrupole-quadrupole and dipole-octupole contributions. Because of the C_{3v} symmetry of the molecule, there is only one non-zero component of the dipole, one independent component of the quadrupole and two independent components of the octupole moment. By orienting the molecular reference system with the z axis coinciding with the three-fold symmetry axis, these components can be chosen as μ_z, Θ_{zz}, Ω_{zzz} and Ω_{xxx}. The electrostatic potential between two molecules has thus the form

$$\begin{aligned} V_{el}^{AB} = & - \sum_{\alpha\beta} \mu_z^2 \Gamma_{\alpha z}^A \Gamma_{\beta z}^B T_{\alpha\beta}^{AB} - \frac{1}{3} \sum_{\alpha\beta\gamma} \sum_{\sigma} \mu_z \Theta_{\sigma\sigma} (\Gamma_{\alpha z}^A \Gamma_{\beta\sigma}^B \Gamma_{\gamma\sigma}^B - \Gamma_{\alpha z}^B \Gamma_{\beta\sigma}^A \Gamma_{\gamma\sigma}^A) T_{\alpha\beta\gamma}^{AB} \\ & + \frac{1}{9} \sum_{\alpha\beta\gamma\delta} \sum_{\tau\sigma} \Theta_{\sigma\sigma} \Theta_{\tau\tau} \Gamma_{\alpha\sigma}^A \Gamma_{\beta\sigma}^A \Gamma_{\gamma\tau}^B \Gamma_{\delta\tau}^B T_{\alpha\beta\gamma\delta}^{AB} + \\ & - \sum_{\alpha\beta\gamma\delta} \{ \frac{1}{6} \mu_z \Omega_{zzz} \Gamma_{\alpha z}^A \Gamma_{\beta z}^B (\Gamma_{\gamma z}^B \Gamma_{\delta z}^B + \Gamma_{\gamma z}^B \Gamma_{\delta z}^B) + \\ & + \frac{1}{15} \mu_z \Omega_{xxx} [\Gamma_{\alpha z}^B \Gamma_{\beta x}^A (\Gamma_{\gamma x}^A \Gamma_{\delta x}^A - 3 \Gamma_{\gamma y}^A \Gamma_{\delta y}^A) \\ & + \Gamma_{\alpha z}^B \Gamma_{\beta x}^A (\Gamma_{\gamma x}^A \Gamma_{\delta x}^A - 3 \Gamma_{\gamma y}^A \Gamma_{\delta y}^A)] \} T_{\alpha\beta\gamma\delta}^{AB} \end{aligned} \qquad 3.6.3$$

where the nomenclature is the same as in Section 3.3.

Higher terms of the electrostatic potential were not included since they depend on the inverse sixth power of the intermolecular distance

Table 3.6 Potential Parameters for Crystalline Ammonia

Atom-atom potential

Contact	A(kcal/mol)	B($Å^{-1}$)	C(kcal/mol×r^6)
H...H	1448	3.798	18.5
H...N	2010	3.150	84.
N...N	12616	2.703	1516.
H...N§	32323	4.563	285

Electrostatic potential

$\mu = 1.53\ 10^{-18}$ e.s.u. $\qquad \Omega_{zzz} = -\ 1.10\ 10^{-34}$ e.s.u.

$\Theta_{zz} = -\ 2.36\ 10^{-26}$ e.s.u. $\qquad \Omega_{xxx} = 1.76\ 10^{-34}$ e.s.u.

§ Hydrogen-bonded contacts only

and thus overlap in the R dependence with the attractive part of the atom-atom potential. Convergence of the calculations was ensured by using the Ewald method described in Section 3.5.

Another calculation [58,98] of the lattice vibrations of crystalline ammonia has been recently made, using a more sophisticated atom-atom potential of the type described by Eq.3.3.9 and a simpler electrostatic potential obtained by localizing effective charges on the atoms. In this case convergence was ensured by extending the crystal sums over different distances and by verifying that the calculated frequencies were not affected by more than 1% by enlarging the sphere of interaction. A third calculation of the lattice vibrations of ammonia was made by computing with an ab-initio SCF-MO method the energy surface of the ammonia dimer for different geometries and fitting this surface with an atom-atom plus an electrostatic potential of the charge-charge type. The Ewald sum was used for the electrostatic part [99] . A fourth calculation was finally made using the multipole expansion potential 3.6.3 plus a polarization potential of the type 3.3.38 in addition to the standard atom-atom potential 3.3.5 [65].

Table 3.7 Observed and Calculated Lattice Frequencies of Solid NH_3

symm.	exp.	I	II	III	IV		
F	533	{533, 522}	{540, 534}	505	{523, 517}	LO TO	R_x, R_y
F	{431, 366}	{435, 373}	{414, 348}	368	{394, 325}	LO TO	R_x, R_y
A	310	304	309	352	366		R_z
E	298	286	302	296	259		R_x, R_y
F	260	272	266	277	258	LO+TO	R_z
F	184	181	180	176	187	LO+TO	T_x, T_y
A		144	137	113	101		T_z
F	138	131	132	137	143	LO+TO	T_x, T_y
E	107	123	117	104	119		T_x, T_y
Lattice energy	-8.5	-8.13	-9.17	-8.7	-8.9		

I - Atom-atom plus multipole-mutipole.Ref.17

II- Potential I plus polarization terms.Ref.65

III- Atom-atom + effective charges on the atoms.Ref.58

IV- Ab-initio SCF-MO potential.Ref.99

In principle all polar modes of F species should show a LO - TO splitting. Experimentally, however,a LO-TO splitting of 65 cm^{-1} is observed only in the case of the F mode at 366-431 cm^{-1}.

The **k** = 0 lattice frequencies of crystalline ammonia calculated with the four intermolecular potential described above are compared with the experimental values in Table 3.7. We notice that using the Ewald method the calculated LO-TO splittings are always within 1 cm^{-1}, except in the case of the two high-frequency modes. Furthermore, the calculated splitting agrees well with the experimental value in all cases for the mode at 366-431 cm^{-1}. The splitting of about 10 cm^{-1} calculated for the band at 533 cm^{-1} is not observed experimentally since this is a very weak band.

The different terms of the multipole-multipole electrostatic potential have different and often very selective effects on the crystal frequencies. The specific contribution of the various terms of the intermolecular potential to the lattice frequencies and to the crystal energy has been investigated in detail in the case of ammonia. In Table 3.8 we list four different calculations of the lattice frequencies and of the crystal energy of NH_3 obtained starting with an atom-atom potential and adding each time a new term of the electrostatic part. The results show very clearly that some lattice modes, for instance ω_3 and ω_6, are practically insensitive to the electrostatic interactions whereas some others are strongly affected by one or more electrostatic terms.

Table 3.8 Contribution to the Lattice Frequencies and to the Crystal Energy of the Terms of the Electrostatic Potential for Crystalline NH_3

Symm.	N	I	II	III	IV	exp.
A	1	95	84	117	143	---
A	2	196	194	203	306	310
E	3	122	122	119	122	107
E	4	192	195	278	299	298
F	5	132	135	125	130	140
F	6	186	183	187	181	183
F	7	128	129	130	273	260
F	8	203	206	327	360	358
F	9	265	316	483	524	533
Crystal Energy		-3.19	-4.40	-6.73	-8.13	-8.50

I = Atom-atom only

II = Potential I + dipole-dipole term

III = Potential II + dipole-quadrupole and quadrupole-quadrupole terms

IV = Potential III + dipole-octupole term

Naphthalene is probably the molecular crystal that has been more studied both experimentally and theoretically. It is the typical representative of the large class of organic crystals which belong to the monoclinic $P2_1/a$ (C_{2h}^5) space group, with two molecules per unit cell located on a C_i site. The crystal structure is known at room temperature and at 120 K. The unit cell constants have been determined also at 4 K. The infrared and Raman spectra are known in a wide range of temperatures including liquid Helium. Very recently the complete set of phonon dispersion curves in the lattice frequency region has been measured at 6 K by means of coherent neutron scattering [100] .

Calculations of the dispersion curves of naphthalene-d_8 have been made by the authors of the experimental neutron scattering experiments using atom-atom potentials of the Kitaigorodsky and William type [100]. These potentials reproduce the gross features of the dispersion curves but give rise to calculated frequencies that are sometime about 20 % off. A drastic improvement of the situation has been obtained[101,102] by using a mixed potential which includes, in addition to the atom-atom part, a quadrupole-quadrupole interaction. The calculated and observed dispersion curves are compared in Fig.3.4. The agreement between the measured and the calculated dispersion curves is extremely satisfactory. The discrepancies still existing between calculations and experiments are essentially due to the fact that the calculated acoustic branches in the (0,**k**,0) and in the the (**k**,0,0) directions are somewhat lower than the observed branches. This is probably due to the fact that a simple atom-atom potential does not correctly account for the anisotropy of the repulsive and dispersion interactions in molecules with π electron inter actions. It is, however, also possible that the representation of the electrostatic potential by means of point quadrupoles in the center of mass of the molecule is not very accurate when one deals with very large molecules. A more sophisticated intermolecular potential in which the anisotropy of the atom-atom part is taken into account and the electrostatic interactions are more realistically localized over the whole molecule should improve the agreement in the acoustic branches region.

Several studies of the dynamics of crystalline ethylene were made in the past. Owing to the fact that the crystal structure was uncertain, these researches were more oriented toward the identification of the

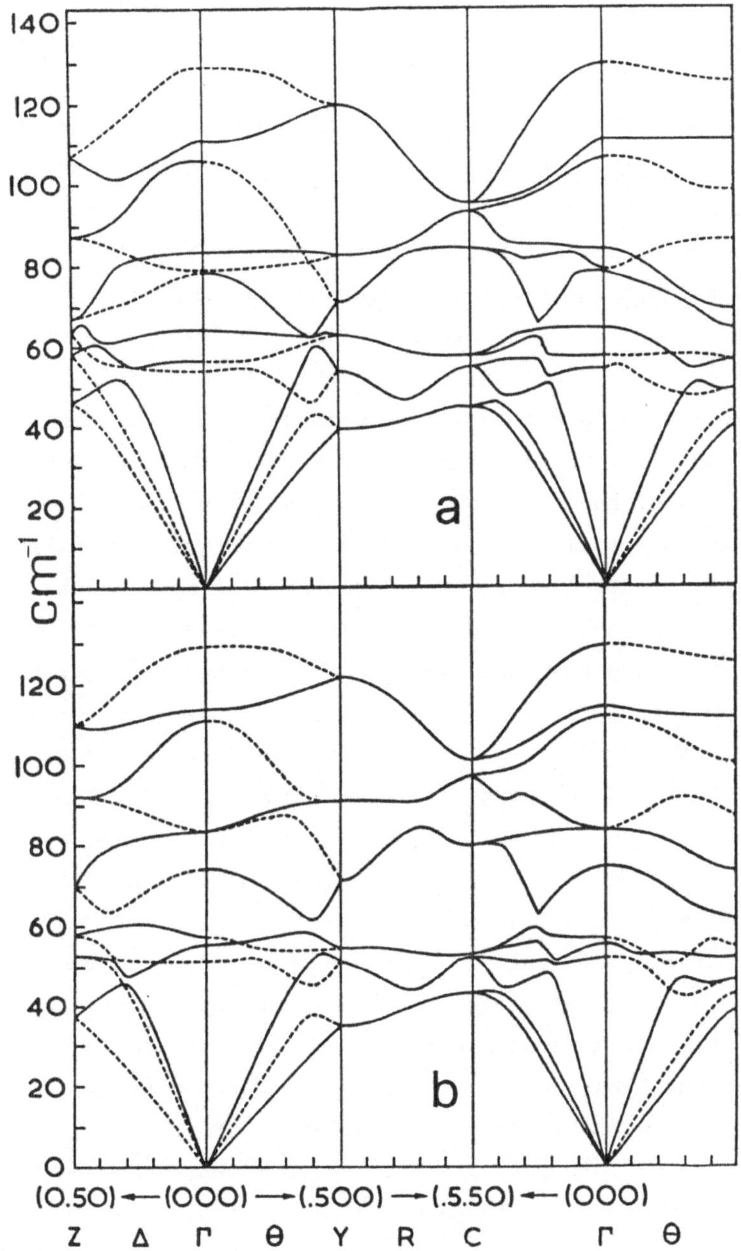

Fig.3.4 Observed (a) and calculated (b) phonon dispersion curves of naphthalene-d_8

correct molecular orientation of the molecules in the unit cell than toward the determination of a convenient intermolecular potential. The crystal structure of ethylene has been determined recently by X-rays and neutron scattering and this has stimulated theoretical researches on the intermolecular interactions and on the dynamical properties of the crystal. An interesting calculation of the intermolecular potential between ethylene molecules has recently been made by Wasiutynski et al. [103]. These authors first calculated, by means of an ab-initio SCF-MO method, the interaction energy of an ethylene dimer for several intermolecular distances and relative orientations of the two molecules. The numerical results so obtained were then fitted with an analytical form of intermolecular potential. The short-range part of the interaction was represented by means of an atom-atom potential and the long-range part by an electrostatic potential. A detailed analysis of the possibilities and of the limits of this fitting procedure was also made by the authors who concluded that the basic deficiencies of the atom-atom model in its simplest form forces the potentials,which give the best fit to the short range part of the ab-initio calculations,to unrealistic forms. The agreement between observed and calculated lattice frequencies and crystal parameters is altogether satisfactory although empirical potentials of the standard type give still rise to a better fit. The comparison between observed and calculated static and dynamical crystal properties for ethylene is given in Table 3.9.

Less satisfactory results have been obtained for acethylene. Two ordered crystal forms of acethylene are known. The first is cubic, space group Pa3 with four molecules per unit cell and the second is orthorhombic, space group Cmca (D_{2h}^{18}) with two molecules per primitive unit cell. Accurate crystallographic data are available for the cubic structure only. The spectroscopic data are also not complete. Some lattice frequencies have been measured in infrared and Raman for both forms but their assignment seems still doubtful. Filippini et al. [104] have recently calculated the lattice frequencies of the two crystal forms. They found that standard atom-atom potentials, normally utilized for hydrocarbons, produce lattice frequencies that are more than 100 % too high and lattice energies that are too low. Inclusion of an electrostatic contribution

Table 3.9 Observed and Calculated Crystal Structure and Lattice Frequencies for Crystalline Ethylene with Ab-initio and Empirical Potentials

		Obs.	I	II	III	IV
Unit Cell Parameters	a(A)	4.626	4.730	4.844	4.856	4.726
	b(A)	6.620	6.205	6.516	6.342	6.435
	c(A)	4.067	4.005	3.886	3.832	4.135
	β(degrees)	94.4	88.5	99.	96.8	94.
Molecular § Orientation (degrees)	ξ	-27.0	-31.7	-34.4	-36.2	-27.0
	η	-14.6	-9.3	-7.4	-5.0	-11.7
	ζ	-34.3	-31.3	-28.7	-27.8	-31.8
Cohesion Energy (kcal/mol)		4.7	5.46	4.28	4.04	3.92
Lattice Frequencies (cm^{-1})		--	75	52	60	63
		73	94	66	67	75
		110	126	111	115	113
		73	50	56	55	38
		90	80	76	80	87
		97	105	82	100	90
		114	157	103	114	131
		167	196	140	183	163
		177	249	142	193	170

§ for the definition of the orientation angles see Ref. 103

I = Ab-initio potential with electrostatic interactions. Ref. 103

II = " " " " without " " " " . Ref. 103

III = Williams potential. Ref. 103

IV = Williams potential. Ref. 103

in the intermolecular potential does not seem to improve the agreement. The authors have refined the potential and have been able to reproduce correctly the Raman active lattice modes of the cubic form but not the lattice energy. Furthermore they found that the same potential gives very unsatisfactory results for the orthorhombic phase.In this latter case imaginary frequencies were obtained for some points of the Brillouin zone.

We consider now in less detail some other calculations of lattice vibrations of molecular crystals which were published in the last few years and which are of some relevance for the determination of the intermolecular potential.

Intermolecular potentials of the atom-atom type are well established for aliphatic hydrocarbons, thanks to the extensive work of Kitaigorodski [49,50] and Williams [47,48]. These potentials have been used with success to calculate the lattice vibrations of n-butane [105,106], n-hexane [106] and adamantane [107] but are inadequate to explain correctly the dynamics and the phase transitions in methane [44]

Atom-atom potentials alone or with the addition of electrostatic interactions have been also utilized successfully for the calculation of the lattice vibrations of molecular crystals made of molecules containing hetero-atoms as for instance CS_2 [108], ethylenediamine [109], 1,3,5-trioxane and 1,3,5,7-tetroxane [110] and succinic anhydride [111]. In this latter case, inclusion of a dipole-dipole interaction term in the intermolecular potential was found to have a negligible effect on the lattice frequencies but to play an important role in the crystal energy and equilibrium structure.

Benzene derivatives has been also extensively investigated, especially by H.Bonadeo and his group,to check the validity of the atom-atom potential and to learn about its transferability among similar molecules. Benzene itself was actually one of the first polyatomic molecules for which the complete optical crystal spectrum was calculated with the aid of a simple atom-atom potential [16]. Later the calculations were extended to the complete Brillouin zone in conjunction with neutron scattering measurements [112]. More recent calculations proved that inclusion of a quadrupole-quadrupole interaction term in the intermolecular potential has almost no influence on the lattice frequencies [101]. The atom-atom potential was then used to calculate the lattice frequencies and other crystal properties of several chloro- and bromo-benzenes including p-dichlorobenzene [114], p-dibromobenzene [115], symm. tribromobenzene [115], 1,2,4,5-tetrachloro- [113] and tetrabromobenzene [115], diiodobenzene [116] and even heterocyclic rings such as pyrazine, triazine and tetrazine [117].

CHAPTER 4

ANHARMONIC INTERACTIONS

4.1 INTRODUCTION

In the previous Chapters we have discussed the vibrational properties of molecular crystals in terms of harmonic motions of the molecules, taking into account only quadratic terms of the crystal potential. This "harmonic" approximation corresponds to the assumption that the vibrational excitation of the crystal can be rigorously separated into a set of independent and non-interacting phonons, each having an infinite lifetime.

The harmonic approximation is undoubtedly of extreme importance in the interpretation of the main features of the infrared, Raman and neutron scattering spectra of molecular solids at low temperatures when displacements from equilibrium positions are small. It is, however, insufficient to explain all the fine details of the vibrational spectra and inadequate to account for many physical properties of these crystals, especially for those which depend on the temperature.

For a more complete treatment we have to consider anharmonic motions of the molecules, allowing for the possibility of interactions among the phonons, through terms of the crystal potential higher than the quadratic ones.

The occurrence of phonon-phonon interactions has several experimental manifestations. Here we are especially concerned with those of interest for solid-state spectroscopy. These can be schematized as follows:
a) Anharmonic phonon frequencies are shifted with respect to the harmonic values. If a sufficient number of overtones and combination bands could be observed in the spectra, the anharmonic shifts could be directly extracted from the vibrational assignment and would constitute an important piece of experimental information on the anharmonic behaviour of the crystal. In practice, however, overtones and combinations are observed only for internal vibrations and very seldom have been identi-

fied for external vibrations. In addition overtones and combinations are frequently observed as broad features that include contributions from different phonons of a given dispersion branch. It is therefore not easy to obtain from experiments the anharmonic shifts of single phonons.

b) Experimental phonons have finite lifetimes even at very low temperatures. The lifetimes are directly connected with the width of the crystal bands and constitute thus an excellent source of information on the anharmonicity of the crystal potential.

c) Phonon frequencies and bandwidths are temperature- and volume-dependent.

d) Multiphonon bands,due to the contemporary excitation of two or more interacting phonons, are often observed in the crystal spectra.

In this chapter we shall present the general treatment of anharmonic crystal vibrations and concentrate our attention on the first three aspects of the problem. Multiphonon bands will be specifically dealt with in the following Chapter.

The theory of anharmonic vibrations of molecular crystals is an extension of the corresponding theory developed for atomic lattices in the last two decades. The main difference is that in the case of molecular crystals molecular coordinates (translations, rotations and internal coordinates) are used instead of atomic displacements. Furthermore in this case one has to deal with two different types of anharmonicity, the one involving internal vibrations of the molecules and the one concerning the lattice vibrations. In principle, both can be treated with the same techniques discussed in this Chapter. It is,however, necessary to consider that the anharmonicity of the lattice vibrations arises from cubic and higher-order terms of the intermolecular potential whereas in the case of internal vibrations is the intramolecular potential that matters.

When the anharmonicity is small, the standard perturbation theory can be used to work out the anharmonic corrections to the phonon energies and to evaluate the lifetimes of the phonon states. When the anharmonicity is large more sophisticated perturbation techniques are necessary.

A convenient parameter for classifying the order of the perturbation terms in these treatments is the ratio ε between the mean square amplitude of vibration and the average next-neighbor distance. If we write the crystal Hamiltonian in orders of magnitude

$$H = H_0 + H_1 + H_2 + \ldots + H_m + \ldots \qquad 4.1.1$$

where H_0 is the Hamiltonian for the harmonic problem and H_1, H_2, H_m represent cubic,quartic and mth order terms of the crystal potential,respectively, we see that H_1 is of the order εH_0,H_2 of the order $\varepsilon^2 H_0$ and, in general,H_m of the order $\varepsilon^m H_0$. Usually $\varepsilon \ll 1$ and thus these terms are small with respect to $H_0$and can be treated as a perturbation to the harmonic Hamiltonian. Their effect is to couple together the unperturbed states, thus leading to a typical many-body problem [119-121]. The formal and powerful techniques of the many-body theory have thus been widely utilized to analyze the physical properties of anharmonic crystals. The great advantage of these approaches is that they are very general and cover not only cases of low anharmonicity,but also cases in which displacements from equilibrium are large and one can reasonably suspect a slow convergence of the ε expansion discussed above.

In this Chapter we discuss two different many-body perturbation methods for the treatment of anharmonicity. The first is a method of renormalization of the crystal Hamiltonian due to Wallace [6,122]. The second is the better known method of the temperature Green's functions [119-121].Both are very general and, to the second order, give exactly the same results for the anharmonic shifts and lifetimes. In addition, we also illustrate a simplified approach, the so called " self-consistent phonon" method that is largely utilized for the calculation of the anharmonic shifts [124].

4.2 THE CRYSTAL HAMILTONIAN

For the anharmonic treatment of the vibrations of a molecular crystal, we need first to express the crystal Hamiltonian in a convenient form. In this Section we will thus describe the general structure of the classical Hamiltonian of the vibrating crystal and introduce some basic simplifications arising from translational symmetry.

In terms of molecular coordinates the kinetic and the potential energy of the crystal are given by 1.6.11 and 1.6.13 respectively. For our purposes it is, however, convenient to express the Hamiltonian in terms of the crystal normal coordinates $Q_{j\mathbf{k}}$ defined in Chapter 1. Using then 1.6.26 and 1.6.36 we have

$$R_{\ell}^{m\mu} = L^{-\frac{1}{2}} \sum_{j\mathbf{k}} E(\ell\mu|j\mathbf{k}) Q_{j\mathbf{k}} \; e^{i\mathbf{k}\cdot\mathbf{R}_m} \qquad 4.2.1$$

where

$$E(\ell\mu|j\mathbf{k}) = \sum_t L_t^{\ell}(\mu) e(t\mu|j\mathbf{k}) \qquad 4.2.2$$

We notice that 4.2.2 is the most general expression for the crystal eigenvectors and covers both external and internal coordinates. Only in this latter case the sum extends over all internal modes. For external coordinates, according to 1.6.26, $L_t^{\ell}(\mu) = \delta_{\ell t}$ and thus $E(\ell\mu|j\mathbf{k}) = e(\ell\mu|j\mathbf{k})$.

By substitution of 4.2.1 in 1.6.11 and 1.6.13 we obtain

$$T = \frac{1}{2} \sum_{j\mathbf{k}} \sum_{j_1\mathbf{k}_1} G(j,\mathbf{k}) \dot{Q}_{j\mathbf{k}} \dot{Q}_{j_1\mathbf{k}_1} \qquad 4.2.3$$

$$V_2 = \frac{1}{2!} \sum_{j\mathbf{k}} \sum_{j_1\mathbf{k}_1} C(jj_1,\mathbf{k}\mathbf{k}_1) Q_{j\mathbf{k}} Q_{j_1\mathbf{k}_1} \qquad 4.2.4a$$

$$V_3 = \frac{1}{3!} \sum_{j\mathbf{k}} \sum_{j_1\mathbf{k}_1} \sum_{j_2\mathbf{k}_2} C(jj_1j_2,\mathbf{k}\mathbf{k}_1\mathbf{k}_2) Q_{j\mathbf{k}} Q_{j_1\mathbf{k}_1} Q_{j_2\mathbf{k}_2} \qquad 4.2.4b$$

$$V_4 = \frac{1}{4!} \sum_{j\mathbf{k}} \sum_{j_1\mathbf{k}_1} \sum_{j_2\mathbf{k}_2} \sum_{j_3\mathbf{k}_3} C(jj_1j_2j_3,\mathbf{k}\mathbf{k}_1\mathbf{k}_2\mathbf{k}_3) Q_{j\mathbf{k}} Q_{j_1\mathbf{k}_1} Q_{j_2\mathbf{k}_2} Q_{j_3\mathbf{k}_3} \qquad 4.2.4c$$

where, from 4.2.2 and 1.6.28a,

$$G(j\mathbf{k}) = L^{-1} \sum_{m\mu} \sum_{\ell} E(\ell\mu|j\mathbf{k}) E(\ell\mu|j_1\mathbf{k}_1) \; e^{i(\mathbf{k}+\mathbf{k}_1)\cdot\mathbf{R}_m}$$

$$C(jj_1,\mathbf{k}\mathbf{k}_1) = L^{-1} \sum_{m\mu} \sum_{n\nu} \sum_{\ell\ell_1} F_{\ell\ell_1} \binom{m\mu}{n\nu} E(\ell\mu|j\mathbf{k}) E(\ell_1\nu_1|j_1\mathbf{k}_1) e^{i(\mathbf{k}\cdot\mathbf{R}_m + \mathbf{k}_1\cdot\mathbf{R}_n)}$$

$$C(jj_1j_2,\mathbf{k}\mathbf{k}_1\mathbf{k}_2) = L^{-3/2} \sum_{m\mu} \sum_{n\nu} \sum_{p\pi} \sum_{\ell\ell_1\ell_2} F_{\ell\ell_1\ell_2} \begin{pmatrix} m\mu \\ n\nu \\ p\pi \end{pmatrix} \times$$

$$\times E(\ell\mu|j\mathbf{k})E(\ell_1\nu|j_1\mathbf{k}_1)E(\ell_2\pi|j_2\mathbf{k}_2)e^{i(\mathbf{k}\cdot\mathbf{R}_m + \mathbf{k}_1\cdot\mathbf{R}_n + \mathbf{k}_2\cdot\mathbf{R}_p)} \qquad 4.2.5$$

$$C(jj_1j_2j_3,\mathbf{k}\mathbf{k}_1\mathbf{k}_2\mathbf{k}_3) = L^{-2}\sum_{m\mu}\sum_{n\nu}\sum_{p\pi}\sum_{r\rho}\sum_{\ell\ell_1\ell_2\ell_3}F_{\ell\ell_1\ell_2\ell_3}\binom{m\mu\ \ p\pi}{n\nu\ \ r\rho}E(\ell\mu|j\mathbf{k})\times$$
$$\times E(\ell_1\nu|j_1\mathbf{k}_1)E(\ell_2\pi|j_2\mathbf{k}_2)E(\ell_3\rho|j_3\mathbf{k}_3)e^{i(\mathbf{k}\cdot\mathbf{R}_m+\mathbf{k}_1\cdot\mathbf{R}_n+\mathbf{k}_2\cdot\mathbf{R}_p+\mathbf{k}_3\cdot\mathbf{R}_r)}$$

The translational symmetry of the crystal helps in simplifying the coefficients C in 4.2.5, since it introduces restrictions on the **k** vectors which give non-zero contributions in 4.2.5. Actually, it is easily seen that **a coefficient** of order **i+1** involves a function $\delta(\mathbf{k}+\mathbf{k}_1+\ldots\mathbf{k}_i)$ such that

$$\delta(\mathbf{k} + \mathbf{k}_1+\ldots+\mathbf{k}_i) = \begin{cases}1 & \text{if } \mathbf{k} + \mathbf{k}_1 + \ldots + \mathbf{k}_i = 0 \text{ or } \mathbf{K}\\ 0 & \text{otherwise}\end{cases} \qquad 4.2.6$$

K being a reciprocal lattice vector. First, as an example, consider the quadratic coefficient. By subtracting $\mathbf{R}_m$ from $\mathbf{R}_n$ and using the symmetry relation 1.6.17 we obtain

$$\sum_{mn}F_{\ell\ell_1}\binom{m\mu}{n\nu}e^{i(\mathbf{k}\cdot\mathbf{R}_m + \mathbf{k}_1\cdot\mathbf{R}_n)} =$$
$$= \sum_{mn}F_{\ell\ell_1}\binom{o\ \ \ \ \mu}{n-m\ \ \nu}e^{i\mathbf{k}_1\cdot(\mathbf{R}_n - \mathbf{R}_m)}\,e^{i(\mathbf{k}+\mathbf{k}_1)\cdot\mathbf{R}_m} \qquad 4.2.7$$

and by putting $\mathbf{R}_n - \mathbf{R}_m = \mathbf{R}_n$ and n - m = n,we can rewrite 4.2.7 as

$$\sum_{mn}F_{\ell\ell_1}\binom{o\mu}{n\nu}e^{i\mathbf{k}_1\cdot\mathbf{R}_n}\,e^{i(\mathbf{k}+\mathbf{k}_1)\cdot\mathbf{R}_m} \qquad 4.2.8$$

Thus , using the property of the lattice sum

$$\sum_m e^{i(\mathbf{k}+\mathbf{k}_1)\cdot\mathbf{R}_m} = L\delta(\mathbf{k}+\mathbf{k}_1) \qquad 4.2.9$$

we obtain

$$\sum_{mn}F_{\ell\ell_1}\binom{m\mu}{n\nu}e^{i(\mathbf{k}\cdot\mathbf{R}_m + \mathbf{k}_1\cdot\mathbf{R}_n)} = L\delta(\mathbf{k}+\mathbf{k}_1)\sum_n F_{\ell\ell_1}\binom{o\mu}{n\nu}e^{i\mathbf{k}\cdot\mathbf{R}_n} \qquad 4.2.10$$

This argument is easily extended to the higher-order coefficients. By subtracting $\mathbf{R}_m$ from all other position vectors of unit cells,we introduce a factor $\exp[(\mathbf{k} + \mathbf{k}_1 + \ldots + \mathbf{k}_i)\cdot\mathbf{R}_m]$. Since the force constants of any order depend only on the relative positions of the molecules, they obey relations of the type

$$F_{\ell\ell_1\ell_2}\begin{pmatrix} m\mu \\ n\nu \\ p\pi \end{pmatrix} = F_{\ell\ell_1\ell_2}\begin{pmatrix} o & \mu \\ n-m & \nu \\ p-m & \pi \end{pmatrix} \quad ; \quad F_{\ell\ell_1\ell_2\ell_3}\begin{pmatrix} m\mu & p\pi \\ n\nu & r\rho \end{pmatrix} = F_{\ell\ell_1\ell_2\ell_3}\begin{pmatrix} o\ \mu & p-m\pi \\ n-m\nu & r-m\rho \end{pmatrix}$$

and therefore, using the condition 4.2.9 in the general form

$$\sum_m e^{i(\mathbf{k} + \mathbf{k}_1 + \ldots + \mathbf{k}_i)\cdot\mathbf{R}_m} = L\delta(\mathbf{k} + \mathbf{k}_1 + \ldots + \mathbf{k}_i) \qquad 4.2.11$$

we obtain

$$\sum_m F_{\ell\ell_1\ell_2\ldots}\begin{pmatrix} m & \mu \\ n & \nu \\ \cdot & \cdot \end{pmatrix} e^{i(\mathbf{k}\cdot\mathbf{R}_m + \mathbf{k}_1\cdot\mathbf{R}_n + \mathbf{k}_2\cdot\mathbf{R}_p + \ldots)} = \qquad 4.2.12$$

$$= L\delta(\mathbf{k} + \mathbf{k}_1 + \mathbf{k}_2 + \ldots)F_{\ell\ell_1\ell_2\ldots}\begin{pmatrix} m & \mu \\ n & \nu \\ \cdot & \cdot \end{pmatrix} e^{i(\mathbf{k}_1\cdot\mathbf{R}_n + \mathbf{k}_2\cdot\mathbf{R}_p + \ldots)}$$

The coefficients C are thus

$$C(jj_1,\mathbf{k}\mathbf{k}_1) = \delta(\mathbf{k} + \mathbf{k}_1)\sum_{\mu\nu}\sum_{\ell\ell} D^{\mu\nu}_{\ell\ell_1}(\mathbf{k}_1)E(\ell\mu|j\mathbf{k})E(\ell_1\nu|j_1\mathbf{k}_1)$$

$$C(jj_1j_2,\mathbf{k}\mathbf{k}_1\mathbf{k}_2) = \delta(\mathbf{k} + \mathbf{k}_1 + \mathbf{k}_2)L^{-\frac{1}{2}}\sum_{\mu\nu\pi}\sum_{\ell\ell_1\ell_2} D^{\mu\nu\ \pi}_{\ell\ell_1\ell_2}(\mathbf{k}_1\mathbf{k}_2)\times \qquad 4.2.13$$

$$\times E(\ell\mu|j\mathbf{k})E(\ell_1\nu|j_1\mathbf{k}_1)E(\ell_2\pi|j_2\mathbf{k}_2)$$

$$C(jj_1j_2j_3,\mathbf{k}\mathbf{k}_1\mathbf{k}_2\mathbf{k}_3) = \delta(\mathbf{k} + \mathbf{k}_1 + \mathbf{k}_2 + \mathbf{k}_3)L^{-1}\sum_{\mu\nu\pi\rho}\sum_{\ell\ell_1\ell_2\ell_3} D^{\mu\nu\ \pi\ \rho}_{\ell\ell_1\ell_2\ell_3}(\mathbf{k}_1\mathbf{k}_2\mathbf{k}_3)\times$$

$$\times E(\ell\mu|j\mathbf{k})E(\ell_1\nu|j_1\mathbf{k}_1)E(\ell_2\pi|j_2\mathbf{k}_2)E(\ell_3\rho|j_3\mathbf{k}_3)$$

where

$$D^{\mu\nu}_{\ell\ell_1}(\mathbf{k}_1) = \sum_n F_{\ell\ell_1}\begin{pmatrix} o\mu \\ n\nu \end{pmatrix} e^{i\mathbf{k}_1\cdot\mathbf{R}_n}$$

$$D^{\mu\nu\ \pi}_{\ell\ell_1\ell_2}(\mathbf{k}_1\mathbf{k}_2) = \sum_{np} F_{\ell\ell_1\ell_2}\begin{pmatrix} o\mu \\ n\nu \\ p\pi \end{pmatrix} e^{i(\mathbf{k}_1\cdot\mathbf{R}_n + \mathbf{k}_2\cdot\mathbf{R}_p)} \qquad 4.2.14$$

$$D^{\mu\nu\ \pi\ \rho}_{\ell\ell_1\ell_2\ell_3}(\mathbf{k}_1\mathbf{k}_2\mathbf{k}_3) = \sum_{npr} F_{\ell\ell_1\ell_2\ell_3}\begin{pmatrix} o\mu & p\pi \\ n\nu & r\rho \end{pmatrix} e^{i(\mathbf{k}_1\cdot\mathbf{R}_n + \mathbf{k}_2\cdot\mathbf{R}_p + \mathbf{k}_3\cdot\mathbf{R}_r)}$$

are generalized dynamical tensors elements. In particular those of the second order are the elements of the dynamical matrix discussed in the first Chapter for the harmonic problem.

The coefficients 4.2.14 can be further simplified when pairwise intermolecular potentials are used. We discuss here the cubic term in detail but the procedure is easily generalized.

When an intermolecular potential of the type

$$V_I = \frac{1}{2}\sum_{m\mu}\sum_{n\nu}{}' V\binom{m\mu}{n\nu} \qquad 4.2.15$$

is used, it is obvious that the cubic force constants, according to their definition 1.6.14, are equal to zero if the three molecules $m\mu$, $n\nu$ and $p\pi$ are different from one another. The only non-zero terms are those of the type

$$\begin{array}{lll} 1) & F_{\ell\ell_1\ell_2}\begin{Bmatrix} o\mu \\ o\mu \\ p\pi \end{Bmatrix} & (o\mu) = (n\nu) \\ 2) & F_{\ell\ell_1\ell_2}\begin{Bmatrix} o\mu \\ n\nu \\ o\mu \end{Bmatrix} & (o\mu) = (p\pi) \\ 3) & F_{\ell\ell_1\ell_2}\begin{Bmatrix} o\mu \\ n\nu \\ n\nu \end{Bmatrix} & (n\nu) = (p\pi) \\ 4) & F_{\ell\ell_1\ell_2}\begin{Bmatrix} o\mu \\ o\mu \\ o\mu \end{Bmatrix} & (o\mu) = (n\nu) = (p\pi) \end{array} \qquad 4.2.16$$

In the first case $\mathbf{R}_n = 0$ and the third order dynamical tensor elements in 4.2.14 become

$$D^{\mu\mu\ \pi}_{\ell\ell_1\ell_2}(\mathbf{k}_1\mathbf{k}_2) = \sum_p{}' F_{\ell\ell_1\ell_2}\begin{Bmatrix} o\mu \\ o\mu \\ p\pi \end{Bmatrix} e^{i\mathbf{k}_2\cdot\mathbf{R}_p} = D^{\mu\mu\ \pi}_{\ell\ell_1\ell_2}(\mathbf{k}_2) \qquad 4.2.17$$

where the prime indicates that $p \neq o$ when $\mu = \pi$.

In the second case, $\mathbf{R}_p = 0$ and thus

$$D^{\mu\nu\ \mu}_{\ell\ell_1\ell_2}(\mathbf{k}_1\mathbf{k}_2) = \sum_n{}' F_{\ell\ell_1\ell_2}\begin{Bmatrix} o\mu \\ n\nu \\ o\mu \end{Bmatrix} e^{i\mathbf{k}_1\cdot\mathbf{R}_n} = D^{\mu\nu\ \mu}_{\ell\ell_1\ell_2}(\mathbf{k}_1) \qquad 4.2.18$$

In the third case, $\mathbf{R}_n = \mathbf{R}_p$ and thus

$$D_{\ell\ell_1\ell_2}(\mathbf{k}_1\mathbf{k}_2) = \sum_n{}' F_{\ell\ell_1\ell_2}\begin{pmatrix}o\mu\\n\nu\\n\nu\end{pmatrix} e^{i(\mathbf{k}_1+\mathbf{k}_2)\cdot\mathbf{R}_n} = \qquad 4.2.19$$

$$= \sum_n{}' F_{\ell\ell_1\ell_2}\begin{pmatrix}m\mu\\o\nu\\o\nu\end{pmatrix} e^{i\mathbf{k}\cdot\mathbf{R}_m} = D_{\ell\ell_1\ell_2}(\mathbf{k})$$

Finally, in the fourth case

$$D_{\ell\ell_1\ell_2}(\mathbf{k}_1\mathbf{k}_2) = F_{\ell\ell_1\ell_2}\begin{pmatrix}o\mu\\o\mu\\o\mu\end{pmatrix} = D_{\ell\ell_1\ell_2}(0) \qquad 4.2.20$$

Substituting these relations in 4.2.13, we obtain

$$C(jj_1j_2,\mathbf{k}\mathbf{k}_1\mathbf{k}_2) = \sum_{\mu\nu}\sum_{\ell\ell_1\ell_2}[\, D^{\mu\mu\;\mu}_{\ell\ell_1\ell_2}(0)E(\ell\mu|j\mathbf{k})E(\ell_1\mu|j_1\mathbf{k}_1)E(\ell_2\mu|j_2\mathbf{k}_2) +$$

$$+ D^{\nu\mu\;\mu}_{\ell\ell_1\ell_2}(\mathbf{k})E(\ell\nu|j\mathbf{k})E(\ell_1\mu|j_1\mathbf{k}_1)E(\ell_2\mu|j_2\mathbf{k}_2) + \qquad 4.2.21$$

$$+ D^{\mu\nu\;\mu}_{\ell\ell_1\ell_2}(\mathbf{k}_1)E(\ell\mu|j\mathbf{k})E(\ell_1\nu|j_1\mathbf{k}_1)E(\ell_2\mu|j_2\mathbf{k}_2) +$$

$$+ D^{\mu\mu\;\nu}_{\ell\ell_1\ell_2}(\mathbf{k}_2)E(\ell\mu|j\mathbf{k})E(\ell_1\mu|j_1\mathbf{k}_1)E(\ell_2\nu|j_2\mathbf{k}_2)]L^{-\frac{1}{2}}\delta(\mathbf{k}+\mathbf{k}_1+\mathbf{k}_2)$$

4.3 QUANTUM FIELD TREATMENT OF PHONONS

In this Section we rewrite the Hamiltonian using the formalism of quantum field theory and introduce some useful expressions for future developments. We first introduce dimensionless normal coordinates $A_{j\mathbf{k}}$ and associate momenta $P_{j\mathbf{k}}$, through the relations [4,123]

$$Q_{j\mathbf{k}} = (\hbar/2\omega_{j\mathbf{k}})^{\frac{1}{2}}A_{j\mathbf{k}} \qquad 4.3.1$$

$$\dot{Q}_{j-\mathbf{k}} = i(\hbar\omega_{j\mathbf{k}}/2)^{\frac{1}{2}}P_{j\mathbf{k}} \qquad 4.3.1a$$

We then express $A_{j\mathbf{k}}$ and $P_{j\mathbf{k}}$ in terms of phonon-creation and -annihilation operators $a^{\dagger}_{j\mathbf{k}}$ and $a_{j\mathbf{k}}$

$$A_{j\mathbf{k}} = a^{\dagger}_{j-\mathbf{k}} + a_{j\mathbf{k}} \qquad 4.3.2$$

$$P_{j\mathbf{k}} = a^{\dagger}_{j\mathbf{k}} - a_{j-\mathbf{k}}$$

The operators $a^{\dagger}_{j\mathbf{k}}$ and $a_{j\mathbf{k}}$ have the property of modifying the state vectors of the harmonic Hamiltonian by changing the number of phonons belonging to the branch j with wave vector **k**. In particular $a^{\dagger}_{j\mathbf{k}}$ will create and $a_{j\mathbf{k}}$ will annihilate a phonon of energy $\hbar\omega_{j\mathbf{k}}$. Correspondingly, they will increase or decrease by one, respectively, the occupation number $n_{j\mathbf{k}}$, according to the relations

$$a^{\dagger}_{j\mathbf{k}}|\ldots n_{j\mathbf{k}}\ \ldots\rangle = (n_{j\mathbf{k}}+1)^{\frac{1}{2}}|\ldots\ n_{j\mathbf{k}}+1\ \ldots\rangle$$
$$a_{j\mathbf{k}}|\ldots n_{j\mathbf{k}}\ \ldots\rangle = n_{j\mathbf{k}}^{\frac{1}{2}}|\ldots\ n_{j\mathbf{k}}-1\ \ldots\rangle \qquad 4.3.3$$

where $|\ldots n_{j\mathbf{k}}\ \ldots\ \rangle$ is a state vector in which, for simplicity, we have specified only the occupation number $n_{j\mathbf{k}}$.

The creation and annihilation operators obey the basic commutation relations

$$\left[a_{j_1\mathbf{k}_1}, a^{\dagger}_{j_2\mathbf{k}_2}\right] = \delta_{j_1 j_2}\delta_{\mathbf{k}_1\mathbf{k}_2}$$
$$\left[a^{\dagger}_{j_1\mathbf{k}_1}, a^{\dagger}_{j_2\mathbf{k}_2}\right] = \left[a_{j_1\mathbf{k}_1}, a_{j_2\mathbf{k}_2}\right] = 0 \qquad 4.3.4$$

and are related to the number operator $n_{j\mathbf{k}}$ by

$$a^{\dagger}_{j\mathbf{k}}a_{j\mathbf{k}} = n_{j\mathbf{k}}$$
$$a_{j\mathbf{k}}a^{\dagger}_{j\mathbf{k}} = n_{j\mathbf{k}} + 1 \qquad 4.3.5$$

We utilize now these relations to express the Hamiltonian of the vibrating crystal in terms of the operators $a^{\dagger}_{j\mathbf{k}}$ and $a_{j\mathbf{k}}$. According to the previous Section, the Hamiltonian can be written in the form

$$H = H_0 + H_1 + H_2 + \ldots \qquad 4.3.6$$

where

$$H_0 = T + V_2 = \frac{1}{2}\sum_{j\mathbf{k}}\left[\dot{Q}_{j\mathbf{k}}\dot{Q}_{j-\mathbf{k}} + \omega^2_{j\mathbf{k}}Q_{j\mathbf{k}}Q_{j-\mathbf{k}}\right] \qquad 4.3.7$$

is the harmonic Hamiltonian and $H_1 = V_3$, $H_2 = V_4$. Higher terms of the potential will not be considered in our treatment since inclusion of the cubic and quartic terms in the perturbation treatment is sufficient to give the second-order correction to the energy.

Using 4.3.1 and 4.3.2 we have then

$$H_0 = \frac{1}{2}\sum_{j\mathbf{k}}\hbar\omega_{j\mathbf{k}}[\,a^{\dagger}_{j\mathbf{k}}a_{j\mathbf{k}} + a_{j\mathbf{k}}a^{\dagger}_{j\mathbf{k}}\,] \qquad 4.3.8$$

where we have used the condition $\omega_{j\mathbf{k}} = \omega_{j-\mathbf{k}}$ and the property [123] that the sum over $\mathbf{k}$ runs symmetrically over positive and negative values, so that

$$\sum_{\mathbf{k}}a^{\dagger}_{j-\mathbf{k}}a_{j-\mathbf{k}} = \sum_{\mathbf{k}}a^{\dagger}_{j\mathbf{k}}a_{j\mathbf{k}} \qquad 4.3.9$$

By means of 4.3.4, H_0 can be written in the more compact form

$$H_0 = \sum_{j\mathbf{k}}\hbar\omega_{j\mathbf{k}}(a^{\dagger}_{j\mathbf{k}}a_{j\mathbf{k}} + \tfrac{1}{2}) = \sum_{j\mathbf{k}}\hbar\omega_{j\mathbf{k}}(n_{j\mathbf{k}} + \tfrac{1}{2}) \qquad 4.3.10$$

In the same way, higher terms of the Hamiltonian assume the form

$$H_1 = \sum_{jj_1j_2}\sum_{\mathbf{k}\mathbf{k}_1\mathbf{k}_2}B\binom{jj_1j_2}{\mathbf{k}\mathbf{k}_1\mathbf{k}_2}A_{j\mathbf{k}}A_{j_1\mathbf{k}_1}A_{j_2\mathbf{k}_2}$$
$$H_2 = \sum_{jj_1j_2j_3}\sum_{\mathbf{k}\mathbf{k}_1\mathbf{k}_2\mathbf{k}_3}B\binom{jj_1j_2j_3}{\mathbf{k}\mathbf{k}_1\mathbf{k}_2\mathbf{k}_3}A_{j\mathbf{k}}A_{j_1\mathbf{k}_1}A_{j_2\mathbf{k}_2}A_{j_3\mathbf{k}_3} \qquad 4.3.11$$

where

$$B\binom{jj_1j_2}{\mathbf{k}\mathbf{k}_1\mathbf{k}_2} = \frac{1}{3!}\left(\frac{\hbar^3}{8\omega_{j\mathbf{k}}\omega_{j_1\mathbf{k}_1}\omega_{j_2\mathbf{k}_2}}\right)^{\frac{1}{2}}C(jj_1j_2,\mathbf{k}\mathbf{k}_1\mathbf{k}_2)$$
$$B\binom{jj_1j_2j_3}{\mathbf{k}\mathbf{k}_1\mathbf{k}_2\mathbf{k}_3} = \frac{1}{4!}\left(\frac{\hbar^4}{16\omega_{j\mathbf{k}}\omega_{j_1\mathbf{k}_1}\omega_{j_2\mathbf{k}_2}\omega_{j_3\mathbf{k}_3}}\right)^{\frac{1}{2}}C(jj_1j_2j_3,\mathbf{k}\mathbf{k}_1\mathbf{k}_2\mathbf{k}_3) \qquad 4.3.12$$

Commutation relations are very useful in working with creation and annihilation operators. The most important are those with the Hamiltonian operator

$$[H_0, a^{\dagger}_{j\mathbf{k}}] = \hbar\omega_{j\mathbf{k}}a^{\dagger}_{j\mathbf{k}} \qquad 4.3.13a$$

$$[H_0, a^{\dagger}_{j\mathbf{k}}] = \hbar\omega_{j\mathbf{k}} a^{\dagger}_{j\mathbf{k}} \qquad 4.3.13b$$

and with the sum operator $A_{j\mathbf{k}}$

$$[A_{j\mathbf{k}}, a^{\dagger}_{j\mathbf{k}}] = 1$$

$$4.3.14$$

$$[A_{j-\mathbf{k}}, a_{j\mathbf{k}}] = -1$$

that are easily obtained using 4.3.4 and 4.3.10.

For the statistical perturbation theory utilized in the next Sections [6,122], it is convenient to summarize here some general expressions concerning the statistical average of phonon operators.

From the commutation relation 4.3.13a it follows

$$H_0 a^{\dagger}_{j\mathbf{k}} = a^{\dagger}_{j\mathbf{k}} (H_0 + \hbar\omega_{j\mathbf{k}}) \qquad 4.3.15$$

and, by operating on it with H_0, we obtain

$$H_0^2 a^{\dagger}_{j\mathbf{k}} = a^{\dagger}_{j\mathbf{k}} (H_0 + \hbar\omega_{j\mathbf{k}})^2 \qquad 4.3.16$$

so that, in general,

$$H_0^n a^{\dagger}_{j\mathbf{k}} = a^{\dagger}_{j\mathbf{k}} (H_0 + \hbar\omega_{j\mathbf{k}})^n \qquad 4.3.17$$

From 4.3.17 it follows that for any constant

$$e^{\alpha H_0} a^{\dagger}_{j\mathbf{k}} = \sum_{0}^{\infty}{}_n (n!)^{-1} \alpha^n H_0^n a^{\dagger}_{j\mathbf{k}} = \sum_{0}^{\infty}{}_n (n!)^{-1} \alpha^n a^{\dagger}_{j\mathbf{k}} (H_0 + \hbar\omega_{j\mathbf{k}})^n$$

$$= a^{\dagger}_{j\mathbf{k}} e^{\alpha(H_0 + \hbar\omega_{j\mathbf{k}})} \qquad 4.3.18$$

The statistical average $<\hat{O}>$ of any operator $\hat{O}$ is defined as [6]

$$<\hat{O}> = Z^{-1} \mathrm{Tr}\{\hat{O} e^{-\beta H_0}\} \qquad 4.3.19$$

where Tr means trace, $\beta = 1/KT$ and the partition function Z is given by

$$Z = Tr\{e^{-\beta H}\} \qquad 4.3.20$$

The statistical average of the number operator $n_{j\mathbf{k}}$ defined by 4.3.5, is then, using the cyclic permutation theorem for traces

$$\langle n_{j\mathbf{k}}\rangle = \langle a^{\dagger}_{j\mathbf{k}}a_{j\mathbf{k}}\rangle = Z^{-1}Tr\{a^{\dagger}_{j\mathbf{k}}a_{j\mathbf{k}}e^{-\beta H_0}\} = Z^{-1}Tr\{a_{j\mathbf{k}}e^{-\beta H_0}a^{\dagger}_{j\mathbf{k}}\} =$$
$$= Z^{-1}Tr\{a_{j\mathbf{k}}a^{\dagger}_{j\mathbf{k}}e^{-\beta(H_0+\hbar\omega_{j\mathbf{k}})}\} = e^{-\beta\hbar\omega_{j\mathbf{k}}}\langle a_{j\mathbf{k}}a^{\dagger}_{j\mathbf{k}}\rangle \qquad 4.3.21$$

from which it follows

$$\langle a_{j\mathbf{k}}a^{\dagger}_{j\mathbf{k}}\rangle = e^{\beta\hbar\omega_{j\mathbf{k}}}\langle a^{\dagger}_{j\mathbf{k}}a_{j\mathbf{k}}\rangle \qquad 4.3.22$$

This expression is easily transformed into one involving commutators by subtracting $a^{\dagger}_{j\mathbf{k}}a_{j\mathbf{k}}$ from both sides

$$(e^{\beta\hbar\omega_{j\mathbf{k}}} - 1)\langle a^{\dagger}_{j\mathbf{k}}a_{j\mathbf{k}}\rangle = \langle a_{j\mathbf{k}}a^{\dagger}_{j\mathbf{k}} - a^{\dagger}_{j\mathbf{k}}a_{j\mathbf{k}}\rangle = \langle[a_{j\mathbf{k}}, a^{\dagger}_{j\mathbf{k}}]\rangle \qquad 4.3.23$$

Using then 4.3.4 and 4.3.5 we obtain the statistical average of the occupation number $n_{j\mathbf{k}}$

$$\bar{n}_{j\mathbf{k}} = \langle a^{\dagger}_{j\mathbf{k}}a_{j\mathbf{k}}\rangle = 1/(e^{\beta\hbar\omega_{j\mathbf{k}}} - 1)$$
$$\bar{n}_{j\mathbf{k}} + 1 = \langle a_{j\mathbf{k}}a^{\dagger}_{j\mathbf{k}}\rangle = 1/(1 - e^{-\beta\hbar\omega_{j\mathbf{k}}}) = e^{\beta\hbar\omega_{j\mathbf{k}}}/(e^{\beta\hbar\omega_{j\mathbf{k}}} - 1) \qquad 4.3.24$$

We now consider, following the treatment of Wallace [6], the more complex case in which the commutation relation 4.3.13a is not exactly satisfied but has the form

$$[H_0, a^{\dagger}_{j\mathbf{k}}] = \hbar\omega_{j\mathbf{k}}a^{\dagger}_{j\mathbf{k}} + R^{\dagger}_{j\mathbf{k}} \qquad 4.3.25$$

where $R^{\dagger}_{j\mathbf{k}}$ is a "small" remainder operator. The contribution of $R^{\dagger}_{j\mathbf{k}}$ to the statistical average can be computed following the same procedure used before for the "exact" case. From 4.3.25 it follows

$$H_0 a^{\dagger}_{j\mathbf{k}} = a^{\dagger}_{j\mathbf{k}}(H_0 + \hbar\omega_{j\mathbf{k}}) + R^{\dagger}_{j\mathbf{k}} \qquad 4.3.26$$

and thus

$$H_0 a^{\dagger}_{j\mathbf{k}} = a^{\dagger}_{j\mathbf{k}}(H_0 + \hbar\omega_{j\mathbf{k}})^2 + R^{\dagger}_{j\mathbf{k}}(H_0 + \hbar\omega_{j\mathbf{k}}) + H_0 R^{\dagger}_{j\mathbf{k}} \qquad 4.3.27$$

In the most general case, we then have

$$H_0 a^{\dagger}_{j\mathbf{k}} = a^{\dagger}_{j\mathbf{k}}(H_0 + \hbar\omega_{j\mathbf{k}})^n + \sum_{0\,p}^{n-1} H_0^p R^{\dagger}_{j\mathbf{k}}(H_0 + \hbar\omega_{j\mathbf{k}})^{n-p-1} \qquad 4.3.28$$

and thus 4.3.18 becomes

$$\begin{aligned} e^{-\beta H_0} a^{\dagger}_{j\mathbf{k}} &= a^{\dagger}_{j\mathbf{k}} e^{-\beta(H_0 + \hbar\omega_{j\mathbf{k}})} + \\ &+ \sum_{1\,n}^{\infty} (n!)^{-1} (-\beta)^n \sum_{0\,p}^{n-1} H_0^p R^{\dagger}_{j\mathbf{k}}(H_0 + \hbar\omega_{j\mathbf{k}})^{n-p-1} \end{aligned} \qquad 4.3.29$$

The statistical average of the number operator now becomes by substitution of 4.3.29 in 4.3.21

$$\begin{aligned} \langle a^{\dagger}_{j\mathbf{k}} a_{j\mathbf{k}} \rangle &= e^{-\beta\hbar\omega_{j\mathbf{k}}} \langle a_{j\mathbf{k}} a^{\dagger}_{j\mathbf{k}} \rangle + \\ &+ Z^{-1} \mathrm{Tr}\{a_{j\mathbf{k}} \sum_{1\,n}^{\infty} (n!)^{-1} (-\beta)^n \sum_{0\,p}^{n-1} H_0^p R^{\dagger}_{j\mathbf{k}}(H_0 + \hbar\omega_{j\mathbf{k}})^{n-p-1}\} \end{aligned} \qquad 4.3.30$$

and, using the cyclic theorem for traces

$$\begin{aligned} \langle a^{\dagger}_{j\mathbf{k}} a_{j\mathbf{k}} \rangle &= e^{-\beta\hbar\omega_{j\mathbf{k}}} \langle a_{j\mathbf{k}} a^{\dagger}_{j\mathbf{k}} \rangle + \\ & Z^{-1} \mathrm{Tr}\{\sum_{n} (n!)^{-1} (-\beta)^n \sum_{0\,p}^{n-1} H_0^p R^{\dagger}_{j\mathbf{k}}(H_0 + \hbar\omega_{j\mathbf{k}})^{n-p-1} a_{j\mathbf{k}}\} \end{aligned} \qquad 4.3.31$$

Since $R_{j\mathbf{k}}$ is small, we can approximate the last term in 4.3.31.For this, we notice that from the Hermitian conjugate of 4.3.26

$$a_{j\mathbf{k}} H_0 = (H_0 + \hbar\omega_{j\mathbf{k}}) a_{j\mathbf{k}} + R_{j\mathbf{k}} \qquad 4.3.32$$

we obtain with the usual procedure

$$a_{j\mathbf{k}}H_0^n = (H_0 + \hbar\omega_{j\mathbf{k}})^n a_{j\mathbf{k}} + O(R_{j\mathbf{k}}) \qquad 4.3.33$$

where we have not specified the form of the last term since we shall neglect it at higher orders. Substituting 4.3.33 in 4.3.31 we obtain

$$\langle a^\dagger_{j\mathbf{k}} a_{j\mathbf{k}}\rangle = e^{-\beta\hbar\omega_{j\mathbf{k}}} \langle a_{j\mathbf{k}} a^\dagger_{j\mathbf{k}}\rangle +$$

$$Z^{-1}\mathrm{Tr}\{\sum_1^\infty{}_n (n!)^{-1}(-\beta)^n \sum_0^{n-1}{}_p H_0^p R^\dagger_{j\mathbf{k}} a_{j\mathbf{k}} H_0^{n-p-1}\} + O(R^\dagger_{j\mathbf{k}} R_{j\mathbf{k}}) \qquad 4.3.34a$$

By permuting H_0^p in the trace, the second term in 4.3.34 becomes

$$Z^{-1}\mathrm{Tr}\{\sum_1^\infty{}_n (n!)^{-1}(-\beta)^n \sum_0^{n-1}{}_p R^\dagger_{j\mathbf{k}} a_{j\mathbf{k}} H_0^{n-1}\} \qquad 4.3.34b$$

and, since none of the terms depends on p, the second sum gives n times the same result. Therefore, using 4.3.18 and 4.3.19

$$Z^{-1}\mathrm{Tr}\{\sum_n [(n-1)!]^{-1}(-\beta)^n R^\dagger_{j\mathbf{k}} a_{j\mathbf{k}} H_0^{n-1}\} =$$

$$= -\beta Z^{-1}\mathrm{Tr}\{R^\dagger_{j\mathbf{k}} a_{j\mathbf{k}} e^{-\beta H_0}\} = -\beta\langle R^\dagger_{j\mathbf{k}} a_{j\mathbf{k}}\rangle \qquad 4.3.34c$$

Thus, neglecting the last term in Eq.(4.3.34), which is of order $R^\dagger_{j\mathbf{k}} R_{j\mathbf{k}}$, we have

$$\langle a^\dagger_{j\mathbf{k}} a_{j\mathbf{k}}\rangle = e^{-\beta\hbar\omega_{j\mathbf{k}}} \langle a_{j\mathbf{k}} a^\dagger_{j\mathbf{k}}\rangle - \beta\langle R^\dagger_{j\mathbf{k}} a_{j\mathbf{k}}\rangle \qquad 4.3.35$$

which is the equivalent of 4.3.21 when the effect of the remainder operator is taken into account to the first order. This expression can be easily transformed to one involving commutators. As in the case of 4.3.23, we obtain

$$[e^{\beta\hbar\omega_{j\mathbf{k}}} - 1]\langle a^\dagger_{j\mathbf{k}} a_{j\mathbf{k}}\rangle = \langle [a_{j\mathbf{k}}, a^\dagger_{j\mathbf{k}}]\rangle - \beta e^{\beta\hbar\omega_{j\mathbf{k}}}\langle R^\dagger_{j\mathbf{k}} a_{j\mathbf{k}}\rangle \qquad 4.3.36$$

or

$$\langle a^\dagger_{j\mathbf{k}} a_{j\mathbf{k}}\rangle = [e^{\beta\hbar\omega_{j\mathbf{k}}} - 1]^{-1}\langle [a_{j\mathbf{k}}, a^\dagger_{j\mathbf{k}}]\rangle - \beta e^{\beta\hbar\omega_{j\mathbf{k}}}[e^{\beta\hbar\omega_{j\mathbf{k}}} - 1]^{-1}\langle R^\dagger_{j\mathbf{k}} a_{j\mathbf{k}}\rangle \qquad 4.3.37$$

We now introduce "renormalized" phonon frequencies, associated to phonon operators which satisfy 4.3.25 instead of 4.3.13, in the form

$$\Omega_{j\mathbf{k}} = \omega_{j\mathbf{k}} + \delta\omega_{j\mathbf{k}} \tag{4.3.38}$$

and we substitute it into 4.3.25 to obtain

$$[H_0, a^\dagger_{j\mathbf{k}}] = \hbar\Omega_{j\mathbf{k}} a^\dagger_{j\mathbf{k}} + (R^\dagger_{j\mathbf{k}} - \hbar\delta\omega_{j\mathbf{k}} a^\dagger_{j\mathbf{k}}) \tag{4.3.39}$$

The procedure used before to derive 4.3.37, now applies equally well if we use as starting equation 4.3.39 instead of 4.3.25. Thus

$$\begin{aligned} \langle a^\dagger_{j\mathbf{k}} a_{j\mathbf{k}} \rangle = {} & [e^{\beta\hbar\Omega_{j\mathbf{k}}} - 1] \; \langle [a_{j\mathbf{k}}, a^\dagger_{j\mathbf{k}}] \rangle \; + \\ & - \beta e^{\beta\hbar\Omega_{j\mathbf{k}}} [e^{\beta\hbar\Omega_{j\mathbf{k}}} - 1]^{-1} \langle (R^\dagger_{j\mathbf{k}} - \hbar\delta\omega_{j\mathbf{k}} a^\dagger_{j\mathbf{k}}) a_{j\mathbf{k}} \rangle \end{aligned} \tag{4.3.40}$$

and the last term in 4.3.40 vanishes if

$$\langle (R^\dagger_{j\mathbf{k}} - \hbar\delta\omega_{j\mathbf{k}} a^\dagger_{j\mathbf{k}}) a_{j\mathbf{k}} \rangle = 0 \tag{4.3.41}$$

i.e. if

$$\hbar\delta\omega_{j\mathbf{k}} = \langle R^\dagger_{j\mathbf{k}} a_{j\mathbf{k}} \rangle / \langle a^\dagger_{j\mathbf{k}} a_{j\mathbf{k}} \rangle \tag{4.3.42}$$

We notice at this point that, although we have derived 4.3.40 in terms of phonon creation and annihilation operators, it is of more general validity. In particular, if we compute the statistical average of the operator $a^\dagger_{j\mathbf{k}} R_{j\mathbf{k}}$, we obtain, following the procedure from 4.3.21 to 4.3.40 and using the condition 4.3.41

$$\langle a^\dagger_{j\mathbf{k}} R_{j\mathbf{k}} \rangle = e^{-\beta\hbar\Omega_{j\mathbf{k}}} \langle R_{j\mathbf{k}} a^\dagger_{j\mathbf{k}} \rangle \tag{4.3.43}$$

which can be rewritten, in terms of commutators, in the form

$$\langle a^\dagger_{j\mathbf{k}} R_{j\mathbf{k}} \rangle = [e^{\beta\hbar\Omega_{j\mathbf{k}}} - 1]^{-1} \langle [R_{j\mathbf{k}}, a^\dagger_{j\mathbf{k}}] \rangle \tag{4.3.44}$$

The Hermitian conjugate of 3.4.44 is

$$<R^{\dagger}_{j\mathbf{k}}a_{j\mathbf{k}}> = [e^{\beta\hbar\Omega_{j\mathbf{k}}} - 1]^{-1}<[a_{j\mathbf{k}},R^{\dagger}_{j\mathbf{k}}]> \tag{4.3.45}$$

In the same way Eq.4.3.40, under the condition 4.3.41, becomes

$$<a^{\dagger}_{j\mathbf{k}}a_{j\mathbf{k}}> = [e^{\beta\hbar\Omega_{j\mathbf{k}}} - 1]^{-1}<[a_{j\mathbf{k}},a^{\dagger}_{j\mathbf{k}}]> \tag{4.3.46}$$

and thus, by substituing the last two equations in 4.3.42 we obtain

$$\hbar\delta\omega_{j\mathbf{k}} = <[a_{j\mathbf{k}},R^{\dagger}_{j\mathbf{k}}]>/<[a_{j\mathbf{k}},a^{\dagger}_{j\mathbf{k}}]> \tag{4.3.47}$$

4.4 THE HAMILTONIAN RENORMALIZATION PROCEDURE

The operator renormalization procedure is illustrated in great detail in the original paper by Wallace[122] as well as in his book [6] entitled "Thermodynamics of crystals". We outline here the essential points.

The basic idea is to produce first-order improved phonon creation and annihilation operators, representing a set of phonons non-interacting to the first order. The procedure takes advantage of the characteristic property of perturbation methods to furnish second-order correct phonon energies from first-order correct operators.

The formalism of Wallace theory makes use of the commutation relation discussed in the previous Section. The single-phonon energies are obtained by evaluation of non-diagonal matrix elements of the commutator of the Hamiltonian with the creation operators.

The theory can be summarized as follows. The usual expansion of the crystal Hamiltonian

$$H = H_0 + H_1 + H_2 + \ldots + H_m \tag{4.4.1}$$

is associated to a corresponding expansion to the mth order of the phonon creation and annihilation operators

$$a^{\dagger}_{\kappa} = a^{0\dagger}_{\kappa} + a^{1\dagger}_{\kappa} + a^{2\dagger}_{\kappa} + \ldots + a^{m\dagger}_{\kappa} \tag{4.4.2a}$$

$$a_{\kappa} = a^{0}_{\kappa} + a^{1}_{\kappa} + a^{2}_{\kappa} + \ldots + a^{m}_{\kappa} \tag{4.4.2b}$$

where, for simplicity, we have used the collective label κ for the two labels j and **k**, owing to the fact that they are always associated in pairs. In what follows, unless otherwise specified, we shall continue to use this simplified nomenclature.

The corrected phonon operators satisfy to the mth-order, the Hamiltonian commutation relations 4.3.13

$$[H,a_\kappa^\dagger] = \hbar\omega_\kappa a_\kappa^\dagger + O(\varepsilon^{m+1})$$
$$[H,a_\kappa] = -\hbar\omega_\kappa a_\kappa + O(\varepsilon^{m+1}) \qquad 4.4.3$$

as well as the appropriate commutators

$$[a_{\kappa_1},a_{\kappa_2}^\dagger] = \delta_{\kappa_1\kappa_2} + O(\varepsilon^{m+1})$$
$$[a_{\kappa_1},a_{\kappa_2}] = [a_{\kappa_1}^\dagger,a_{\kappa_2}^\dagger] = 0 + O(\varepsilon^{m+1}) \qquad 4.4.4$$

where $O(\varepsilon^{m+1})$ means a quantity of the order m+1 that can be neglected at the order m.

Under the conditions specified by 4.4.3 and 4.4.4, the Hamiltonian is diagonal to the mth-order and, in terms of the corrected operators, has the form

$$H = G(m) + \sum_\kappa \hbar\omega_\kappa a_\kappa^\dagger a_\kappa + O(\varepsilon^{m+1}) \qquad 4.4.5$$

where the mth-order corrected phonon frequency ω_κ is also given by an expansion

$$\omega_\kappa = \omega_\kappa^0 + \omega_\kappa^1 + \omega_\kappa^2 + \ldots + \omega_\kappa^m \qquad 4.4.6$$

and where G(m) represents the mth-order correct zero-point energy

$$G(m) = G_0 + G_1 + G_2 + \ldots + G_m \qquad 4.4.7$$

with

$$G = \frac{1}{2}\hbar\sum_{\kappa}\omega_{\kappa} \tag{4.4.8}$$

As pointed out before, it is possible to calculate the correction of order m+1 for the single phonon energies without finding the corresponding correction of order m+1 to the phonon operators. This can be proved in the following way. To the order m+1, the Hamiltonian commutator 4.4.3 is given by

$$[H, a_{\kappa}^{\dagger} + a_{\kappa}^{m+1\dagger}] = \hbar(\omega_{\kappa} + \omega_{\kappa}^{m+1})(a_{\kappa}^{\dagger} + a_{\kappa}^{m+1\dagger}) \tag{4.4.9}$$

where $a_{\kappa}^{\dagger}$ and ω_{κ} are mth-order correct quantities given by 4.4.2 and 4.4.6, respectively. Since the commutators up to the order m are already satisfied, we obtain to the order m+1

$$[H_0, a_{\kappa}^{m+1\dagger}] + [H_1, a_{\kappa}^{m\dagger}] + \ldots + [H_{m+1}, a_{\kappa}^{0\dagger}] =$$

$$\hbar\omega_{\kappa}^{0} a_{\kappa}^{m+1\dagger} + \hbar\omega_{\kappa}^{1} a_{\kappa}^{m\dagger} + \ldots + \hbar\omega_{\kappa}^{m+1} a_{\kappa}^{0\dagger} \tag{4.4.10}$$

In order to find the correction for the single-phonon energy to the order m+1, we can evaluate upper-diagonal matrix elements of the commutator sum 4.4.10. We obtain

$$\langle \ldots n_{\kappa}+1 \ldots | \, [H_0, a_{\kappa}^{m+1\dagger}] - \hbar\omega_{\kappa}^{0} a_{\kappa}^{m+1\dagger} \, | \ldots n_{\kappa} \ldots \rangle \; +$$

$$\langle \ldots n_{\kappa}+1 \ldots | \, [H_1, a_{\kappa}^{m\dagger}] - \hbar\omega_{\kappa}^{1} a_{\kappa}^{m\dagger} \, | \ldots n_{\kappa} \ldots \rangle \; + \ldots \tag{4.4.11}$$

$$\langle \ldots n_{\kappa}+1 \ldots | \, [H_{m+1}, a_{\kappa}^{0\dagger}] \, | \ldots n_{\kappa} \ldots \rangle = \hbar\omega_{\kappa}^{m+1}(n_{\kappa}+1)^{\frac{1}{2}}$$

Using the obvious relations

$$\langle \ldots n_{\kappa}+1 \ldots | H_0 a_{\kappa}^{m+1\dagger} = (n_{\kappa}+1)\hbar\omega_{\kappa}^{0} \langle \ldots n_{\kappa}+1 \ldots | a_{\kappa}^{m+1\dagger} \tag{4.4.12a}$$

and

$$a_{\kappa}^{m+1\dagger} H_0 \, | \ldots n_{\kappa} \ldots \rangle = n_{\kappa}\hbar\omega_{\kappa}^{0} a_{\kappa}^{m+1\dagger} | \ldots n_{\kappa} \ldots \rangle \tag{4.4.12b}$$

it is easily seen that the first term in 4.4.11 is equal to zero. Therefore the correction of order m+1 for the single-phonon energy can be obtained if the mth-order correction for the operators is known.

We now consider the application of the formalism described above to the different orders.

1) Zero Order

The commutator sum 4.4.10 reduces, to the zero-order, to

$$[H_0, a_\kappa^{0\dagger}] = \hbar\omega_\kappa^0 a_\kappa^{0\dagger} \qquad 4.4.13$$

and 4.4.11 then yields the zero-order single-phonon energy

$$\langle \ldots n_\kappa+1 \ldots | [H_0, a_\kappa^{0\dagger}] | \ldots n_\kappa \ldots \rangle = \hbar\omega_\kappa^0 (n_\kappa + 1) \qquad 4.4.14$$

Furthermore the zero-order operators satisfy the commutation relations

$$\begin{aligned} &[a_\kappa^0, a_{\kappa_1}^{0\dagger}] = \delta_{\kappa\kappa_1} \\ &[a_\kappa^{0\dagger}, a_{\kappa_1}^{0\dagger}] = [a_\kappa^0, a_{\kappa_1}^0] = 0 \end{aligned} \qquad 4.4.15$$

2) First Order

To the first order, Eq.4.4.10 becomes

$$[H_0, a_\kappa^{1\dagger}] + [H_1, a_\kappa^{0\dagger}] = \hbar\omega_\kappa^0 a_\kappa^{1\dagger} + \hbar\omega_\kappa^1 a_\kappa^{0\dagger} \qquad 4.4.16$$

and 4.4.11 gives

$$\langle \ldots n_\kappa+1 \ldots | [H_1, a_\kappa^{0\dagger}] | \ldots n_\kappa \ldots \rangle = \hbar\omega_\kappa^1 (n_\kappa + 1)^{\frac{1}{2}} \qquad 4.4.17$$

The commutator in 4.4.17 is easily obtained from 4.3.11

$$[H_1, a_\kappa^{0\dagger}] = \sum_{\kappa_1\kappa_2\kappa_3} B(\kappa_1\kappa_2\kappa_3) [A_{\kappa_1}^0 A_{\kappa_2}^0 A_{\kappa_3}^0, a_\kappa^{0\dagger}] \qquad 4.4.18$$

$$[H_1, a_\kappa^{0\dagger}] = 3\sum_{\kappa_1\kappa_2} B(\kappa\kappa_1\kappa_2)\,[A_\kappa^0, a_\kappa^{0\dagger}]\,A_{\kappa_1}^0 A_{\kappa_2}^0$$

where

$$A_\kappa^0 = a_{-\kappa}^{0\dagger} + a_\kappa^0 \qquad 4.4.19$$

and where the factor 3 originates from the symmetry of the indices. Using 4.3.14 , the commutator 4.4.18 reduces then to

$$[H_1, a_\kappa^{0\dagger}] = 3\sum_{\kappa_1\kappa_2} B(\kappa\kappa_1\kappa_2)A_{\kappa_1}^0 A_{\kappa_2}^0 \qquad 4.4.20$$

and 4.4.17 gives

$$3\sum_{\kappa_1\kappa_2} B(\kappa\kappa_1\kappa_2) < \ldots\ n_\kappa+1\ \ldots|A_{\kappa_1}^0 A_{\kappa_2}^0|\ldots\ n_\kappa\ \ldots> = 0 \qquad 4.4.21$$

since the operator $A_{\kappa_1}^0 A_{\kappa_2}^0$ can couple only states in which, if $\kappa_1 = \kappa_2 = \kappa$, the occupation number changes by 0 or 2. We then obtain the well-known result of perturbation theory

$$\hbar\omega_\kappa^1 = 0 \qquad 4.4.22$$

and thus 4.4.16 becomes

$$[H_0, a_\kappa^{1\dagger}] + [H_1, a_\kappa^{0\dagger}] = \hbar\omega_\kappa^0 a_\kappa^{1\dagger} \qquad 4.4.23$$

We can now use this relation to find the first-order correction $a_\kappa^{1\dagger}$ to the phonon-creation operator. Since the commutator $[H_1, a_\kappa^{0\dagger}]$ is quadratic in the phonon operators (see 4.4.20), we expect that both $a_\kappa^{1\dagger}$ and $[H_0, a_\kappa^{1\dagger}]$ should be quadratic in the same operators. Therefore, $a_\kappa^{1\dagger}$ should have the general form

$$a_\kappa^{1\dagger} = \sum_{\kappa_1\kappa_2} B(\kappa\kappa_1\kappa_2)\,[\ \alpha(\kappa\kappa_1\kappa_2)a_{\kappa_1}^0 a_{\kappa_2}^0 + \beta(\kappa\kappa_1\kappa_2)a_{\kappa_1}^0 a_{-\kappa_2}^{0\dagger} + \gamma(\kappa\kappa_1\kappa_2)a_{-\kappa_1}^{0\dagger} a_{\kappa_2}^0 + \delta(\kappa\kappa_1\kappa_2)a_{-\kappa_1}^{0\dagger} a_{-\kappa_2}^{0\dagger}\ ] \qquad 4.4.24$$

where the coefficients $\alpha(\kappa\kappa_1\kappa_2)$, $\beta(\kappa\kappa_1\kappa_2)$, $\gamma(\kappa\kappa_1\kappa_2)$ and $\delta(\kappa\kappa_1\kappa_2)$ are determined by the condition that $a^{1\dagger}_{\kappa}$ should satisfy 4.4.23. We thus obtain

$$[H_0,a^{1\dagger}_{\kappa}] = \sum_{\kappa_1\kappa_2} B(\kappa\kappa_1\kappa_2)\{\alpha(\kappa\kappa_1\kappa_2)[H_0,a^0_{\kappa_1}a^0_{\kappa_2}] + \beta(\kappa\kappa_1\kappa_2)[H_0,a^0_{\kappa_1}a^{0\dagger}_{-\kappa_2}]$$

$$+ \gamma(\kappa\kappa_1\kappa_2)[H_0,a^{0\dagger}_{-\kappa_1}a^0_{\kappa_2}] + \delta(\kappa\kappa_1\kappa_2)[H_0,a^{0\dagger}_{-\kappa_1}a^{0\dagger}_{-\kappa_2}]\} \qquad 4.4.25$$

The four commutators in 4.4.25 are easily evaluated using 4.4.3 and the relation

$$[A,BC] = B[A,C] + [A,B]C \qquad 4.4.26$$

In this way we obtain

$$[H_0,a^{1\dagger}_{\kappa}] = \hbar\sum_{\kappa_1\kappa_2} B(\kappa\kappa_1\kappa_2)[\, - \alpha(\kappa\kappa_1\kappa_2)(\omega^0_{\kappa_1} + \omega^0_{\kappa_2})a^0_{\kappa_1}a^0_{\kappa_2} \quad +$$

$$- \beta(\kappa\kappa_1\kappa_2)(\omega^0_{\kappa_1} - \omega^0_{\kappa_2})a^0_{\kappa_1}a^{0\dagger}_{-\kappa_2} + \gamma(\kappa\kappa_1\kappa_2)(\omega^0_{\kappa_1} - \omega^0_{\kappa_2})a^{0\dagger}_{-\kappa_1}a^0_{\kappa_2} +$$

$$+ \delta(\kappa\kappa_1\kappa_2)(\omega^0_{\kappa_1} + \omega^0_{\kappa_2})a^{0\dagger}_{-\kappa_1}a^{0\dagger}_{-\kappa_2}] \qquad 4.4.27$$

We could now substitute 4.4.27 in 4.4.23 to obtain the four coefficients. If this is done, one finds, under resonance conditions, coefficients with vanishing denominators. To avoid these divergencies, a stand ard device of perturbation theory is utilized. This consists in replacin $\hbar\omega^0_{\kappa}$ in 4.4.23 by $(\hbar\omega^0_{\kappa} + i\lambda_{\kappa})$ where λ_{κ} is an infinitesimally small quantity such that $\lambda_{\kappa} << \hbar\omega^0_{\kappa}$. We notice that, since the three translations of the crystal with $\omega^0_{\kappa} = 0$ are not included in the calculations, the procedure is perfectly legitimate. After having taken the necessary statistical averages and having completed the operator calculations, we can take the limit for $\lambda_{\kappa}\to 0$ of our expressions.

We therefore require that the first order commutators 4.4.20 and 4.4.27 satisfy, instead of 4.4.23, the more general relation

$$[H_0,a^{1\dagger}_{\kappa}] + [H_1,a^{0\dagger}_{\kappa}] = (\hbar\omega^0_{\kappa} + i\lambda_{\kappa})a^{1\dagger}_{\kappa} \qquad 4.4.28$$

and thus, by substituing 4.4.20 and 4.4.27 in this expression and equating coefficients of the same operators, we find

$$\alpha(\kappa\kappa_1\kappa_2) = 3\left[\hbar(\omega^0_\kappa + \omega^0_{\kappa_1} + \omega^0_{\kappa_2}) + i\lambda_\kappa\right]^{-1}$$

$$\beta(\kappa\kappa_1\kappa_2) = 3\left[\hbar(\omega^0_\kappa + \omega^0_{\kappa_1} - \omega^0_{\kappa_2}) + i\lambda_\kappa\right]^{-1}$$

$$\gamma(\kappa\kappa_1\kappa_2) = 3\left[\hbar(\omega^0_\kappa - \omega^0_{\kappa_1} + \omega^0_{\kappa_2}) + i\lambda_\kappa\right]^{-1} \qquad 4.4.29$$

$$\delta(\kappa\kappa_1\kappa_2) = 3\left[\hbar(\omega^0_\kappa - \omega^0_{\kappa_1} - \omega^0_{\kappa_2}) + i\lambda_\kappa\right]^{-1}$$

It is now easy,although time-consuming, to show that the first-order corrected operators

$$a^{0\dagger}_\kappa + a^{1\dagger}_\kappa$$

$$a^0_\kappa + a^1_\kappa \qquad 4.4.30$$

describe phonons independent to the first order. For this, we need to show that they satisfy the relations

$$[a^0_\kappa + a^1_\kappa, a^{0\dagger}_{\kappa_1} + a^{1\dagger}_{\kappa_1}] = \delta_{\kappa\kappa_1} + O(\varepsilon^2)$$

$$[a^0_\kappa + a^1_\kappa, a^0_{\kappa_1} + a^1_{\kappa_1}] = O(\varepsilon^2) \qquad 4.4.31$$

Since the zero-order phonon operators already satisfy the zero-order commutators 4.4.15, these relations reduce to

$$[a^0_\kappa, a^{1\dagger}_{\kappa_1}] + [a^1_\kappa, a^{0\dagger}_{\kappa_1}] = 0$$

$$[a^0_\kappa, a^1_{\kappa_1}] + [a^1_\kappa, a^0_{\kappa_1}] = 0 \qquad 4.4.32$$

In order to prove these expressions we recall that a^1_κ is the Hermitian conjugate of $a^{1\dagger}_\kappa$ and thus, according to 4.4.24, has the form

$$a^1_\kappa = \sum_{\kappa_1\kappa_2} B(-\kappa-\kappa_1-\kappa_2)\left[\alpha(\kappa\kappa_1\kappa_2)^* a^{0\dagger}_{\kappa_2} a^{0\dagger}_{\kappa_1} + \beta(\kappa\kappa_1\kappa_2)^* a^0_{-\kappa_2} a^{0\dagger}_{\kappa_1} \right. +$$

$$+ \quad \gamma(\kappa\kappa_1\kappa_2)^* a^{0\dagger}_{\kappa_2} a^0_{-\kappa_1} \quad + \quad \delta(\kappa\kappa_1\kappa_2)^* a^{0\dagger}_{-\kappa_2} a^{0\dagger}_{-\kappa_1}] \qquad 4.4.33$$

where we have used the condition $B(\kappa\kappa_1\kappa_2)^* = B(-\kappa-\kappa_1-\kappa_2)$.

Since $\omega_\kappa = \omega_{-\kappa}$, λ_κ in 4.4.28 can be taken without loss of generality such that $\lambda_\kappa = \lambda_{-\kappa}$. In this way, the coefficients 4.4.29 do not change when the sign of one of the κ indices is changed.

Using then 4.4.24,4.4.33 and 4.4.26, we obtain

$$[a^0_\kappa, a^{1\dagger}_{\kappa_1}] + [a^1_\kappa, a^{0\dagger}_{\kappa_1}] = \sum_{\kappa_2} B(-\kappa\kappa_1\kappa_2)\{[\beta(\kappa\kappa_2\kappa_1)^* + \gamma(\kappa\kappa_1\kappa_2)^* +$$

$$+ \delta(\kappa_1\kappa_2\kappa) + \delta(\kappa_1\kappa\kappa_2)]a^{0\dagger}_{-\kappa_2} + \qquad 4.4.34$$

$$+ [\beta(\kappa_1\kappa_2\kappa) + \gamma(\kappa_1\kappa\kappa_2) + \delta(\kappa\kappa_1\kappa_2)^* + \delta(\kappa\kappa_2\kappa_1)^*]a^0_{\kappa_2}\} = 0$$

$$[a^0_\kappa, a^1_{\kappa_1}] + [a^1_\kappa, a^0_{\kappa_1}] = \sum_{\kappa_2} B(-\kappa-\kappa_1\kappa_2)\{[\alpha(\kappa_1\kappa\kappa_2)^* + \alpha(\kappa_1\kappa_2\kappa)^* +$$

$$- \alpha(\kappa\kappa_1\kappa_2)^* - \alpha(\kappa\kappa_2\kappa_1)^*]a^{0\dagger}_{-\kappa_2} +$$

$$+ [\beta(\kappa_1\kappa\kappa_2)^* + \gamma(\kappa_1\kappa_2\kappa)^* - \beta(\kappa\kappa_1\kappa_2)^* - \gamma(\kappa\kappa_2\kappa_1)^*]a^0_{\kappa_2}\} = 0$$

and thus

$$\beta(\kappa\kappa_2\kappa_1)^* + \gamma(\kappa\kappa_1\kappa_2)^* + \delta(\kappa_1\kappa_2\kappa) + \delta(\kappa_1\kappa\kappa_2) = 0$$

$$\beta(\kappa_1\kappa_2\kappa) + \gamma(\kappa_1\kappa\kappa_2) + \delta(\kappa\kappa_1\kappa_2)^* + \delta(\kappa\kappa_2\kappa_1)^* = 0$$

$$\qquad 4.4.35$$

$$\alpha(\kappa_1\kappa\kappa_2)^* + \alpha(\kappa_1\kappa_2\kappa)^* - \alpha(\kappa\kappa_1\kappa_2)^* - \alpha(\kappa\kappa_2\kappa_1)^* = 0$$

$$\beta(\kappa_1\kappa\kappa_2)^* + \gamma(\kappa_1\kappa_2\kappa)^* - \beta(\kappa\kappa_1\kappa_2)^* - \gamma(\kappa\kappa_2\kappa_1)^* = 0$$

These equations are satisfied only if λ_κ is independent of κ, i.e. only if $\lambda_\kappa = \lambda$.

3) Second Order

To the second order Eq.4.4.10 gives

$$[H_0,a_\kappa^{2\dagger}] + [H_1,a_\kappa^{1\dagger}] + [H_2,a_\kappa^{0\dagger}] = \hbar\omega_\kappa^0 a_\kappa^{2\dagger} + \hbar\omega_\kappa^2 a_\kappa^{0\dagger} \qquad 4.4.36$$

and correspondingly 4.4.11 gives

$$\langle \ldots\; n_\kappa+1 \;\ldots | \; [H_1,a_\kappa^{1\dagger}] + [H_2,a_\kappa^{0\dagger}] \; | \ldots\; n_\kappa \;\ldots \rangle = \hbar\omega_\kappa^2 (n_\kappa+1)^{\frac{1}{2}} \qquad 4.4.37$$

The calculation of these matrix elements is straightforward. We summarize here the essential steps.

The commutator $[H_2,a_\kappa^{0\dagger}]$, using 4.3.11 and the symmetry of the four indices is given by

$$[H_2,a_\kappa^{0\dagger}] = \sum_{\kappa\kappa_1\kappa_2\kappa_3} B(\kappa\kappa_1\kappa_2\kappa_3)[A_\kappa^0 A_{\kappa_1}^0 A_{\kappa_2}^0 A_{\kappa_3}^0, a_\kappa^{0\dagger}] =$$

$$= 4 \sum_{\kappa_1\kappa_2\kappa_3} B(\kappa\kappa_1\kappa_2\kappa_3) A_{\kappa_1}^0 A_{\kappa_2}^0 A_{\kappa_3}^0 \qquad 4.4.38$$

The only terms of 4.4.38 which will give rise to non-zero matrix elements in 4.4.37 are those involving a creation operator $a_\kappa^{0\dagger}$ and products of the type $a_{\kappa_1}^{0\dagger} a_{\kappa_1}^0$ or $a_{\kappa_1}^0 a_{\kappa_1}^{0\dagger}$ for any κ_1. The "effective" operator is thus

$$\begin{aligned}[H_2,a_\kappa^{0\dagger}]_{eff.} = 4\sum_{\kappa_1} \{ & B(\kappa-\kappa\kappa_1-\kappa_1) a_\kappa^{0\dagger}(a_{\kappa_1}^0 a_{\kappa_1}^{0\dagger} + a_{-\kappa_1}^{0\dagger} a_{-\kappa_1}^0) + \\ & B(\kappa\kappa_1-\kappa_1-\kappa)(a_{\kappa_1}^0 a_{\kappa_1}^{0\dagger} + a_{-\kappa_1}^{0\dagger} a_{-\kappa_1}^0) a_\kappa^{0\dagger} + \qquad 4.4.39 \\ & B(\kappa\kappa_1-\kappa-\kappa_1)(a_{\kappa_1}^0 a_\kappa^{0\dagger} a_{\kappa_1}^{0\dagger} + a_{-\kappa_1}^{0\dagger} a_\kappa^{0\dagger} a_{-\kappa_1}^0)\}\end{aligned}$$

The matrix elements of 4.4.39 are conveniently evaluated using 4.4.26 to commute them to normal order. In particular, we have

$$a_{\kappa_1}^0 a_\kappa^{0\dagger} a_{\kappa_1}^{0\dagger} = a_\kappa^{0\dagger} a_{\kappa_1}^{0\dagger} a_{\kappa_1}^0 + a_\kappa^{0\dagger} + a_{\kappa_1}^{0\dagger} \delta_{\kappa\kappa_1} \qquad 4.4.40$$

and thus, neglecting the last term which removes the sum over κ_1 and is of relative order 1/L with respect to other terms, we have

$$<\dots n_\kappa+1 \dots| \, [H_2, a^{0\dagger}_\kappa] \, | \dots n_\kappa \dots> = \qquad 4.4.41$$

$$= 12(n_\kappa + 1)^{\frac{1}{2}} \sum_{\kappa_1} B(\kappa-\kappa\kappa_1-\kappa_1)(n_{\kappa_1} + n_{-\kappa_1} + 1)$$

The calculation of the matrix element of $|H_1, a^{1\dagger}_\kappa|$ is considerably more time-consuming but nevertheless involves no special difficulties. According to Eq.4.4.24, the commutator has the form

$$[H_1, a^{1\dagger}_\kappa] = \sum_{\kappa_1\kappa_2} B(\kappa\kappa_1\kappa_2)\{ \alpha(\kappa\kappa_1\kappa_2)[H_1, a^0_{\kappa_1} a^0_{\kappa_2}] + \beta(\kappa\kappa_1\kappa_2)[H_1, a^0_{\kappa_1} a^{0\dagger}_{-\kappa_2}] +$$

$$+ \gamma(\kappa\kappa_1\kappa_2)[H_1, a^{0\dagger}_{-\kappa_1} a^0_{\kappa_2}] + \delta(\kappa\kappa_1\kappa_2)[H_1, a^{0\dagger}_{-\kappa_1} a^{0\dagger}_{-\kappa_2}]\} \qquad 4.4.42$$

where H_1 is given by 4.3.11. It is easily seen that here we have products of five phonon operators and these will give rise to non-zero matrix elements only if $a^{0\dagger}_\kappa$ and products of the type $a^0_{\kappa_1} a^{0\dagger}_{-\kappa_1} a^0_{\kappa_2} a^{0\dagger}_{-\kappa_2}$ are involved, with all possible permutations of the operators.

Consider, for instance, the first commutator in 4.4.42. From 4.3.11 we have

$$[H_1, a^0_{\kappa_1} a^0_{\kappa_2}] = \sum_{\kappa_3\kappa_4\kappa_5} B(\kappa_3\kappa_4\kappa_5)[A^0_{\kappa_3} A^0_{\kappa_4} A^0_{\kappa_5}, a^0_{\kappa_1} a^0_{\kappa_2}] \qquad 4.4.43$$

and the effective part of it, taking into account the symmetry of the indices, will be

$$6B(-\kappa-\kappa_1-\kappa_2) a^{0\dagger}_\kappa [a^{0\dagger}_{\kappa_1} a^{0\dagger}_{\kappa_2}, a^0_{\kappa_1} a^0_{\kappa_2}] \qquad 4.4.44$$

using then 4.4.26 and neglecting terms of order 1/L, the first term in 4.4.42 becomes

$$-6\sum_{\kappa_1\kappa_2} B(\kappa\kappa_1\kappa_2) B(-\kappa-\kappa_1-\kappa_2) \alpha(\kappa\kappa_1\kappa_2)(n_{\kappa_1} + n_{\kappa_2} + 1) a^{0\dagger}_\kappa \qquad 4.4.45$$

Proceeding in the same way for all other commutators, we obtain for the matrix element of $[H_1, a^{1\dagger}_\kappa]$

$$<\ldots n_\kappa+1 \ldots | \, [H_1, a^{1\dagger}_\kappa] \, | \ldots n_\kappa \ldots> = \tag{4.4.46}$$

$$-6(n_\kappa + 1)^{\frac{1}{2}} \sum_{\kappa_1\kappa_2} \{B(\kappa\kappa_1\kappa_2)B(-\kappa-\kappa_1-\kappa_2) \left[\alpha(\kappa\kappa_1\kappa_2)(n_{\kappa_1} + n_{\kappa_2} + 1) + \right.$$

$$\left. \beta(\kappa\kappa_1\kappa_2)(n_{-\kappa_2} - n_{\kappa_1}) + \gamma(\kappa\kappa_1\kappa_2)(n_{-\kappa_1} - n_{\kappa_2}) - \delta(\kappa\kappa_1\kappa_2)(n_{-\kappa_1} + n_{-\kappa_2} + 1)\right]$$

$$+ B(\kappa-\kappa\kappa_2)B(\kappa_1-\kappa_1-\kappa_2)\left[\gamma(\kappa\kappa_1\kappa_2) - \delta(\kappa\kappa_1\kappa_2)\right](n_{\kappa_1} + n_{-\kappa_1} + 1) \}$$

where we have used the conditions

$$\beta(\kappa\kappa_1\kappa_2) = \gamma(\kappa\kappa_2\kappa_1)$$
$$\delta(\kappa\kappa_1\kappa_2) = \delta(\kappa\kappa_2\kappa_1) \tag{4.4.47}$$

Substituing 4.4.41 and 4.4.46 in 4.4.37, we finally obtain the second-order correction to the single phonon energy

$$\hbar\omega^2_\kappa = 12\sum_{\kappa_1} B(\kappa-\kappa\kappa_1-\kappa_1)(n_{\kappa_1} + n_{\kappa_1} + 1) +$$

$$-6\sum_{\kappa_1\kappa_2} \{B(\kappa\kappa_1\kappa_2)B(-\kappa-\kappa_1-\kappa_2)\left[\alpha(\kappa\kappa_1\kappa_2)(n_{\kappa_1} + n_{\kappa_2} + 1) + \right.$$

$$+ \beta(\kappa\kappa_1\kappa_2)(n_{-\kappa_2} - n_{\kappa_1}) + \gamma(\kappa\kappa_1\kappa_2)(n_{-\kappa_1} - n_{\kappa_2}) + \tag{4.4.48}$$

$$\left. - \delta(\kappa\kappa_1\kappa_2)(n_{-\kappa_1} + n_{-\kappa_2} + 1)\right] +$$

$$+ B(\kappa-\kappa\kappa_2)B(\kappa_1-\kappa_1-\kappa_2)\left[\gamma(\kappa\kappa_1\kappa_2) - \delta(\kappa\kappa_1\kappa_2)\right](n_{\kappa_1} + n_{-\kappa_1} + 1) \}$$

All spectroscopic measurements furnish, however, not directly the single-phonon energy but rather a statistical average over a large number of them. To obtain the statistical average of 4.4.48, we use 4.3.24 and the resulting condition $\bar{n}_\kappa = \bar{n}_{-\kappa}$. We obtain

$$\hbar\omega^2_\kappa = 12\sum_{\kappa_1} B(\kappa-\kappa\kappa_1-\kappa_1)(2\bar{n}_{\kappa_1} + 1) +$$

$$- 6\sum_{\kappa_1\kappa_2} \{B(\kappa\kappa_1\kappa_2)B(-\kappa-\kappa_1-\kappa_2)\{\left[\alpha(\kappa\kappa_1\kappa_2) - \delta(\kappa\kappa_1\kappa_2)\right](\bar{n}_{\kappa_1} + \bar{n}_{\kappa_2} + 1) +$$

$$+ \left[\beta(\kappa\kappa_1\kappa_2) - \gamma(\kappa\kappa_1\kappa_2)\right](\bar{n}_{\kappa_2} - \bar{n}_{\kappa_1}) \quad + \tag{4.4.49}$$

$$+ \quad B(\kappa-\kappa\kappa_2)B(\kappa_1-\kappa_1-\kappa_2)\left[\gamma(\kappa\kappa_1\kappa_2) - \delta(\kappa\kappa_1\kappa_2)\right](2\bar{n}_{\kappa_1} + 1) \}$$

where

$$\bar{n}_\kappa = \langle n_\kappa \rangle = 1/(e^{\beta\hbar\omega_\kappa} - 1) \tag{4.4.50}$$

The coefficients α, β, γ and δ occurring in 4.4.49 are complex according to 4.4.29. After the single occupation numbers have been replaced by their statistical averages, we can take the limit for $\lambda \to 0$ of these coefficients. We use the general relation

$$\lim_{\lambda\to 0} \sum_\kappa \frac{f(\kappa)}{(A_\kappa \pm i\lambda)} = \sum_\kappa \frac{f(\kappa)}{(A_\kappa)_p} \mp i\pi \sum_\kappa f(\kappa)\,\delta(A_\kappa) \tag{4.4.51}$$

where the suffix p denotes the principal part of the sum.

Using 4.4.51 to compute the limit for $\lambda \to 0$ of the coefficients 4.4.29, expression 4.4.49 can be put in the form

$$\hbar\omega_\kappa^2 = \hbar\Delta_\kappa - i\hbar\Gamma_\kappa \tag{4.4.52}$$

where

$$\begin{aligned}
\hbar\Delta_\kappa = {} & 12\sum_{\kappa_1} B(\kappa-\kappa\kappa_1-\kappa_1)(2\bar{n}_{\kappa_1} + 1) \quad + \\
& - 18\hbar^{-1}\sum_{\kappa_1\kappa_2} \{B(\kappa\kappa_1\kappa_2)B(-\kappa-\kappa_1-\kappa_2)\Big[\frac{(\bar{n}_{\kappa_1} + \bar{n}_{\kappa_2} + 1)}{(\omega_\kappa^0 + \omega_{\kappa_1}^0 + \omega_{\kappa_2}^0)_p} \quad + \\
& + \frac{(\bar{n}_{\kappa_1} - \bar{n}_{\kappa_2})}{(\omega_\kappa^0 + \omega_{\kappa_1}^0 - \omega_{\kappa_2}^0)_p} + \frac{(\bar{n}_\kappa - \bar{n}_\kappa)}{(\omega_\kappa^0 - \omega_{\kappa_1}^0 + \omega_{\kappa_2}^0)_p} \quad + \\
& - \frac{(\bar{n}_{\kappa_1} + \bar{n}_{\kappa_2} + 1)}{(\omega_\kappa^0 - \omega_{\kappa_1}^0 - \omega_{\kappa_2}^0)_p}\Big] + 2B(\kappa-\kappa\kappa_1)B(-\kappa_1\kappa_2-\kappa_2)\frac{(2\bar{n}_{\kappa_2} + 1)}{(\omega_{\kappa_1}^0)_p} \}
\end{aligned} \tag{4.4.53}$$

and

$$\hbar\Gamma_\kappa = 18\hbar^{-1}\pi\sum_{\kappa_1\kappa_2} B(\kappa\kappa_1\kappa_2)B(-\kappa-\kappa_1-\kappa_2)\{(\bar{n}_{\kappa_1} + \bar{n}_{\kappa_2} + 1)\left[\delta(\omega_\kappa^0 - \omega_{\kappa_1}^0 - \omega_{\kappa_2}^0) + \right.$$

$$- \delta(\omega^0_\kappa + \omega^0_{\kappa_1} + \omega^0_{\kappa_2})] + (\bar{n}_{\kappa_1} - \bar{n}_{\kappa_2})[\delta(\omega^0_\kappa + \omega^0_{\kappa_1} - \omega^0_{\kappa_2}) -$$

$$\delta(\omega^0_\kappa - \omega^0_{\kappa_1} + \omega^0_{\kappa_2})]\} \qquad 4.4.54$$

According to 4.4.6, the renormalized single-phonon energy is given by

$$\hbar\omega_\kappa = \hbar(\omega^0_\kappa + \omega^2_\kappa) = \hbar(\omega^0_\kappa + \Delta_\kappa - i\Gamma_\kappa) \qquad 4.4.55$$

since the first-order correction ω^1_κ is equal to zero.

It is easily seen that the real part of 4.4.55 represents the frequency of the renormalized phonon whereas the imaginary part is the inverse of the phonon lifetime. For this we recall that the time-dependence of the renormalized phonon frequency is given by

$$Q_\kappa(t) = Q_\kappa(0)e^{-i\omega_\kappa t} = Q_\kappa(0)e^{-i(\omega^0_\kappa+\Delta_\kappa)t}e^{-\Gamma_\kappa t} \qquad 4.4.56$$

where Q_κ is the corresponding normal coordinate. Eq.4.4.56 therefore describes a Lorentian phonon band with a frequency peak at

$$\omega_\kappa = \omega^0_\kappa + \Delta_\kappa \qquad 4.4.57$$

and a band-width at half maximum equal to $2\Gamma_\kappa$.

4.5 THE SELF-CONSISTENT PHONON METHOD

When the anharmonicity is large, it is not appropriate to use perturbation methods in anharmonic frequency calculations since it may occur that the anharmonic corrections are as large or even larger than the bare harmonic frequencies. In these cases, if one whishes to avoid the more elaborate and complex Green's functions method described in the next Section, it may be convenient to use the so-called "self-consistent phonon method" (SCP) originally developed for quantum crystals [123-126] like solid He and later extended, because of its simplifying features, to less anharmonic crystals.

The basic idea is to include anharmonic interactions in the treatment right at the beginning, by averaging the total potential acting on a molecule, over all vibrational modes of the interacting molecules in the crystal. The SCP method is thus a kind of harmonic calculation in which the harmonic force constants are substituted by "effective" force constants which include the anharmonic interactions averaged over all vibrational modes of the interacting molecules.

There are several possible approaches to the SCP theory. The most popular one is due to Werthamer[124] who formulated the SCP theory on the basis of the cumulant expansion. The method was later improved by Klein et al [127] and extended to molecular crystals by several authors[128-130]. Different derivations of the SCP theory are due to Plakida and Siclos[131] using the Green's function technique, to Samathiyakanit and Glyde [132] by the path integral method and to Wallace [6] by an extension of the statistical perturbation theory.

Although these derivations are all equivalent, we believe that the reader who has been exposed to the operator formalism of the previous Sections can follow better the statistical perturbation approach of Wallace. In addition, this method is better adapted than others to the use of molecular rotation coordinates of basic interest in molecular crystal dynamics.

Consider the general crystal Hamiltonian

$$H = T + V \qquad 4.5.1$$

where V represents the complete crystal potential, including all higher order terms. The SCP method consists in replacing the true anharmonic potential V with an "effective" harmonic potential of the form

$$\Phi = \frac{1}{2}\sum_{m\mu}\sum_{n\nu}\sum_{\ell_1\ell_2}\Phi_{\ell_1\ell_2}\binom{m\mu}{n\nu}R^{m\mu}_{\ell_1}R^{n\nu}_{\ell_2} \qquad 4.5.2$$

such that the effective force constants $\Phi_{\ell_1\ell_2}\binom{m\mu}{n\nu}$ include all the average statistical influence of the anharmonic terms of V.

By adding and subtracting Φ from 4.5.1, we can actually rewrite the Hamiltonian in the form

$$H = T + \Phi + (V - \Phi) = H_{eff} + (V - \Phi) \qquad 4.5.3$$

and the effective Hamiltonian H_{eff}, can be then diagonalized in the standard way as long as the difference $(V - \Phi)$ is small in the statistical sense.

Under this condition, the Hamiltonian is diagonalized by means of a transformation analogous to 4.2.1

$$R_\ell^{m\mu} = L^{-\frac{1}{2}} \sum_\kappa Y(\ell\mu|\kappa) e^{i\mathbf{k}\cdot\mathbf{R}_m} Q_\kappa \qquad 4.5.4$$

with

$$Q_\kappa = (\hbar/2\Omega_\kappa)^{\frac{1}{2}} (b_{-\kappa}^\dagger + b_\kappa) \qquad 4.5.5$$

where $Y(\ell\mu|\kappa)$ and Ω_κ are effective eigenvectors and phonon frequencies, respectively,and $b_{-\kappa}^\dagger$ and b_κ are effective phonon operators, which obey the standard commutation relations

$$[H_{eff}, b_\kappa^\dagger] = \hbar\Omega_\kappa b_\kappa^\dagger \qquad 4.5.6$$

$$[b_\kappa, b_{\kappa_1}^\dagger] = \delta_{\kappa\kappa_1} \qquad 4.5.7a$$

$$[b_\kappa, b_{\kappa_1}] = [b_\kappa^\dagger, b_{\kappa_1}^\dagger] = 0 \qquad 4.5.7b$$

We then obtain from 4.5.3 and 4.5.6

$$[H, b_\kappa^\dagger] = \hbar\Omega_\kappa b_\kappa^\dagger + [(V - \Phi), b_\kappa^\dagger] \qquad 4.5.8$$

and, to the extent to which the last term cannot be neglected but is small in the statistical sense, Eq.4.5.8 is identical to 4.3.25 with

$$R_\kappa^\dagger = [(V - \Phi), b_\kappa^\dagger] \qquad 4.5.9$$

From Section 4.3 we obtain then that the shift in the phonon energy is given by

$$\hbar\delta\Omega_\kappa = \frac{<[b_\kappa, R_\kappa^\dagger]>}{<[b_\kappa, b_\kappa^\dagger]>} = -<[[(V-\Phi), b_\kappa^\dagger], b_\kappa]> \tag{4.5.10}$$

since the denominator is equal to 1, according to 4.5.7a.

Our objective is then to choose Φ so that for every κ, the energy shift 4.5.10 vanishes. This means that for every κ, the relation

$$<[[(V-\Phi), b_\kappa^\dagger], b_\kappa]> = 0 \tag{4.5.11}$$

must be fulfilled. In order to evaluate the double commutator in 4.5.11, we proceed as follows. By inversion of 4.5.5 and of its time derivative we obtain (see 4.3.1)

$$\begin{aligned} b_\kappa^\dagger &= (2\hbar\Omega_\kappa)^{-\frac{1}{2}}(\Omega_\kappa Q_{-\kappa} - i\dot{Q}_{-\kappa}) \\ b_\kappa &= (2\hbar\Omega_\kappa)^{-\frac{1}{2}}(\Omega_\kappa Q_\kappa + i\dot{Q}_\kappa) \end{aligned} \tag{4.5.12}$$

In turn, we express Q_κ and $\dot{Q}_\kappa$ in terms of the molecular coordinates $R_\ell^{m\mu}$. By inversion of 4.5.4 and its time derivative we obtain

$$\begin{aligned} Q_\kappa &= L^{-\frac{1}{2}}\sum_{m\mu}\sum_\ell Y(\ell\mu|-\kappa)e^{-i\mathbf{k}\cdot\mathbf{R}_m}R_\ell^{m\mu} \\ \dot{Q}_\kappa &= L^{-\frac{1}{2}}\sum_{m\mu}\sum_\ell Y(\ell\mu|-\kappa)e^{-i\mathbf{k}\cdot\mathbf{R}_m}\dot{R}_\ell^{m\mu} \end{aligned} \tag{4.5.13}$$

We now substitute 4.5.13 into 4.5.12 and the resulting expressions into 4.5.11, disregarding all terms in $R_\ell^{m\mu}$ since they commute with both V and Φ given by 1.6.13 and 4.5.2, respectively. The result is

$$\begin{aligned} <[(V-\Phi), b_\kappa^\dagger]> &= -i(2\hbar\Omega_\kappa)^{-\frac{1}{2}} <[(V-\Phi), \dot{Q}_{-\kappa}]> = \\ &= -i(2\hbar\Omega_\kappa)^{-\frac{1}{2}}L^{-\frac{1}{2}} \sum_{\mu}\sum_\ell Y(\ell\mu|\kappa)e^{i\mathbf{k}\cdot\mathbf{R}_m}<[(V-\Phi), \dot{R}_\ell^{m\mu}]> = 0 \end{aligned} \tag{4.5.14}$$

and

$$<[[(V-\Phi), b_\kappa^\dagger], b_\kappa]> =$$

$$= (2\hbar\Omega_\kappa L)^{-1}\sum_{m\mu}\sum_{n\nu}\sum_{\ell\ell_1} Y(\ell\mu|\kappa)Y(\ell_1\nu|-\kappa)e^{i\mathbf{k}\cdot(\mathbf{R}_m+\mathbf{R}_n)}\langle[[(V - \Phi),\dot{R}_\ell^{m\mu}],\dot{R}_{\ell_1}^{n\nu}]\rangle = 0$$

The commutators in 4.5.15 are easily calculated. From 4.3.1 and 4.3.4 we obtain

$$[Q_\kappa,\dot{Q}_{\kappa_1}] = i\hbar\delta_{\kappa-\kappa_1} \qquad 4.5.16$$

and, by substitution of 4.5.13 into 4.5.16

$$[R_\ell^{m\mu},\dot{R}_{\ell_1}^{n\nu}] = i\hbar\delta_{mn}\delta_{\mu\nu}\delta_{\ell\ell_1} \qquad 4.5.17$$

The commutator of Φ with $\dot{R}_\ell^{m\mu}$ is then

$$[\Phi,\dot{R}_\ell^{m\mu}] = \frac{1}{2}\sum_{n\nu p\pi}\sum_{\ell_1\ell_2}\Phi_{\ell_1\ell_2}\binom{n\nu}{p\pi}[R_{\ell_1}^{n\nu}R_{\ell_2}^{p\pi},\dot{R}_\ell^{m\mu}] =$$

$$= i\hbar\sum_{n\nu}\sum_{\ell_1}\Phi_{\ell\ell_1}\binom{m\mu}{n\nu}R_\ell^{n\nu} = i\hbar(\partial\Phi/\partial R_\ell^{m\mu}) \qquad 4.5.18$$

and that of V, according to 4.2.4

$$[V,\dot{R}_\ell^{m\mu}] = [V_2,\dot{R}_\ell^{m\mu}] + [V_3,\dot{R}_\ell^{m\mu}] + [V_4,\dot{R}_\ell^{m\mu}] + \ldots = i\hbar(\partial V/\partial R_\ell^{m\mu}) \qquad 4.5.19$$

In the same way, the relations

$$[[\Phi,\dot{R}_\ell^{m\mu}],\dot{R}_{\ell_1}^{n\nu}] = (i\hbar)^2(\partial^2\Phi/\partial R_\ell^{m\mu}\partial R_{\ell_1}^{n\nu})$$

$$[[V,\dot{R}_\ell^{m\mu}],\dot{R}_{\ell_1}^{n\nu}] = (i\hbar)^2(\partial^2 V/\partial R_\ell^{m\mu}\partial R_{\ell_1}^{n\nu}) \qquad 4.5.20$$

are easily obtained. From 4.5.2 and 1.6.13 we obtain then

$$(\partial^2\Phi/\partial R_\ell^{m\mu}\partial R_{\ell_1}^{n\nu}) = \Phi_{\ell\ell_1}\binom{m\mu}{n\nu} \qquad 4.5.21$$

$$(\partial^2 V/\partial R_\ell^{m\mu}\partial R_{\ell_1}^{n\nu}) = F_{\ell\ell_1}\binom{m\mu}{n\nu} + \sum_{p\pi}\sum_{\ell_2}F_{\ell\ell_1\ell_2}\binom{m\mu}{n\nu}{}_{p\pi}R_{\ell_2}^{p\pi} +$$

$$+ \frac{1}{2}\sum_{p\pi}\sum_{r\rho}\sum_{\ell_2\ell_3}F_{\ell\ell_1\ell_2\ell_3}\binom{m\mu\ p\pi}{n\nu\ r\rho}R_{\ell_2}^{p\pi}R_{\ell_3}^{r\rho} \qquad 4.5.22$$

Substituing 4.5.20 into 4.5.15 we have

$$<\left[\left[(V - \Phi), b_{\kappa}^{\dagger}\right], b_{\kappa}\right]> = (i\hbar)^2 (2\hbar\Omega_{\kappa} L)^{-1} \sum_{m\mu}\sum_{n\nu}\sum_{\ell\ell_1} Y(\ell\mu|\kappa) Y(\ell_1\nu|-\kappa) \times$$

$$\times e^{i\mathbf{k}\cdot(\mathbf{R}_m + \mathbf{R}_n)} <\partial^2(V - \Phi)/\partial R_{\ell}^{m\mu}\partial R_{\ell_1}^{n\nu}> = 0 \qquad 4.5.23$$

This relation is fulfilled for every value of κ only if each term of the triple sum vanishes separately, i.e. only if

$$<\partial^2(V - \Phi)/\partial R_{\ell}^{m\mu}\partial R_{\ell_1}^{n\nu}> = 0 \qquad 4.5.24$$

from which, using 4.5.21, results

$$\Phi_{\ell\ell_1}\binom{m\mu}{n\nu} = <\partial^2 V/\partial R_{\ell}^{m\mu}\partial R_{\ell_1}^{n\nu}> \qquad 4.5.25$$

i.e.

$$\Phi_{\ell\ell_1}\binom{m\mu}{n\nu} = F_{\ell\ell_1}\binom{m\mu}{n\nu} + \sum_{p\pi}\sum_{\ell_2} F_{\ell\ell_1\ell_2}\begin{pmatrix}m\mu\\n\nu\\p\pi\end{pmatrix} <R_{\ell_2}^{p\pi}> +$$

$$+ \frac{1}{2} \sum_{p\pi}\sum_{r\rho}\sum_{\ell_2\ell_3} F_{\ell\ell_1\ell_2\ell_3}\begin{pmatrix}m\mu & p\pi\\n\nu & r\rho\end{pmatrix} <R_{\ell_2}^{p\pi} R_{\ell_3}^{r\rho}> + \ldots \qquad 4.5.26$$

The statistical averages in 4.5.26 can be taken using 4.3.19. We recall that the trace of an operator is obtained by summing over all its diagonal matrix elements. In our case therefore the statistical averages can be evaluated by summing over all the diagonal matrix elements of the operators $R_{\ell}^{m\mu}$ and $R_{\ell}^{m\mu}R_{\ell_1}^{n\nu}$ in the harmonic representation, i.e. using the state vectors $Y(\ell\mu|\kappa)$ of the harmonic Hamiltonian. It is easily seen from 4.5.4 and 4.5.5 that only products of an even number of molecular coordinates have non-zero statistical averages. Therefore, the effective force constants are given by

$$\Phi_{\ell\ell_1}\binom{m\mu}{n\nu} = F_{\ell\ell_1}\binom{m\mu}{n\nu} + \frac{1}{2} \sum_{p\pi}\sum_{r\rho}\sum_{\ell_2\ell_3} F_{\ell\ell_1\ell_2\ell_3}\begin{pmatrix}m\mu & p\pi\\n\nu & r\rho\end{pmatrix} <R_{\ell_2}^{p\pi} R_{\ell_3}^{r\rho}> \qquad 4.5.27$$

and differ from the corresponding harmonic force constants because they include a correction due to the quartic terms and, more generally, to the even terms of the potential averaged over the molecular motions.

The renormalized phonons obtained in this lowest-order SCP theory are still harmonic since the cubic terms which contribute to the damping are not effective. Higher order SCP approximations[133], which take into account the phonon damping have been developed, but they have not yet been utilized for molecular crystals.

4.6 THE METHOD OF THE GREEN'S FUNCTIONS

One of the most powerful techniques for the study of the physical properties of a system of interacting particles is that of the Green's functions[119-121]. The standard formalism has been first developed in applications of quantum field theory, in conjunction with the diagrammatic representation of the perturbation expansion of the interaction Hamiltonian. The method has then been extended to cover complex systems at finite temperature which are of interest to solid-state physics. Different types of Green's functions exist for different applications. Using these it is possible to describe the response of the system to an external perturbation, to characterize its microscopic properties, to derive thermodynamic functions, to follow the time evolution of the particles,etc. In the discussion of the anharmonic interactions in molecular solids, we are interested essentially in a particular type of such functions, the so-called "temperature Green's functions". Other important functions of this type, which are of interest in response theory, will not be discussed here.

Excellent reviews, articles and books are available on the general theory of the Green's functions[119-121] and in particular on their applications to solid-state physics [134-139]. The interested reader should consult them for a more complete understanding of the mathematical procedure and of the physical interpretation.

The Green's functions are appropriate generalizations of the time-dependent correlation functions, widely utilized in statistical mechanics. These are defined as the statistical average of the product of time-dependent operators written in the Heisenberg representation. We discuss them briefly before introducing the temperature Green's functions.

Correlation Functions

We recall that in the usual Schrödinger representation, the state of a system is described by time-dependent wave functions Ψ_s which satisfy the well-known equation

$$i\hbar\partial\Psi_s/\partial t = H_s\Psi_s \qquad 4.6.1$$

In this representation all operators $\hat{O}_s$ are time-independent and their expectation value is

$$\langle\hat{O}_s\rangle = \langle\Psi_s|\hat{O}_s|\Psi_s\rangle \qquad 4.6.2$$

The Heisenberg representation uses instead time-independent wave functions and time-dependent operators. Since the expectation value 4.6.2 corresponds to a physical observable, it must be the same in both representations. This condition is fulfilled, together with 4.6.1 if wave functions and operators in the two representations are related by the unitary transformation

$$\Psi_s = e^{-iHt/\hbar}\Psi_H \qquad 4.6.3$$

$$\hat{O}_H = e^{iHt/\hbar}\hat{O}_s e^{-iHt/\hbar} \qquad 4.6.4$$

By direct substitution of these transformations in 4.6.2, the invariance of the expectation value is easily verified. Differentiation of these equations with respect to time gives the equations of motion in the two representations. From 4.6.3 we obtain obviously 4.6.1. From 4.6.4 we have instead

$$i\hbar\partial\hat{O}_H/\partial t = e^{iHt/\hbar}(\hat{O}_s\hat{H} - \hat{H}\hat{O}_s)e^{-iHt/\hbar} = [\hat{O}_H,\hat{H}] \qquad 4.6.5$$

where $[\ldots,\ldots]$ means the commutator of the bracketed operators.

We then define the correlation function of two time-dependent operators $A(t)$ and $B(t')$ as the statistical average of their product. By generalization of 4.3.19, the correlation function is then

$$\langle A(t)B(t')\rangle = Z^{-1}\mathrm{Tr}\{A(t)B(t')e^{-\beta H}\} = \\ = Z^{-1}\mathrm{Tr}\{e^{iHt/\hbar}A_0 e^{-iHt/\hbar} e^{iHt'/\hbar} B_0 e^{-iHt'/\hbar} e^{-\beta H}\} \quad 4.6.6$$

Using the cyclic theorem for traces it is easily seen that the correlation function 4.6.6 depends only on the difference t-t' and not separately on t or t'

$$\langle A(t)B(t')\rangle = Z^{-1}\mathrm{Tr}\{e^{iH(t-t')/\hbar}A_0 e^{-iH(t-t')/\hbar}B_0 e^{-\beta H}\} = \\ = \langle A(t-t')B(0)\rangle \quad 4.6.7$$

Without loss of generality we can therefore take always t'= 0 and write the correlation functions in the form

$$F_{AB}(t) = \langle A(t)B(0)\rangle \\ F_{BA}(t) = \langle B(0)A(t)\rangle \quad 4.6.8$$

In our treatment of anharmonic effects in crystals we shall essentially use products of phonon operators in which one is the conjugate of the other. For generality, however, in the following discussion we shall continue to use the symbol B(0) and only when dealing explicitly with the phonon problem, we shall replace B(0) by A(0)*. It is nevertheless understood that B(0) is in any case a phonon operator like A(t), with all the simplifications of the general formalism that this implies.

The equations of motion for the time-dependent correlation functions 4.6.8 are easily obtained from 4.6.5

$$i\hbar\frac{\partial}{\partial t}\langle A(t)B(0)\rangle = \langle [A(t),H(t)]B(0)\rangle \\ i\hbar\frac{\partial}{\partial t}\langle B(0)A(t)\rangle = \langle B(0)[A(t),H(t)]\rangle \quad 4.6.9$$

The correlation functions can, in principle, be evaluated by direct integration[119] of these equations with appropriate boundary conditions. It is however simpler to obtain them from the correlated Green's functions using their spectral representation for which well-developed analytical methods exist[119-121].

We recall that the Fourier representation of a function f(t) is

$$f(t) = \int_{-\infty}^{+\infty} F(\omega) e^{-i\omega t} d\omega \qquad 4.6.10$$

where the coefficient $F(\omega)$ is called the Fourier transform of f(t) and is given by

$$F(\omega) = \frac{1}{2\pi} \int_{-\infty}^{+\infty} f(t) e^{i\omega t} d\omega \qquad 4.6.11$$

The correlation functions 4.6.8 can be then written in terms of their Fourier transforms as

$$\begin{aligned} F_{AB}(t) &= \int_{-\infty}^{+\infty} F_{AB}(\omega) e^{-i\omega t} d\omega \\ F_{BA}(t) &= \int_{-\infty}^{+\infty} F_{BA}(\omega) e^{-i\omega t} d\omega \end{aligned} \qquad 4.6.12$$

An important property of the Fourier transforms in 4.6.12 is that they are not independent but can be both expressed in terms of the same spectral function. To show this, let us consider the complete set of eigenfunctions $|n\rangle$ and eigenvalues E_n of the Hamiltonian and write in full expressions 4.6.12 using their definition 4.6.6. We have for $F_{AB}(t)$

$$\begin{aligned} F_{AB}(t) &= Z^{-1} \mathrm{Tr}\{A(t)B(0)e^{-\beta H}\} = Z^{-1} \sum_n \langle n|A(t)B(0)e^{-\beta H}|n\rangle = \\ &= Z^{-1} \sum_{mn} \langle n|e^{iHt/\hbar} A_0 e^{-iHt/\hbar}|m\rangle\langle m|B_0|n\rangle e^{-\beta E_n} \\ &= Z^{-1} \sum_{mn} \langle n|A_0|m\rangle\langle m|B_0|n\rangle e^{-\beta E_n} e^{-i(E_m-E_n)t/\hbar} \end{aligned} \qquad 4.6.13$$

In the same way we obtain

$$F_{BA}(t) = Z^{-1} \sum_{mn} \langle n|B_0|m\rangle\langle m|A_0|n\rangle e^{-\beta E_n} e^{-i(E_n-E_m)t/\hbar} \qquad 4.6.14$$

or, interchanging the indices in the double sum

$$F_{BA}(t) = Z^{-1}\sum_{mn} \langle m|B_0|n\rangle\langle n|A_0|m\rangle e^{-\beta E_m} e^{-i(E_m-E_n)t/\hbar} \qquad 4.6.15$$

Equations 4.6.13 and 4.6.15 are very similar and differ only with respect to an exponential factor. We can thus define a function $J_{AB}(\omega)$

$$J_{AB}(\omega) = Z^{-1}\sum_{mn} \langle n|A_0|m\rangle\langle m|B_0|n\rangle e^{-\beta E_n}\delta(E_m-E_n-\hbar\omega) \qquad 4.6.16$$

such that the correlation functions $F_{AB}(t)$ and $F_{BA}(t)$ can be expressed in integral form as

$$F_{AB}(t) = \int_{-\infty}^{+\infty} J_{AB}(\omega)e^{-i\omega t}d\omega \qquad 4.6.17$$

$$F_{BA}(t) = \int_{-\infty}^{+\infty} J_{AB}(\omega)e^{-i\omega t}e^{-\hbar\beta\omega}d\omega \qquad 4.6.18$$

By direct substitution of 4.6.16 in these relations and using the definition of the δ function

$$\int_{-\infty}^{+\infty} f(x)\delta(x-x_0)dx = f(x_0) \qquad 4.6.19$$

expressions 4.6.14 and 4.6.15 are easily obtained. By comparison with 4.6.12 we see that

$$F_{AB}(\omega) = J_{AB}(\omega)$$

$$\qquad 4.6.20$$

$$F_{BA}(\omega) = J_{AB}(\omega)e^{-\hbar\beta\omega} = F_{AB}(\omega)e^{-\hbar\beta\omega}$$

The function $J_{AB}(\omega)$ is called "spectral function" and is non-zero only if $\hbar\omega = E_m - E_n$. The density of $J_{AB}(\omega)$ values therefore reproduces that of all possible transitions between eigenstates of the Hamiltonian. The importance of the spectral representation of the correlation functions is thus evident. It is also important to observe that 4.6.18 gives directly 4.6.17 if the variable t is replaced by $t + i\hbar\beta$. The correlation functions thus satisfy the condition

$$\langle B(0)A(t+i\hbar\beta)\rangle = \langle A(t)B(0)\rangle \qquad 4.6.21$$

We shall reconsider later this similarity between time and temperature ($\beta = 1/KT$) when dealing with equations of the type 4.6.21.

The Temperature Green's Functions

The Green's functions are closely connected to correlation function They are however simpler to use and well-developed analytical methods are available for their calculation.

The important Green's function for our purpose is the so-called "temperature" Green's function, defined in terms of a fictitious imaginary time τ rather than by the real time variable t used previously. To arrive at its definition, we begin with the definition of the " causal" Green's function

$$G_{AB}(t) = - i\Theta(t)\langle A(t)B(0)\rangle - i\Theta(-t)\langle A(t)B(0)\rangle \qquad 4.6.22$$

where $\Theta(t)$ is the Heaviside step function

$$\Theta(t) = \begin{matrix} 0 \text{ for } t<0 \\ 1 \text{ for } t>0 \end{matrix} \qquad 4.6.23$$

It is normal practice to write $G_{AB}(t)$ in the more compact form

$$G_{AB}(t) = - i\langle T_t A(t)B(0)\rangle \qquad 4.6.24$$

where T_t is a time-ordering operator which orders the operators that follow him in the time-scale so that the one with earlier time operates first. The two definitions are equivalent and simply mean that

$$G_{AB}(t) = - i\langle A(t)B(0)\rangle \qquad \text{for } t > 0$$

$$4.6.25$$

$$G_{AB}(t) = - i\langle B(0)A(t)\rangle \qquad \text{for } t < 0$$

The causal Green's function is thus defined in terms of correlation functions. In contrast to them, however, the Green's function is not defined at t = 0, due to the presence of the step function 4.6.23.

The equation of motion for the causal Green's function can be obtained from 4.6.9, using the relation

$$\frac{d}{dt}\,\Theta(t) = -\frac{d}{dt}\,\Theta(-t) = \delta(t) \qquad 4.6.26$$

and is

$$i\hbar\,\frac{dG_{AB}(t)}{dt} = \hbar\langle[A(t),B(0)]\rangle\delta(t) - i\Theta(t)\left\langle\Big[[A(t),H(t)],B(0)\Big]\right\rangle \qquad 4.6.27$$

The last term in 4.6.27 is, by definition, a higher order Green's function involving the operator $[A(t),H(t)]$ in place of A(t). If we write the equation of motion for this new Green's function $G'_{AB}(t)$, we obtain with the same procedure

$$i\hbar\,\frac{dG_{AB}(t)}{dt} = \hbar\langle[[A(t),H(t)],B(0)]\rangle\delta(t) + \qquad 4.6.28$$

$$- i\Theta(t)\left\langle\Big[\big[[A(t),H(t)],H(t)\big],B(0)\Big]\right\rangle$$

Continuing this process a chain of coupled equations with a hierarchy of Green's functions results. To solve this system of coupled equations, we must uncouple the chain at a given stage and supplement a set of boundary conditions. As no general method exists for finding the uncoupling procedure, this is normally done by introducing specific approximations into the higher order Green's functions. A more convenient approach is that of utilizing the powerful diagrammatic techniques developed in quantum field theory for perturbation procedures[119,120]. We shall consider only this second method.

In the previous discussion of correlation functions we have noticed that $F_{AB}(t)$ can be obtained from $F_{BA}(t)$ by replacing in the latter the variable t by t+iħβ. This suggests that, on the imaginary time axis, properly defined Green's functions could show important periodic properties. On the basis of these considerations we introduce then the temperature Green's functions, defined in terms of the imaginary time variable. From 4.6.24 we define the temperature Green's function

$$G_{AB}(\tau) = \langle T_\tau A(\tau)B(0)\rangle \tag{4.6.29}$$

where T_τ is now the time-ordering operator in the imaginary time so that

$$G_{AB}(\tau) = \langle A(\tau)B(0)\rangle \qquad \text{for} \quad \tau > 0$$
$$\tag{4.6.30}$$
$$G_{AB}(\tau) = \langle B(0)A(\tau)\rangle \qquad \text{for} \quad \tau < 0$$

Using the cyclic theorem for traces, we then obtain

$$\langle B(0)A(\tau)\rangle = Z^{-1}\mathrm{Tr}\{B(0)e^{H\tau}A_0e^{-H\tau}e^{-\beta H}\} =$$

$$= Z^{-1}\mathrm{Tr}\{B_0e^{-\beta H}e^{\beta H}e^{H\tau}A_0e^{-(\tau+\beta)H}\} = \tag{4.6.31}$$

$$= Z^{-1}\mathrm{Tr}\{e^{(\tau+\beta)H}A_0e^{-(\tau+\beta)H}B_0e^{-\beta H}\} = \langle A(\tau+\beta)B(0)\rangle$$

Therefore, as long as $\tau < 0$ but $\tau+\beta > 0$, the temperature Green's function $G_{AB}(\tau)$ is periodic with periodicity β

$$G_{AB}(\tau) = G_{AB}(\tau+\beta) \tag{4.6.32}$$

and can thus be expanded in a Fourier series

$$G_{AB}(\tau) = \sum_{-\infty}^{+\infty}{}_n G_{AB}(i\omega_n)e^{-i\hbar\omega_n\tau} \tag{4.6.33}$$

with

$$\omega_n = 2\pi n/\beta\hbar \tag{4.6.34}$$

The Fourier coefficients in the series are given by

$$G_{AB}(i\omega_n) = \frac{1}{\beta}\int_0^\beta G_{AB}(\tau)e^{i\hbar\omega_n\tau}d\tau \tag{4.6.35}$$

and, by means of 4.6.29 and 4.6.17, can be written in the form

$$G_{AB}(i\omega_n) = \frac{1}{\beta}\int_0^\beta <A(\tau)B(0)>e^{i\hbar\omega_n\tau}d\tau$$

$$= \frac{1}{\beta}\int_{-\infty}^{+\infty}d\omega J_{AB}(\omega)\int_0^\beta e^{\hbar\tau(i\omega_n-\omega)}d\tau \qquad 4.6.36$$

By integration of 4.6.36 on τ we obtain

$$G_{AB}(i\omega_n) = \frac{1}{\beta}\int_{-\infty}^{+\infty}d\omega J_{AB}(\omega)\frac{(1 - e^{-\beta\hbar\omega})}{\hbar(\bar{\omega} - i\omega_n)} \qquad 4.6.37$$

since $e^{i\beta\hbar\omega_n} = 1$ for all values of n.

The Fourier coefficients 4.6.37 are defined only at the infinite set of points $i\omega_n$,where ω_n takes the values given by 4.6.34. The extension of 4.6.37 to the whole complex plane, necessary for many applications of the theory, is possible by defining a continuous complex variable w such that

$$G_{AB}(w) \equiv G_{AB}(i\omega_n) \qquad \text{when } w = i\omega_n \qquad 4.6.38$$

In terms of the variable w the coefficients 4.6..37 become

$$G_{AB}(w) = \frac{1}{\beta\hbar}\int_{-\infty}^{+\infty}d\omega J_{AB}(\omega)\frac{(1 - e^{-\beta\hbar\omega})}{\omega - w} \qquad 4.6.39$$

This function is now analytical except on the real axis. Near the real axis we can write $w = \Omega \pm i\varepsilon$ where Ω is real and ε a small quantity. Using then 4.4.51, we obtain, by approaching the real axis from the positive direction

$$\lim_{\varepsilon\to 0} G_{AB}(\Omega\pm i\varepsilon) = \frac{1}{\beta\hbar}\Big[\int_{-\infty}^{+\infty}d\omega J_{AB}(\omega)\frac{(1 - e^{-\beta\hbar\omega})}{(w - \Omega)_p} + $$

$$\mp i\pi J_{AB}(\Omega)(1 - e^{-\beta\hbar\Omega})\Big] \qquad 4.6.40$$

where p denotes the principal part of the integral. From this relation, by taking the difference between the limit of $G_{AB}(\Omega+i\varepsilon)$ and the limit of $G_{AB}(\Omega-i\varepsilon)$, we obtain directly the spectral function

$$J_{AB}(\Omega) = -\frac{i\beta\hbar}{2\pi}(1 - e^{-\beta\hbar\Omega})^{-1}\lim_{\varepsilon\to 0}\big[G_{AB}(\Omega+i\varepsilon) - G_{AB}(\Omega-i\varepsilon)\big] \qquad 4.6.41$$

The Harmonic Phonon Propagator

For a harmonic lattice the one-particle Green's function 4.6.29 assumes a very simple form. In this case, it is convenient[134-137] to take as phonon operators the sum operators introduced in Section 4.3

$$A(\tau) = A_{\kappa}(\tau) = a^{\dagger}_{-\kappa}(\tau) + a_{\kappa}(\tau)$$
$$B(\tau) = A^{*}_{\kappa}(\tau) = a^{\dagger}_{\kappa}(\tau) + a_{-\kappa}(\tau) \qquad 4.6.42$$

since these operators are directly connected to the crystal normal coordinates and are thus the basis for the expansion of the crystal Hamiltonian. We notice that the operators 4.6.42 differ from the corresponding operators 4.3.2 only for the fact that the latter are time-independent. The operators 4.6.42 are instead written in the Heisenberg representation and are thus time-dependent.

Using the operators 4.6.42, the temperature Green's function 4.6.29 becomes

$$G^{0}_{\kappa\kappa_1}(\tau) = \langle T_{\tau} A_{\kappa}(\tau) A^{*}_{\kappa_1}(0)\rangle =$$
$$= \langle T_{\tau}(a^{\dagger}_{-\kappa}(\tau) + a_{\kappa}(\tau))(a^{\dagger}_{\kappa_1}(0) + a_{-\kappa_1}(0))\rangle \qquad 4.6.43$$

where the superscript zero denotes the harmonic case.

The thermal averages in 4.6.43 are different from zero only if $\kappa=\kappa_1$ and only for operator products involving one creation and one annihilation operator of the same type, as discussed in Section 4.4 when dealing with matrix elements. Equation 4.6.43 then simplifies to

$$G^{0}_{\kappa\kappa_1}(\tau) = \delta_{\kappa\kappa_1} g^{0}_{\kappa}(\tau) \qquad 4.6.44$$

where

$$g^{0}_{\kappa}(\tau) = \langle T_{\tau}(a^{\dagger}_{-\kappa}(\tau) a_{-\kappa}(0) + a_{\kappa}(\tau) a^{\dagger}_{\kappa}(0))\rangle \qquad 4.6.45$$

$g^0_\kappa(\tau)$ is often called the "harmonic phonon propagator" and is defined for $-\beta < \tau < \beta$. From 4.3.18 and 4.3.24 we have

$$\begin{aligned}<a^\dagger_{-\kappa}(\tau)a_{-\kappa}(0)> &= Z^{-1}\mathrm{Tr}\{e^{H_0\tau}a^\dagger_{-\kappa}e^{-H_0\tau}a_{-\kappa}e^{-\beta H_0}\} = \\ &= Z^{-1}\mathrm{Tr}\{e^{H_0\tau}e^{-H_0\tau}e^{\hbar\omega^0_\kappa\tau}a^\dagger_{-\kappa}a_{-\kappa}e^{-\beta H_0}\} = \\ &= e^{\hbar\omega^0_\kappa\tau}<a^\dagger_{-\kappa}a_{-\kappa}> = \bar{n}_{-\kappa}e^{\hbar\omega^0_\kappa\tau}\end{aligned} \qquad 4.6.46$$

In the same way we obtain

$$<a_\kappa(\tau)a^\dagger_\kappa(0)> = e^{-\hbar\omega^0_\kappa\tau}<a_\kappa a^\dagger_\kappa> = (\bar{n}_\kappa + 1)e^{-\hbar\omega^0_\kappa\tau} \qquad 4.6.47$$

Considering now the two possible time-orderings in 4.6.45, Expression 4.6.48 results:

$$g^0_\kappa(\tau) = \bar{n}_\kappa e^{\hbar\omega^0_\kappa|\tau|} + (\bar{n}_\kappa + 1)e^{-\hbar\omega^0_\kappa|\tau|} \qquad 4.6.48$$

From 4.6.35 the Fourier transform of $g^0_\kappa(\tau)$ is

$$\begin{aligned}g^0_\kappa(i\omega_n) &= \frac{1}{\beta}\int_0^\beta[\,\bar{n}_\kappa e^{\hbar\omega^0_\kappa\tau} + (\bar{n}_\kappa + 1)e^{-\hbar\omega^0_\kappa\tau}\,]e^{i\hbar\omega_n\tau}d\tau \\ &= \frac{1}{\beta\hbar}\left[\frac{\bar{n}_\kappa(e^{\hbar\omega^0_\kappa\beta} - 1)}{\omega^0_\kappa + i\omega_n} - \frac{(\bar{n}_\kappa + 1)(e^{-\hbar\omega^0_\kappa\beta} - 1)}{\omega^0_\kappa - i\omega_n}\right]\end{aligned} \qquad 4.6.49$$

since $e^{i\hbar\omega_n\beta} = 1$ for $\omega_n = 2\pi n/\beta\hbar$. From 4.3.24 it then follows

$$g^0_\kappa(i\omega_n) = \frac{1}{\beta\hbar}\left(\frac{1}{\omega_\kappa + i\omega_n} + \frac{1}{\omega_\kappa - i\omega_n}\right) = \frac{2\omega^0_\kappa}{\beta\hbar(\omega^{0\,2}_\kappa + \omega^2_n)} \qquad 4.6.50$$

As discussed in the previous Section, $g^0_\kappa(i\omega_n)$ can be analytically continued in the complex plane. Thus, from 4.6.39 we obtain

$$g^0_\kappa(w) = \frac{1}{\beta\hbar}\left(\frac{1}{\omega^0_\kappa + w} + \frac{1}{\omega^0_\kappa - w}\right) = \frac{2\omega^0_\kappa}{\beta\hbar(\omega^{0\,2}_\kappa + w^2)} \qquad 4.6.51$$

and from this relation, using 4.6.40 and 4.6.41, we can obtain the spectral function $J^0_\kappa(w)$

$$\mathcal{J}^0_\kappa(\hat{w}) = \bar{n}_\kappa\ (w+\omega_\kappa) + (\bar{n}_\kappa + 1)\ (w-\omega_\kappa) \qquad 4.6.52$$

The Anharmonic Temperature Green's Function

We now consider the more complex case of anharmonic crystals and discuss the corresponding temperature Green's function. As customary in perturbation theory, we expand the complete Hamiltonian in orders of magnitude as in 4.1.1. For simplicity we write H in the form

$$H = H_0 + H_i \qquad 4.6.53$$

where H_i includes all the non-harmonic terms

$$H_i = H_1 + H_2 + \ldots + H_m \qquad 4.6.54$$

H_1, H_2, H_m being of the type 4.3.11.

Using this expansion, the temperature Green's function has the form

$$\begin{aligned} G^{>0}_{AB}(\tau) &= \langle A(\tau)B(0)\rangle = Z^{-1}\mathrm{Tr}\{e^{(H_0+H_i)\tau}A_0 e^{-(H_0+H_i)\tau}B_0 e^{-\beta(H_0+H_i)}\} \\ G^{<0}_{AB}(\tau) &= \langle B(0)A(\tau)\rangle = Z^{-1}\mathrm{Tr}\{B_0 e^{(H_0+H_i)\tau}A_0 e^{-(H_0+H_i)\tau}e^{-\beta(H_0+H_i)}\} \end{aligned} \qquad 4.6.55$$

where the upper expression is valid for $\tau>0$ and the lower for $\tau<0$.

Since only the anharmonic terms of the Hamiltonian are time-dependent, it is convenient to use a third representation for operators and wave functions when dealing with the perturbation expansion 4.6.53. This is called the "interaction representation" and is intermediate between the Schrödinger and the Heisenberg representations. In the interaction representation the wave functions are taken as time-independent under the action of H_0 only, i.e. they can vary with time under the perturbation H_i. Operators are then defined by

$$\hat{O}_I = e^{iH_0t/\hbar}\hat{O}_s e^{-iH_0t/\hbar} \qquad 4.6.56$$

and wave functions are related to those in the Schrödinger picture by the transformation

$$\Psi_s = e^{-iH_0t/\hbar}\Psi_s \qquad 4.6.57$$

so that the expectation value 4.6.2 remains unchanged. In the following discussion we shall always use, unless otherwise specified, operators in the interaction representation. These will be indicated by the symbol $\tilde{O}$.

From 4.6.4 and 4.6.56 we find that operators in the Heisenberg and in the interaction representations are related by

$$\hat{O}_H = e^{H\tau}e^{-H_0\tau}\tilde{O}e^{H_0\tau}e^{-H\tau} \qquad 4.6.58$$

where we have used as previously the complex time variable $\tau = it/\hbar$. This relation is more conveniently written in the form

$$\hat{O}_H = S(0,\tau)\tilde{O}S(\tau,0) \qquad 4.6.59$$

where the operator $S(\tau,0)$ is defined, in the most general case, by

$$S(\tau,\tau') = e^{H_0\tau}e^{-H(\tau-\tau')}e^{-H_0\tau'} \qquad 4.6.60$$

and satisfies the conditions

$$S(\tau',\tau) = S(\tau,\tau')^{-1}$$

$$S(\tau,\tau'')S(\tau'',\tau') = S(\tau,\tau') \qquad 4.6.61$$

$$S(\tau,\tau) = 1$$

so that, in particular

$$S(\tau,0) = e^{H_0\tau}e^{-H\tau}$$

$$4.6.62$$

$$S(0,\tau) = e^{H\tau}e^{-H_0\tau}$$

From these relations it follows, using 4.6.53

$$e^{-(H_0+H_i)\tau} = e^{-H_0\tau}S(\tau,0)$$
$$e^{(H_0+H_i)\tau} = S(0,\tau)e^{H_0\tau} \qquad 4.6.63$$

The $S(\tau,\tau')$ operator is very useful for perturbation calculations since it is easy to find for it an integral equation that can be solved by iterations. For this purpose, we differentiate 4.6.63a with respect to τ

$$-(H_0 + H_i)e^{-(H_0+H_i)\tau} = e^{-H_0\tau}\{[dS(\tau,0)/d\tau] - H_0S(\tau,0)\}$$
$$-(H_0 + H_i)e^{-H_0\tau}S(\tau,0) = e^{-H_0\tau}\{[dS(\tau,0)/d\tau] - H_0S(\tau,0)\} \qquad 4.6.64$$
$$- H_ie^{-H_0\tau}S(\tau,0) = e^{-H_0\tau}[dS(\tau,0)/d\tau]$$

and from this it follows

$$dS(\tau,0)/d\tau = - e^{H_0\tau}H_ie^{-H_0\tau}S(\tau,0) = - \tilde{H}_iS(\tau,0) \qquad 4.6.65$$

This equation can be solved by iterations[119] with the condition $S(0,0) = 1$ giving

$$S(\tau,0) = 1 + \sum_{-\infty n}^{+\infty}(-1)^n\frac{1}{n!}\int_0^\tau d\tau_1\int_0^\tau d\tau_2\ldots\int_0^\tau d\tau_n T_\tau\tilde{H}_i(\tau_1)\tilde{H}_i(\tau_2)\ldots\tilde{H}_i(\tau_n) \qquad 4.6.66$$

where T_τ is the time-ordering operator which orders operators in 4.6.66 so that the value of τ increases from the right to the left. A detailed discussion of the derivation of 4.6.66 can be found in Refs. 119,120.

Using now 4.6.61 and 4.6.63, the temperature Green's function 4.6.55 becomes for $\tau > 0$

$$G_{AB}^{>0}(\tau) = Z^{-1}\mathrm{Tr}\{S(0,\tau)e^{H_0\tau}A_0e^{-H_0\tau}S(\tau,0)B_0e^{-\beta H_0}S(\beta,0)\}$$
$$= Z^{-1}\mathrm{Tr}\{S(\beta,0)S(0,\tau)\tilde{A}(\tau)S(\tau,0)\tilde{B}(0)e^{-\beta H_0}\} \quad (\tau > 0) \qquad 4.6.67$$

$$G^{>0}_{AB}(\tau) = Z^{-1}\mathrm{Tr}\{S(\beta,\tau)\tilde{A}(\tau)S(\tau,0)\tilde{B}(0)e^{-\beta H_0}\}$$

where $S(\beta,0)$ is obtained from 4.6.66 by replacing τ with β.

In the same way, we obtain for $\tau < 0$

$$G^{<0}_{AB}(\tau) = Z^{-1}\mathrm{Tr}\{B_0 S(0,\tau)e^{H_0\tau}A_0 e^{-H_0\tau}S(\tau,0)e^{-\beta H_0}S(\beta,0)\}$$

$$= Z^{-1}\mathrm{Tr}\{\tilde{B}(0)S(0,\tau)\tilde{A}(\tau)S(\tau,0)e^{-\beta H_0}S(\beta,0)\} \qquad (\tau < 0) \qquad 4.6.68$$

and we can combine $G^{>0}_{AB}(\tau)$ and $G^{<0}_{AB}(\tau)$ to the single, more compact expression

$$G_{AB}(\tau) = Z^{-1}\mathrm{Tr}\{T_\tau\tilde{A}(\tau)\tilde{B}(0)S(\beta,0)e^{-\beta H_0}\} \qquad 4.6.69$$

where now the time-ordering operator orders the operators $\tilde{H}_i(\tau)$ in the expansion of $S(\beta,0)$ simultaneusly with $\tilde{A}(\tau)$ and $\tilde{B}(0)$. Using 4.3.20, we obtain for the partition function of the anharmonic system

$$Z = \mathrm{Tr}\{e^{-\beta(H_0+H_i)}\} = \mathrm{Tr}\{e^{-\beta H_0}S(\beta,0)\} = Z_0\langle S(\beta,0)\rangle \qquad 4.6.70$$

The temperature Green's function becomes then

$$G_{AB}(\tau) = \mathrm{Tr}\{T_\tau\tilde{A}(\tau)\tilde{B}(0)S(\beta,0)e^{-\beta H_0}\}/Z_0\langle S(\beta,0)\rangle_0 =$$

$$= \langle T_\tau\tilde{A}(\tau)\tilde{B}(0)S(\beta,0)\rangle_0/\langle S(\beta,0)\rangle_0 \qquad 4.6.71$$

where the symbol $\langle\ldots\rangle_0$ means that matrix elements are taken in the harmonic representation and where, according to 4.6.66

$$\langle S(\beta,0)\rangle = 1 + \sum_{1n}^{\infty}(-1)^n\frac{1}{n!}\int_0^\beta d\tau_1\int_0^\beta d\tau_2\ldots\int_0^\beta d\tau_n\langle T_\tau\tilde{H}_i(\tau_1)\tilde{H}_i(\tau_2)\ldots\tilde{H}_i(\tau_n)\rangle \qquad 4.6.72$$

The anharmonic perturbation Hamiltonian $\tilde{H}_i(\tau)$ is given by 4.6.54 expressed in the interaction representation

$$\tilde{H}_i(\tau) = \tilde{H}_1(\tau) + \tilde{H}_2(\tau) + \ldots + \tilde{H}_m(\tau) + \ldots \qquad 4.6.73$$

where, for the specific problem of anharmonic effects in crystals, the terms of the Hamiltonian are

$$\tilde{H}_1(\tau) = \sum_{\kappa_1\kappa_2\kappa_3} B(\kappa_1\kappa_2\kappa_3)\tilde{A}_{\kappa_1}(\tau)\tilde{A}_{\kappa_2}(\tau)\tilde{A}_{\kappa_3}(\tau)$$

$$\tilde{H}_2(\tau) = \sum_{\kappa_1\kappa_2\kappa_3\kappa_4} B(\kappa_1\kappa_2\kappa_3\kappa_4)\tilde{A}_{\kappa_1}(\tau)\tilde{A}_{\kappa_2}(\tau)\tilde{A}_{\kappa_3}(\tau)\tilde{A}_{\kappa_4}(\tau) \qquad 4.6.74$$

$$\tilde{H}_m(\tau) = \sum_{\kappa_1\ldots\kappa_{m+2}} B(\kappa_1\ldots\kappa_{m+2})\tilde{A}_{\kappa_1}(\tau)\ldots\tilde{A}_{\kappa_{m+2}}(\tau)$$

with

$$\tilde{A}_\kappa(\tau) = e^{H_0\tau}(a^\dagger_{-\kappa} + a_\kappa)e^{-H_0\tau} = e^{\hbar\omega_\kappa\tau}a^\dagger_{-\kappa} + e^{-\hbar\omega_\kappa\tau}a_\kappa \qquad 4.6.75$$

As discussed before in the harmonic case, thermal averages of the type involved in 4.6.71 are different from zero only if there is an ever number of phonon operators $\tilde{A}_\kappa(\tau)$ and only if these occur in conjugate pairs so that for each operator $\tilde{A}_\kappa(\tau)$ there is always a corresponding operator $\tilde{A}_{-\kappa}(\tau)$. For an even number of operators, say 2n, we have then $(2n!)/2^n(n!)$ different possibilities of pairing. It can be shown[119] by means of Wick's theorem that

$$\langle T_\tau\tilde{A}_{\kappa_1}(\tau_1)\tilde{A}_{\kappa_2}(\tau_2)\tilde{A}_{\kappa_3}(\tau_3)\ldots\tilde{A}_{\kappa_{2n}}(\tau_{2n})\rangle =$$

$$= \langle T_\tau\tilde{A}_{\kappa_1}(\tau_1)\tilde{A}_{\kappa_2}(\tau_2)\rangle\langle T_\tau\tilde{A}_{\kappa_3}(\tau_3)\tilde{A}_{\kappa_4}(\tau_4)\rangle\ldots\langle T_\tau\tilde{A}_{\kappa_{2n-1}}(\tau_{2n-1})\tilde{A}_{\kappa_{2n}}(\tau_{2n})\rangle \qquad 4.6.76$$

$$+ \left[(2n!)/2^n(n!) - 1\right] \text{ equivalent pairing schemes}$$

The thermal averages in 4.6.71 are thus sums of products of harmonic Green's functions of the type 4.6.43. These depend only on the difference $\tau_i - \tau_j$ and can in general be written as

$$\langle T_\tau\tilde{A}_{\kappa_1}(\tau_i)\tilde{A}_{\kappa_2}(\tau_j)\rangle = \delta_{\kappa_1-\kappa_2}g^0_{\kappa_1}(\tau_i - \tau_j) =$$

$$= \delta_{\kappa_1-\kappa_2}\{n_{\kappa_1}e^{\hbar\omega_{\kappa_1}(\tau_i - \tau_j)} + (n_{\kappa_1} + 1)e^{\hbar\omega_{\kappa_1}(\tau_i - \tau_j)}\} \qquad 4.6.77$$

by simple extension of 4.6.44 and 4.6.48.

The operator pairing schemes occurring in 4.6.68 are most conveniently represented by means of diagrams of the type of Figs. 4.1 and 4.2 . These can be utilized to evaluate, by means of simple rules, the contribution of each scheme to the one-phonon Green's function.

To construct the diagrams we represent phonons by lines and anharmonic coupling coefficients by vertices joining lines. For instance, a triple vertex joining three phonon lines represents a third-order coefficient, a vertex with four lines represents a fourth-order coefficient and so on. The lines are paired together according to the scheme to be represented. Some simple examples will clarify the procedure.

Consider Eq. 4.6.72 for $\langle S(\beta,0)\rangle$. If we limit the expansion of the perturbation Hamiltonian 4.6.73 to the first two terms, we see that the only contribution for n = 1 to $\langle S(\beta,0)\rangle$ arises from $\tilde{H}_2(\tau)$ which includes four $\tilde{A}_\kappa(\tau)$ operators. These can be paired in three different ways as shown below

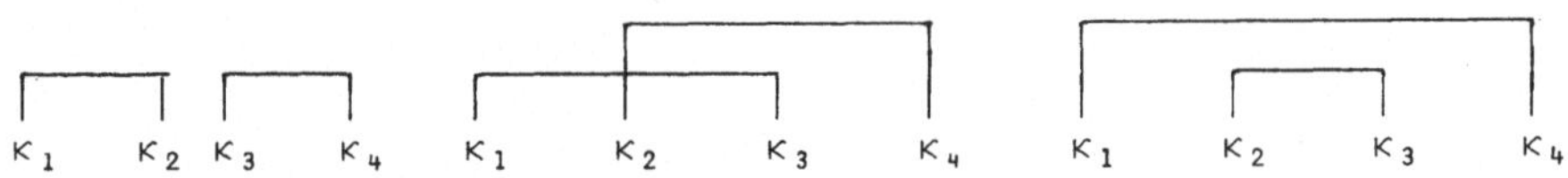

(1) (2) (3)

where the line joining two indices means that these should be equal but with opposite sign. Owing to the symmetry of the indices, these three pairing schemes are equivalent in the sense that they give the same contribution to $\langle S(\beta,0)\rangle$. They can be thus concisely represented by the diagram a of Fig.4.1. The contribution of this diagram must be obviously multiplied by three.

For n = 2 we have two possible products with an even number of operators in 4.6.72, i.e. $\tilde{H}_1(\tau_i)\tilde{H}_1(\tau_j)$ and $\tilde{H}_2(\tau_i)\tilde{H}_2(\tau_j)$. These are of different order, the first being of order two and the second of order four in the expansion parameter(see Section 4.1). We shall thus consider only the product $\tilde{H}_1(\tau_i)\tilde{H}_1(\tau_j)$. The six phonon operators can be paired in this case either as

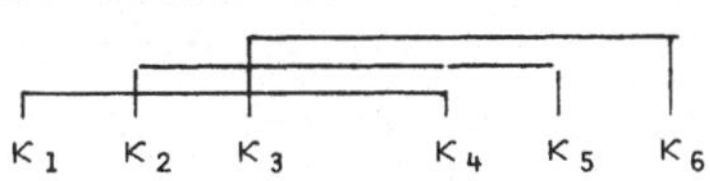

+ 5 equivalent pairing schemes

or as

+ 8 equivalent pairing schemes

The first six equivalent pairings are condensed in the diagram b and the other nine equivalent schemes in the diagram c of Fig.4.1.

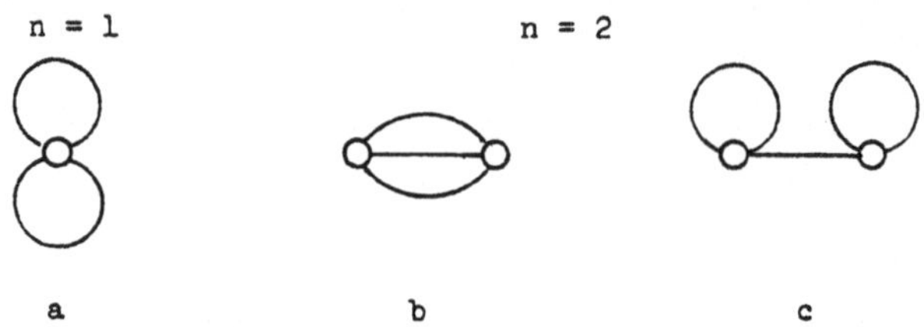

Fig.4.1 The simplest diagrams occurring in the expansion of $\langle S(\beta,0)\rangle$.

We now consider the numerator of 4.6.71. There we have two additional phonon operators to pair to each product in the expansion of $S(\beta,0)$. Until now we have used for them the generic symbols $\tilde{A}(\tau)$ and $\tilde{B}(0)$ but from now on we shall be more specific and denote them by the symbols $\tilde{A}_\kappa(\tau)$ and $\tilde{A}^*_\kappa(0)$ as done in 4.6.42 when dealing with the harmonic phonon propagator. If the operator $\tilde{A}^*_\kappa(0)$ is represented by a line entering the figure from the right and the operator $\tilde{A}_\kappa(\tau)$ by a line leaving the figure from the left, then the new diagrams represented in Fig.4.2 result.

Since these latter diagrams are of greater interest for our treatment we shall concentrate our attention on them. Those of Fig.4.1, without external lines, are important for the calculation of thermodynamic functions, but as we shall see, do not enter directly in the calculation of spectroscopic anharmonic effects.

We notice that also in this case each diagram corresponds to a set of equivalent coupling schemes. Consider, for instance, the diagram b of Fig.4.2. The six different phonon operators can be paired, as shown below, in 12 different ways since the operator on the right can be paired to the fourth-order vertex in four different ways and for each one of these there are three possible couplings of the vertex with the operator on the left side

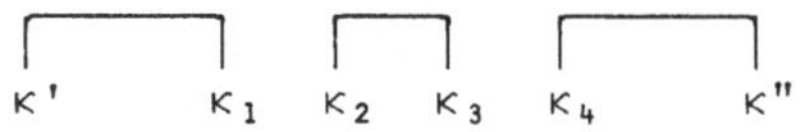

+ 11 equivalent pairing schemes

In the same way it is easily seen that the 8 κ values of diagrams d and e in Fig.4.2 can be coupled in 18 different ways

diagram d

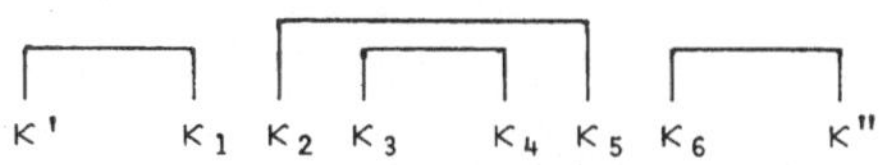

+ 17 equivalent pairing schemes

diagram e

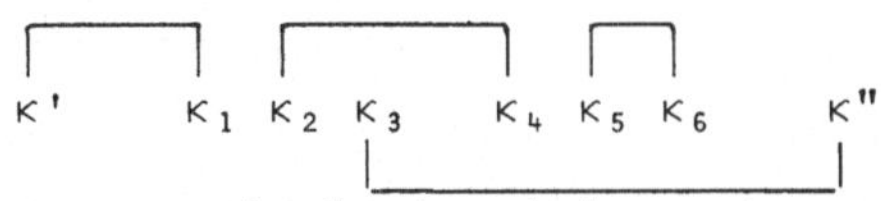

+ 17 equivalent pairing schemes

The phonon diagrams of Fig.4.2 can be divided into two classes. Diagrams such as b,d and e , in which each vertex is connected by lines to the external operators, are called "connected". Diagrams such as c,f, g and h in which one or more vertices are not paired with the external lines, are called "disconnected".

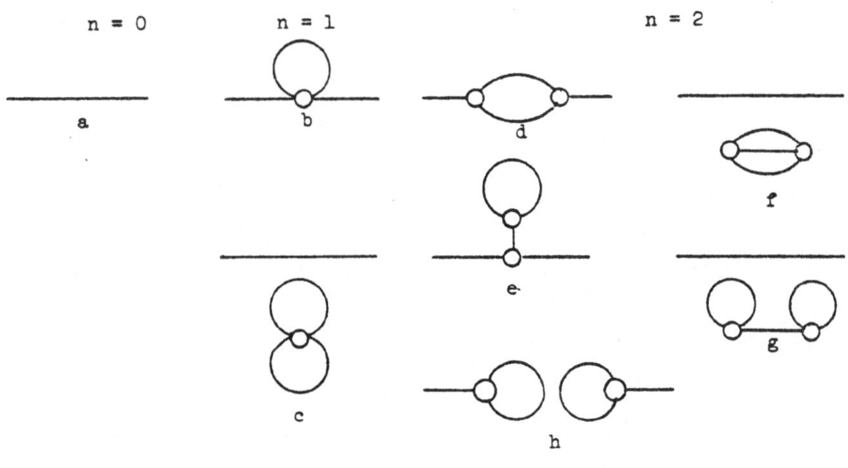

Fig.4.2 Phonon diagrams for n = 0, 1, 2. Diagrams a, b, d and e are connected diagrams. The diagrams c, f, g and h are disconnected.

The contribution of a disconnected diagram to the Green's function consists of the product of two factors. The first includes all the $\tilde{H}_i(\tau)$ operators connected in a chain to $\tilde{A}_\kappa(\tau)$ and $\tilde{A}^*_\kappa(0)$ whereas the second includes the remaining part of the operator product. The contribution to the numerator of 4.6.71 is then

$$(-1)\frac{1}{n!}\int d\tau_1 \ldots d\tau_m \langle T_\tau \tilde{A}_\kappa(\tau)\tilde{A}^*_{\kappa'}(0)\tilde{H}_i(\tau_1)\tilde{H}_i(\tau_2)\ldots\tilde{H}_i(\tau_m)\rangle_{conn.} \times$$
$$\times\int d\tau_{m+1}\ldots d\tau_n \langle T_\tau \tilde{H}_i(\tau_{m+1})\ldots\tilde{H}_i(\tau_n)\rangle \qquad 4.6.78$$

where $\langle\ldots\rangle_{conn.}$ denotes the connected part. All diagrams with m vertices in the connected and n - m vertices in the disconnected part, give rise to the same contribution since they differ only in the relative ordering of the operators inside the brackets. The number of such diagrams is $n!/m!(n-m)!$ and their total contribution is

$$(-1)^m \frac{1}{m!}\int d\tau_1 \ldots d\tau_m \langle T_\tau \tilde{A}_\kappa(\tau)\tilde{A}^*_{\kappa'}(0)\tilde{H}_i(\tau_1)\ldots\tilde{H}_i(\tau_m)\rangle_{conn.} \times$$
$$\times\frac{(-1)^{n-m}}{(n-m)!}\int d\tau_{m+1}\ldots d\tau_n \langle T_\tau \tilde{H}_i(\tau_{m+1})\ldots\tilde{H}_i(\tau_n)\rangle \qquad 4.6.79$$

If we sum all contributions from all diagrams of any order, containing the same connected part and an arbitrary disconnected part, we obtain

$$(-1)^m \frac{1}{m!}\int d\tau_1 \ldots d\tau_m \langle T_\tau \tilde{A}_\kappa(\tau)\tilde{A}^*_{\kappa'}(0)\tilde{H}_i(\tau\ )\ldots\tilde{H}_i(\tau_m)\rangle_{conn.}\Big[1 +$$
$$-\int d\tau_{m+1}\langle\tilde{H}_i(\tau_{m+1})\rangle + \frac{1}{2}\int d\tau_{m+1}d\tau_{m+2}\langle T_\tau\tilde{H}_i(\tau_{m+1})\tilde{H}(\tau_{m+2})\rangle + \qquad 4.6.80$$
$$+\ldots+\frac{(-1)^k}{k!}\int d\tau_{m+1}\ldots d\tau_{m+k}\langle T_\tau\tilde{H}_i(\tau_{m+1})\ldots\tilde{H}_i(\tau_{m+k})\rangle\Big]$$

Comparison with 4.6.72 shows that the expression in square brackets is nothing else that the expansion of $S(\beta,0)$. Thus, 4.6.71 becomes

$$\begin{aligned} G_{\kappa\kappa'}(\tau) &= \langle T_\tau\tilde{A}_\kappa(\tau)\tilde{A}^*_{\kappa'}(0)S(\beta,0)\rangle_{conn.}\langle S(\beta,0)\rangle/\langle S(\beta,0)\rangle \\ &= \langle T_\tau\tilde{A}_\kappa(\tau)\tilde{A}^*_{\kappa'}(0)S(\beta,0)\rangle_{conn.} \end{aligned} \qquad 4.6.81$$

and thus, for the calculation of the Green's function, we need to consider only the connected diagrams.

The one-phonon Green's function 4.6.81 is called the " phonon propagator". We now consider all possible time orderings in it. Clearly, all pairings which differ only in the relative time ordering of the operators $\tilde{H}_i(\tau_1)\ldots\tilde{H}_i(\tau_n)$ give the same contribution. Since there are n! different orderings of the n operators, we can limit the sum to topologically distinct diagrams and neglect the factor 1/n! in 4.6.72.

To evaluate the phonon propagator 4.6.81, we can use the pairing theorem 4.6.76 summing all the $(2n!)/2^n(n!)$ products of harmonic phonon propagators of the type 4.6.77. Owing to the fact that we are more interested in the Fourier transform $G_{\kappa\kappa'}(i\omega_n)$ of 4.6.81 rather than in $G_{\kappa\kappa'}(\tau)$ itself, we can use Eq. 4.6.33 to replace each $g^0_\kappa(\tau_i-\tau_j)$ (see Eq.4.6.77) by its expansion in the Fourier series

$$g^0_\kappa(\tau_i-\tau_j) = \sum_{-\infty n}^{+\infty} g^0_\kappa(i\omega_n)e^{-i\hbar\omega_n(\tau_i-\tau_j)} \qquad 4.6.82$$

where $g^0_\kappa(i\omega_n)$ is given by 4.6.50. For a particular time τ_i,all phonons coming from an earlier time to a vertex have the factor $\exp(-i\hbar\omega_n\tau_i)$ whereas those going to a later time have the factor $\exp(i\hbar\omega_n\tau_i)$. For each vertex, we then have, from 4.6.72, an integral of the type

$$\int_0^\beta d\tau_i e^{-i\hbar\tau_i[(\omega_{n'}+\omega_{n''}+\ldots)_\leftarrow - (\omega_{n'}+\omega_{n''}+\ldots)_\rightarrow]} = \qquad 4.6.83$$

$$= \beta\delta[(\sum\omega_n)_\leftarrow - (\sum\omega_n)_\rightarrow]$$

where the arrows indicate the time direction "in" and "out". This is simply the energy conservation law at each vertex and the contribution of each time integral is thus equal to β.

We can now summarize the previous discussion into a set of simple rules for the utilization of the diagrams to obtain the phonon propagator 4.6.81. The rules are the following

1) For each value of n, draw all topologically distinct connected diagrams. The phonon associated to the operator $\tilde{A}^*_{\kappa'}(\tau_j)$ enters from the right the diagram whereas the phonon associated to $\tilde{A}_\kappa(\tau_i)$ leaves from

the left. Each phonon is represented by a line. In the diagram mark n interaction vertices representing the interaction coefficients and to each vertex associate a number of lines equal to its order.

2) At every vertex energy and momentum are conserved

$$\sum \omega_n = 0$$

$$\sum \mathbf{k} = 0 \text{ or a reciprocal lattice vector}$$

3) Each phonon line contributes a factor

$$g^0_\kappa(i\omega_n) = 2\omega^0_\kappa/\beta\hbar(\omega^{0\,2}_\kappa + \omega^2_n)$$

4) Multiply all the phonon contributions, one for each line.
5) Multiply the result by the number of equivalent pairing schemes.
6) Multiply by $(-\beta)^n$
7) Multiply by the appropriate coefficient characteristic of each diagram. This is the product of n anharmonic coefficients, one for each vertex.
8) Sum over all independent intermediate ω_n, j and **k** values.

We have already discussed the classification of diagrams in Fig.4.2 into connected and disconnected diagrams. The connected diagrams can be further divided into "proper" and "improper" diagrams. A diagram is called proper if it cannot be broken into two parts by cutting a single phonon line. Improper diagrams can be instead devided into two parts in this way. For instance, the diagrams a, b and d in Fig.4.2 are proper diagrams. The diagram e is improper. Other proper and improper diagrams are shown in Fig.4.3. We shall see that the total contribution to the phonon propagator can be expressed in terms of proper diagrams only.

In applying the rules discussed above to the evaluation of the phonon propagator, we shall distinguish between different modes with the same wave vector. For this we shall replace the collective index κ by the indices j and **k** used previously. The phonon propagator will be then written

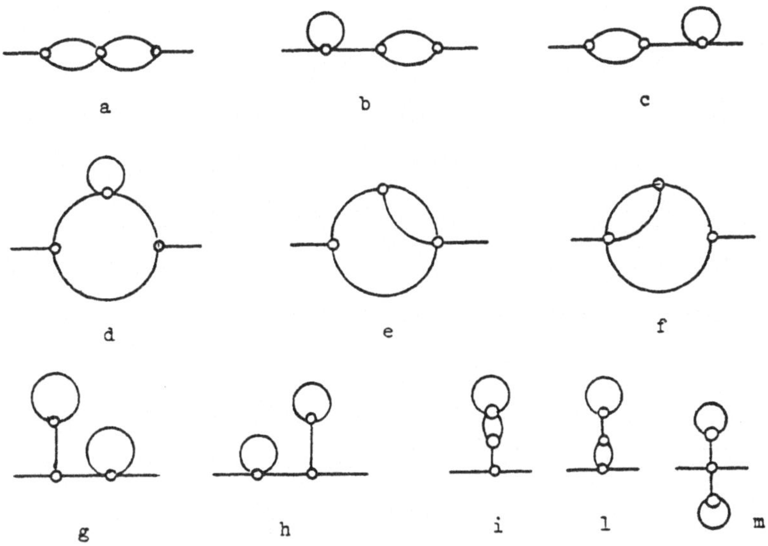

Fig.4.3 Some diagrams for n = 3. Diagrams a, d, e, f are proper diagrams.

$$G_{\mathbf{k}jj'}(\tau) = < T_\tau \tilde{A}_{\mathbf{k}j}(\tau)\tilde{A}_{-\mathbf{k}j'}(0)S(\beta,0)>_{conn.} \qquad 4.6.86$$

and the corresponding Fourier transform will be indicated by the symbol $G_{\mathbf{k}jj'}(i\omega_n)$. We now consider the contribution of proper diagrams to the phonon propagator 4.6.86. Each of them contributes a term

$$g^0_{\mathbf{k}j}(i\omega_n)\Sigma^p_{\mathbf{k}jj'}(i\omega_n)g^0_{\mathbf{k}j'}(i\omega_n) \qquad 4.6.87$$

where the side terms originate from the external phonons and the factor $\Sigma^p_{\mathbf{k}jj'}(i\omega_n)$ from the core of the diagram. The contribution of all proper diagrams is thus

$$G_{\mathbf{k}jj'}(i\omega_n) = \delta_{jj'}g^0_{\mathbf{k}j}(i\omega_n) + g^0_{\mathbf{k}j}(i\omega_n)S_{\mathbf{k}jj'}(i\omega_n)g^0_{\mathbf{k}j'}(i\omega_n) \qquad 4.6.88$$

in which the first term is the contribution of the harmonic phonon (n = 0, diagram a of Fig.4.2) and the second term collects all contributions of the proper diagrams, with any number of vertices. The function

$$S_{\mathbf{k}jj'}(i\omega_n) = \sum_p \Sigma^p_{\mathbf{k}jj'}(i\omega_n) \qquad 4.6.89$$

is called "self-energy".

It is normal practice to represent these contributions by means of very schematic "bubble" diagrams such as those of Fig.4.4. The shaded bubble in diagram a represents the self-energy.

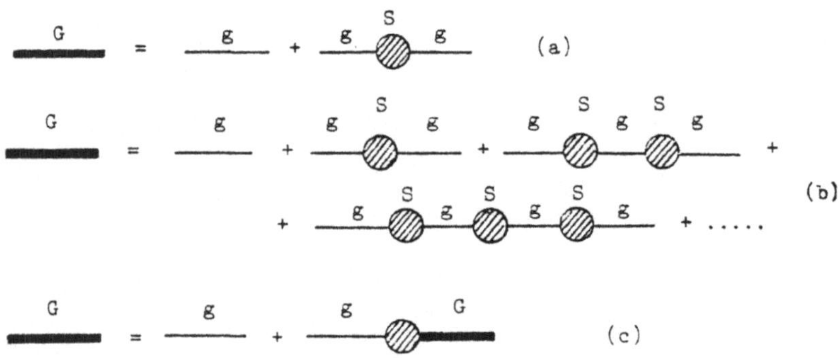

Fig.4.4 The bubble diagram representations of the phonon propagator.

Consider now improper diagrams such as diagram b of Fig.4.3 which splits into two proper diagrams by cutting the internal phonon line. Because of momentum conservation at each vertex, the wave vector of the internal phonon is the same as that of the external phonons. The contribution of all improper diagrams of this type is thus expressed in terms of proper diagram contributions in the form

$$g^0_{\mathbf{k}j}(i\omega_n)\sum_{j''} S_{\mathbf{k}jj''}(i\omega_n)\, g^0_{\mathbf{k}j''}(i\omega_n)\, S_{\mathbf{k}j''j'}(i\omega_n)\, g^0_{\mathbf{k}j'}(i\omega_n) \qquad 4.6.90$$

and is schematically represented by the two-bubble diagram of Fig.4.4.

In addition to these, we have improper diagrams which split into three proper diagrams by cutting two phonon lines, improper diagrams which split into four proper diagrams by cutting three phonon lines etc.

Expression 4.6.90 is thus the first member of an infinite series of terms, as shown in diagram b of Fig.4.4. The total contribution to the phonon propagator is thus the sum of the harmonic term, of the contribution of all proper diagrams and of the series of contributions of all improper diagrams. This is schematically represented in Fig. 4.2 by diagram b where, for simplicity, only the first two terms of the series of the improper diagrams are shown. We then have

$$G_{\mathbf{k}jj'}(i\omega_n) = g^0_{\mathbf{k}j}(i\omega_n)\delta_{jj'} + g^0_{\mathbf{k}j}(i\omega_n)S_{\mathbf{k}jj'}(i\omega_n)g^0_{\mathbf{k}j'}(i\omega_n) + \qquad 4.6.91$$

$$g^0_{\mathbf{k}j}(i\omega_n)\sum_{j''}S_{\mathbf{k}jj''}(i\omega_n)g^0_{\mathbf{k}j''}(i\omega_n)S_{\mathbf{k}j''j'}(i\omega_n)g^0_{\mathbf{k}j'}(i\omega_n) + \ldots$$

or in a more concise form (see diagram c in Fig.4.4)

$$G_{\mathbf{k}jj'}(i\omega_n) = g^0_{\mathbf{k}j}(i\omega_n)\delta_{jj'} + \sum_{j''}g^0_{\mathbf{k}j}(i\omega_n)S_{\mathbf{k}jj''}(i\omega_n)G_{\mathbf{k}j''j'}(i\omega_n) \qquad 4.6.92$$

which can be rewritten in the form

$$g^0_{\mathbf{k}j}(i\omega_n)\delta_{jj'} = \sum_{j''}\{\delta_{jj''} - g^0_{\mathbf{k}j}(i\omega_n)S_{\mathbf{k}jj''}(i\omega_n)\}G_{\mathbf{k}j''j'}(i\omega_n) \qquad 4.6.93$$

This equation is known as the Dyson equation for the phonon propagator and is often written in matrix notation

$$\mathbb{G} = (\mathbb{I} - \mathbb{g}\mathbb{S})^{-1}\mathbb{g} \qquad 4.6.94$$

where $\mathbb{I}$ is the unity matrix.

Equation 4.6.93 is the basic equation for the description of the anharmonic phonon propagator in crystals. The self-energy matrix is Hermitian and plays the role of mixing modes with different j values, as long as they belong to the same representation of the space group.

For some applications it is convenient to consider an approximate solution of 4.6.93 in which non-diagonal terms are neglected as they are normally small. In this approximation, we have, using 4.6.51

$$2\omega_{\mathbf{k}j} = \sum_{j''}\{\delta_{jj''}\beta\hbar(\omega^2_{\mathbf{k}j} + \omega^2_n) - 2\omega_{\mathbf{k}j}S_{\mathbf{k}jj''}(i\omega_n)\}G_{\mathbf{k}j''j}(i\omega_n) \qquad 4.6.95$$

and thus

$$G_{\mathbf{k}jj}(i\omega_n) = \frac{2\omega_{\mathbf{k}j}}{\beta\hbar(\omega_{\mathbf{k}j}^2 + \omega_n^2) - 2\omega_{\mathbf{k}j}S_{\mathbf{k}jj}(i\omega_n)} \qquad 4.6.96$$

We notice that in the harmonic approximation, $S_{\mathbf{k}jj}(i\omega_n)$ is zero and Eq. 4.6.96 reduces to expression 4.6.51 for the harmonic phonon propagator. The meaning of the self-energy is thus evident. It has the effect of changing the energy of the phonon with respect to the harmonic value. Later, we shall see that it has also the effect of leading to a finite phonon lifetime.

The Dyson equation 4.6.93 is very general and, in principle, includes the effect of all terms of the Hamiltonian expansion. In actual calculations, however, only the first few contributions to the self-energy are considered. We shall see in the following discussions that it is sufficient to consider the contribution of the simpler diagrams of Fig.4.2 to obtain the same anharmonic shifts and dampings found in Sec.6.4 by means of the Hamiltonian perturbation method.

We therefore evaluate the self-energy considering only the contributions of diagrams b, d and e of Fig.6.2. These are the first three connected diagrams of the series and account for perturbation terms up to the second order.

To evaluate the self-energy we use the rules discussed before for the phonon propagator. According to Eq.4.6.87 the only difference between the phonon propagator and the self-energy is that in this latter the contribution due to the two side-phonon lines is excluded.

The contribution to the self-energy of diagram b is

$$S_{\mathbf{k}jj'}(i\omega_n) = -12\beta\sum_{j_1\mathbf{k}_1} B\begin{pmatrix} j & j & j_1 & j_1 \\ \mathbf{k} & -\mathbf{k} & \mathbf{k}_1 & -\mathbf{k}_1 \end{pmatrix} \sum_m g^0_{\mathbf{k}_1 j_1}(i\omega_m) \qquad 4.6.97$$

where the factor 12 arises from the number of equivalent schemes (rule 5), the factor $-\beta$ from rule 6, the fourth order coefficient from rule 7 and the sum from rule 8. Phonons of the type $\mathbf{k}_1 j_1$ in diagram b formed and destroyed at the same vertex are called "instantaneous phonons". The sum over m in this case can be evaluated in a very simple way. From 4.6.33 and 4.6.48 we have

$$g^0_{\mathbf{k}_1 j_1}(0) = \sum_m g^0_{\mathbf{k}_1 j_1}(i\omega_m) = 2\bar{n}_{\mathbf{k}_1 j_1} + 1 \qquad 4.6.98$$

This is a general result for instantaneous phonons since they always contribute a factor $\sum_m g^0_{\mathbf{k}j}(i\omega_m)$.

A more general method of evaluating summations of the type 4.6.98, encountered in the applications of the diagrammatic technique, is by contour integration. The method [135] is based on the fact that, provided F(z) is a rapidly decreasing function for $z \to \infty$, the contour integral

$$\int_c (e^{\beta\hbar z} - 1)^{-1} F(z)dz = 0 \qquad 4.6.99$$

The integral is also equal to $2\pi i$ the sum of the residues of the poles. The function $1/|\exp(\beta\hbar z) - 1|$ has poles on the imaginary axis z for all values of $z = 2\pi i n/\beta\hbar = i\omega_n$, all with residue $1/\beta\hbar$. If the poles of F(z) occur at $z = i\omega_p \neq i\omega_n$, then

$$\sum_n F(i\omega_n) + \beta\hbar \sum_p (e^{\beta\hbar i\omega_p} - 1)^{-1} R_p(i\omega_p) = 0 \qquad 4.6.100$$

where $R_p(i\omega_p)$ is the residue of $F(i\omega)$ at ω_p.

We shall now utilize this relation to evaluate the sum 4.6.98. The function $g^0_{\mathbf{k}j}(i\omega_n)$ has poles of $i\omega_n$ at $\omega_{j\mathbf{k}}$ and $-\omega_{j\mathbf{k}}$ (see Eq.4.6.50), with residues $-1/\beta\hbar$ and $1/\beta\hbar$ respectively. We then obtain from 4.6.100

$$\sum_n g^0_{\mathbf{k}_1 j_1}(i\omega_n) = -\beta\hbar\Big[-\frac{1}{\beta\hbar}(e^{\beta\hbar\omega_{j_1\mathbf{k}_1}} - 1)^{-1} + \frac{1}{\beta\hbar}(e^{-\beta\hbar\omega_{j_1\mathbf{k}_1}} - 1)^{-1}\Big] = 2\bar{n}_{j_1\mathbf{k}_1} + 1 \qquad 4.6.101$$

which is the same result obtained by 4.6.98.

The contribution of diagram b to the self-energy is thus

$$S^b_{\mathbf{k}jj'}(i\omega_n) = -12\beta \sum_{j_1\mathbf{k}_1} B\begin{pmatrix} j & j & j_1 & j_1 \\ \mathbf{k} & -\mathbf{k} & \mathbf{k}_1 & -\mathbf{k}_1 \end{pmatrix}(2\bar{n}_{j_1\mathbf{k}_1} + 1) \qquad 4.6.102$$

We now consider the diagram d of Fig.4.2. The contribution to the self-energy is

$$S^d_{\mathbf{k}jj'}(i\omega_n) = 18\beta^2 \sum_{j_1k_1j_2k_2} B\begin{pmatrix} j & j_1 & j_2 \\ \mathbf{k} & -\mathbf{k}_1 & -\mathbf{k}_2 \end{pmatrix} B\begin{pmatrix} j & j_1 & j_2 \\ -\mathbf{k} & \mathbf{k}_1 & \mathbf{k}_2 \end{pmatrix} \sum_m g^0_{\mathbf{k}_1 j_1}(i\omega_n) g^0_{\mathbf{k}_2 j_2}(i\omega_n - i\omega_m) \qquad 4.6.103$$

Again the factor 18 comes from the number of equivalent schemes, the factor β^2 from rule 6, the product of the two third-order coefficients from the two triple vertices (rule 7) and the sums from rule 8.

The sum over m can be performed with the help of 4.6.100. We obtain from 4.6.50 and using again the collective label κ for **jk**, whenever it is not necessary to specify both labels separately,

$$\sum_m g^0_{\kappa_1}(i\omega_m) g^0_{\kappa_2}(i\omega_n - i\omega_m) = \qquad 4.6.104$$

$$= -\hbar\beta \sum_m \left(\frac{1}{\beta\hbar}\right)^2 \left(\frac{1}{\omega_{\kappa_1} + i\omega_m} + \frac{1}{\omega_{\kappa_1} - i\omega_m}\right)\left(\frac{1}{\omega_{\kappa_2} + i(\omega_n - \omega_m)} + \frac{1}{\omega_{\kappa_2} - i(\omega_n - \omega_m)}\right)$$

$$= -\frac{1}{\beta\hbar}\sum_m \left[\frac{1}{(\omega_{\kappa_1} + i\omega_m)(\omega_{\kappa_2} + i\omega_n - i\omega_m)} + \frac{1}{(\omega_{\kappa_1} + i\omega_m)(\omega_{\kappa_2} - i\omega_n + i\omega_m)} + \right.$$

$$\left. + \frac{1}{(\omega_{\kappa_1} - i\omega_m)(\omega_{\kappa_2} + i\omega_n - i\omega_m)} + \frac{1}{(\omega_{\kappa_1} - i\omega_m)(\omega_{\kappa_2} - i\omega_n + i\omega_m)}\right]$$

The first term in 4.6.104 has poles at $i\omega_m = -\omega_{\kappa_1}$ and at $i\omega_m = i\omega_n + \omega_{\kappa_2}$. Their contribution is then

$$-\frac{1}{\beta\hbar}\left[\frac{1}{\omega_{\kappa_2} + i\omega_n + \omega_{\kappa_1}}(e^{-\beta\hbar\omega_{\kappa_1}} - 1)^{-1} - \frac{1}{\omega_{\kappa_1} + i\omega_n + \omega_{\kappa_2}}(e^{\beta\hbar(\omega_{\kappa_2} + i\omega_n)} - 1)^{-1}\right]$$

$$= \frac{1}{\beta\hbar}\left[\frac{1}{\omega_{\kappa_1} + \omega_{\kappa_2} + i\omega_n}(\bar{n}_{\kappa_1} + 1) + \frac{1}{\omega_{\kappa_1} + \omega_{\kappa_2} + i\omega_n}(\bar{n}_{\kappa_2})\right] = \qquad 4.6.105$$

$$= \frac{1}{\beta\hbar}\,\frac{1}{\omega_{\kappa_1} + \omega_{\kappa_2} + i\omega_n}(\bar{n}_{\kappa_1} + \bar{n}_{\kappa_2} + 1)$$

since $\exp(\beta\hbar i\omega_n) = 1$ for all ω_n.

The second term has poles at $i\omega_m = -\omega_{\kappa_1}$ and at $i\omega_m = -\omega_{\kappa_2} + i\omega_n$. The contribution is then

$$\frac{1}{\beta\hbar}\left[\frac{1}{\omega_{\kappa_2} - \omega_{\kappa_1} - i\omega_n}(\bar{n}_{\kappa_1} + 1) + \frac{1}{\omega_{\kappa_1} - \omega_{\kappa_2} + i\omega_n}(\bar{n}_{\kappa_2} + 1)\right] =$$

$$= \frac{1}{\beta\hbar}\,\frac{1}{\omega_{\kappa_1} - \omega_{\kappa_2} + i\omega_n}(\bar{n}_{\kappa_2} - \bar{n}_{\kappa_1}) \qquad 4.6.106$$

In the same way, it is easily seen that the contribution of the third term with poles at ω_{κ_1} and $\omega_{\kappa_2} + i\omega_n$ and that of the fourth term with poles at ω_{κ_1} and $-\omega_{\kappa_2} + i\omega_n$ are respectively

$$\frac{1}{\beta\hbar(\omega_{\kappa_1} - \omega_{\kappa_2} - i\omega_n)}(\bar{n}_{\kappa_2} - \bar{n}_{\kappa_1})$$

$$\frac{1}{\beta\hbar(\omega_{\kappa_1} + \omega_{\kappa_2} - i\omega_n)}(\bar{n}_{\kappa_1} + \bar{n}_{\kappa_2} + 1) \qquad 4.6.107$$

Therefore, the contribution of diagram d to the self-energy is

$$S^{d}_{\mathbf{k}jj'}(i\omega_n) = \frac{18\beta}{\hbar}\sum_{\kappa_1\kappa_2} B(\kappa-\kappa_1-\kappa_2)B(-\kappa\kappa_1\kappa_2)\left[\frac{(\bar{n}_{\kappa_1} + \bar{n}_{\kappa_2} + 1)}{(\omega_{\kappa_1} + \omega_{\kappa_2} + i\omega_n)} + \right.$$

$$\left. + \frac{(\bar{n}_{\kappa_2} - \bar{n}_{\kappa_1})}{(\omega_{\kappa_1} - \omega_{\kappa_2} + i\omega_n)} + \frac{(\bar{n}_{\kappa_1} - \bar{n}_{\kappa_2})}{(\omega_{\kappa_2} - \omega_{\kappa_1} + i\omega_n)} + \frac{(\bar{n}_{\kappa_1} + \bar{n}_{\kappa_2} + 1)}{(\omega_{\kappa_1} + \omega_{\kappa_2} + i\omega_n)}\right] \qquad 4.6.108$$

Finally, the contribution of diagram e is

$$S^{e}_{\mathbf{k}jj'}(i\omega_n) = 18\beta^2\sum_{\kappa_1\kappa_2} B(\kappa-\kappa-\kappa_1)B(\kappa_1-\kappa_2\kappa_2)\sum_m g^0_{\kappa_1}(0)g^0_{\kappa_2}(i\omega_m) \qquad 4.6.109$$

and, using 4.6.98 and 4.6.50

$$S^{e}_{\mathbf{k}jj'}(i\omega_n) = \frac{18\beta}{\hbar}\sum_{\kappa_1\kappa_2} B(\kappa-\kappa-\kappa_1)B(\kappa_1-\kappa_2\kappa_2)\frac{2(2\bar{n}_{\kappa_2} + 1)}{(\omega_{\kappa_1})} \qquad 4.6.110$$

The total contribution of diagrams b, d and e is then

$$S_{kjj'}(i\omega_n) = S^{b}_{kjj'}(i\omega_n) + S^{d}_{kjj'}(i\omega_n) + S^{e}_{kjj'}(i\omega_n) \qquad 4.6.111$$

As discussed before (Eqs. 4.6.38 to 4.6.41) we shall analytically continue the function 4.6.111 in the complex plane by defining a continuous complex variable w. Near the real axis the variable w can be written in the form $w = \Omega + i\varepsilon$ and thus, using 4.4.51, we obtain

$$\lim_{\varepsilon\to 0} S_{\mathbf{k}jj'}(\Omega+i\varepsilon) = -\beta\hbar\{\Delta_{\mathbf{k}jj'}(\Omega) - i\Gamma_{\mathbf{k}jj'}(\Omega)\} \qquad 4.6.112$$

where the real part gives the anharmonic shift of the phonon frequency and the imaginary part the inverse lifetime of the phonon. Using this relation we then obtain from 4.6.102,108 and 110

$$\Delta^{b}_{\mathbf{k}jj'}(\Omega) = \frac{12}{\hbar}\sum_{\kappa} B(\kappa-\kappa\kappa_1-\kappa_1)(2\bar{n}_{\kappa} + 1)$$

$$\Gamma^{b}_{\mathbf{k}jj'}(\Omega) = 0 \qquad 4.6.113$$

$$\Delta^{d}_{\mathbf{k}jj'}(\Omega) = -\frac{18}{\hbar^2}\sum_{\kappa_1\kappa_2} B(\kappa-\kappa_1-\kappa_2)B(-\kappa\kappa_1\kappa_2)\Bigg[\frac{(\bar{n}_{\kappa_1} + \bar{n}_{\kappa_2} + 1)}{(\omega_{\kappa_1} + \omega_{\kappa_2} + \Omega)_p} +$$

$$+ \frac{(\bar{n}_{\kappa_2} - \bar{n}_{\kappa_1})}{(\omega_{\kappa_1} - \omega_{\kappa_2} + \Omega)_p} + \frac{(\bar{n}_{\kappa_1} - \bar{n}_{\kappa_2})}{(\omega_{\kappa_2} - \omega_{\kappa_1} + \Omega)_p} + \frac{(\bar{n}_{\kappa_1} + \bar{n}_{\kappa_2} + 1)}{(\omega_{\kappa_1} + \omega_{\kappa_2} - \Omega)_p}\Bigg]$$

$$\Gamma^{d}_{\mathbf{k}jj'}(\Omega) = \frac{18\pi}{\hbar^2}\sum_{\kappa_1\kappa_2} B(\kappa-\kappa_1-\kappa_2)B(-\kappa\kappa_1\kappa_2)\{(\bar{n}_{\kappa_1} + \bar{n}_{\kappa_2} + 1)[\delta(\omega_{\kappa_1} + \omega_{\kappa_2} - \Omega) +$$

$$- \delta(\omega_{\kappa_1} + \omega_{\kappa_2} + \Omega)] + (\bar{n}_{\kappa_1} - \bar{n}_{\kappa_2})[\delta(\omega_{\kappa_2} - \omega_{\kappa_1} - \Omega) - \delta(\omega_{\kappa_2} - \omega_{\kappa_1} + \Omega)]\}$$

$$\Delta^{e}_{\mathbf{k}jj'}(\Omega) = -\frac{18}{\hbar^2}\sum_{\kappa_1\kappa_2} B(\kappa-\kappa-\kappa_1)B(\kappa_1-\kappa_2\kappa_2)\frac{2(2\bar{n}_{\kappa_2} + 1)}{(\omega_{\kappa_1})_p}$$

$$\Gamma^{e}_{\mathbf{k}jj'}(\Omega) = 0$$

It is easily seen that the sum

$$\Delta_{\mathbf{k}jj'}(\Omega) = \Delta^{b}_{\mathbf{k}jj'}(\Omega) + \Delta^{d}_{\mathbf{k}jj'}(\Omega) + \Delta^{e}_{\mathbf{k}jj'}(\Omega) \qquad 4.6.114$$

is identical with the anharmonic frequency shift given by 4.4.53 and that the sum

$$\Gamma_{\mathbf{k}jj'}(\Omega) = \Gamma^{b}_{\mathbf{k}jj'}(\Omega) + \Gamma^{d}_{\mathbf{k}jj'}(\Omega) + \Gamma^{e}_{\mathbf{k}jj'}(\Omega) \qquad 4.6.115$$

is the same inverse phonon lifetime of Eq.4.4.54.

In order to utilize the relations 4.4.113 for the calculation of the anharmonic shifts or inverse phonon lifetimes, we need a representation of the Dirac delta function and of the principal part of a sum. Following Maradudin and Fein [134], it is common practice to use the expressions

$$\delta(x) = \lim_{\varepsilon \to 0} \frac{1}{\pi} \frac{\varepsilon}{x^2 + \varepsilon^2}$$
$$\frac{1}{(x)_p} = \lim_{\varepsilon \to 0} \frac{x}{x^2 + \varepsilon^2} \qquad 4.6.116$$

where ε is a small but finite quantity whose value is chosen according to the type of spectral resolution required.

4.7 THERMAL STRAIN

An important consequence of anharmonicity is the thermal expansion of crystals. A temperature variation always produces an elastic strain in the solid since all intermolecular distances change and the molecules readjust themselves to new equilibrium positions. This leads to a variation of the phonon frequencies and therefore introduces an additional shift that must be taken into account in calculating anharmonic shifts as a function of the temperature.

The theory of thermal strain is rather complex and is beyond the scope of this book. The detailed treatment for atomic lattices can be found in references 3,123,137 and 138. Here we shall limit ourselves to a very simplified treatment adapted to the specific case of molecular crystals and intended to furnish only a basic understanding of the problem.

Let R_ℓ^a represent a molecular displacement coordinate (rotation or translation) of molecule a in the unstrained lattice and $\bar{R}_\ell^a$ the corresponding coordinate for the strained lattice at a different temperature. In principle we should take as unstrained the lattice at T = 0. It is, however, convenient to consider often as unstrained the lattice in a reference state of strain for which the crystal structure is known. The coordinate $\bar{R}_\ell^a$ of the strained lattice can be written as

$$\bar{R}_\ell^a = R_\ell^a + \sum_t \varepsilon_{t\ell}^a Z_t^a \qquad 4.7.1$$

where $\varepsilon_{t\ell}^a$ are the components of the strain tensor in the molecular coordinate space and Z_t^a represent finite molecular displacements (translations and rotations) whose combined effects convert the unstrained into the strained lattice. Substituting 4.7.1 in 1.6.13, we can express the crystal potential of the strained lattice in the form

$$V = \frac{1}{2} \sum_{ab}\sum_{\ell\ell_1} [\, F_{\ell\ell_1}(\begin{smallmatrix}a\\b\end{smallmatrix}) + \sum_c \sum_{\ell_2} (\sum_{t_2} \varepsilon^c_{\ell_2 t_2} Z^c_{t_2}) F_{\ell\ell_1\ell_2}(\begin{smallmatrix}a\\b\\c\end{smallmatrix}) +$$

$$+ \frac{1}{2} \sum_{cd}\sum_{\ell_2\ell_3} (\sum_{t_2 t_3} \varepsilon^c_{\ell_2 t_2} \varepsilon^d_{\ell_3 t_3} Z^c_{t_2} Z^d_{t_3}) F_{\ell\ell_1\ell_2\ell_3}(\begin{smallmatrix}a & c\\b & d\end{smallmatrix}) + \dots\,] R^a_\ell R^b_{\ell_1} +$$

$$+ \frac{1}{3!} \sum_{abc}\sum_{\ell\ell_1\ell_2} [\, F_{\ell\ell_1\ell_2}(\begin{smallmatrix}a\\b\\c\end{smallmatrix}) + \sum_d \sum_{\ell_3} (\sum_{t_3} \varepsilon^d_{\ell_3 t_3} Z^d_{t_3}) F_{\ell\ell_1\ell_2\ell_3}(\begin{smallmatrix}a & c\\b & d\end{smallmatrix}) +$$

$$+ \dots\,] R^a_\ell R^b_{\ell_1} R^c_{\ell_2} + \dots \qquad 4.7.2$$

where the terms linear in R^a_ℓ have been omitted, assuming that the new equilibrium positions of the molecules also correspond to a minimum of the crystal potential.

The crystal Hamiltonian can be then written in the form

$$H = (H_0 + H^s_0) + (H_{anh} + H^s_{anh}) \qquad 4.7.3$$

where H_0 and H_{anh} are the usual harmonic and anharmonic Hamiltonians of the unstrained lattice and

$$H^s_0 = \frac{1}{2} \sum_{abc}\sum_{\ell\ell_1\ell_2} [(\sum_{t_2} \varepsilon^c_{\ell_2 t_2} Z^c_{t_2}) F_{\ell\ell_1\ell_2}(\begin{smallmatrix}a\\b\\c\end{smallmatrix}) +$$

$$+ \frac{1}{2} \sum_d \sum_{\ell_3} (\sum_{t_2 t_3} \varepsilon^c_{\ell_2 t_2} \varepsilon^d_{\ell_3 t_3} Z^c_{t_2} Z^d_{t_3}) F_{\ell\ell_1\ell_2\ell_3}(\begin{smallmatrix}a & c\\b & d\end{smallmatrix})\,] R^a_\ell R^b_{\ell_1} \qquad 4.7.4$$

$$H^s_{anh} = \frac{1}{3!} \sum_{abcd}\sum_{\ell\ell_1\ell_2\ell_3} [\, (\sum_{t_3} \varepsilon^d_{\ell_3 t_3} Z^d_{t_3}) F_{\ell\ell_1\ell_2\ell_3}(\begin{smallmatrix}a & c\\b & d\end{smallmatrix}) + \dots\,] R^a_\ell R^b_{\ell_1} R^c_{\ell_2} + \dots$$

Since H^s_0 is quadratic in the molecular displacement coordinates as H_0, the transformation 4.2.1 that diagonalizes H_0, also diagonalizes the Hamiltonian $H_0 + H^s_0$. We thus have

$$H_0 + H^s_0 = \sum_{\mathbf{k}j} \hbar\omega'_{\mathbf{k}j} (a^\dagger_{\mathbf{k}j} a_{\mathbf{k}j} + \frac{1}{2}) \qquad 4.7.5$$

where

$$\omega'_{\mathbf{k}j} = \omega_{\mathbf{k}j} + \Delta^s_{\mathbf{k}j} \qquad 4.7.6$$

The lowest-order frequency shift $\Delta_{\mathbf{k}j}$, due to the thermal strain, can thus be computed in the harmonic approximation, using the Hamiltonian $H_0 + H_0^s$ instead of H_0. This is called the"quasi-harmonic"approximation. It can be conveniently utilized as long as calculations of the frequency of phonon modes at different temperatures are concerned. In principle, the contribution of the thermal strain to the harmonic Hamiltonian can be calculated by means of 4.7.4. This requires a knowledge of the strain tensor coefficients and of the finite molecular displacements Z_t^a. These quantities are normally obtained, following a standard procedure, by equating to zero the derivatives of the thermal average of the potential with respect to the strain parameters. The calculations are, however, rather complex and time-consuming and shall not be discussed here.

Alternative approaches of a more heuristic type are often utilized in actual calculations. For instance, when the crystal structure is known at different temperatures, the harmonic frequencies can be directly computed for these different structures without the complications associated to the use of the Hamiltonian 4.7.5. Another convenient approach is to derive optical Grüneisen parameters for the different phonon modes through the relation [140]

$$\gamma_{\mathbf{k}j} = -(\partial \ln \omega_{\mathbf{k}j} / \partial \ln \upsilon) \qquad 4.7.7$$

where υ is the volume of the unit cell. The strain shift is then given by

$$\Delta^s_{\mathbf{k}j} = \omega_{\mathbf{k}j} [\, e^{-\gamma_{\mathbf{k}j} \int_0^T \alpha(T)\, dT} \,]^{-1} \qquad 4.7.8$$

For a detailed discussion of the derivation of 4.7.8 see Reference 140.

4.8 ANHARMONIC CALCULATIONS

A very limited number of anharmonic calculations of phonon frequencies and lifetimes has been made until now for molecular crystals. Researches in this field are,however,in rapid progress and significant developments can be easily foreseen in the near future.

Most of the available calculations deal with solid N_2 in the different phases. These calculations are collected in Table 4.1 for the α- and in Table 4.2 for the γ-phase. The comparison of these data shows marked differences between calculations made with different methods and using different intermolecular potentials.

Table 4.1 Anharmonic lattice frequencies of crystalline N_2. α-phase

	Exp.		I		II		III			
	a	b	har.	anhar.	har.	anhar.	har.	anhar.	$2\Gamma_{exp}$	$2\Gamma_{obs}$
E_g	32.3	35.6	33.5	36.7	40.8	39.5	33.6	40.2	0.8	0.8
T_g^-	36.3	39.7	37.7	40.8	50.7	48.5	37.8	42.5	0.8	1.2
T_g^+	59.7	65.0	45.7	47.6	74.3	70.3	45.8	51.8	5.0	4.7
T_u^-	48.4	49.0	46.0	48.8	52.0	48.4	45.5	50.9	---	1.3
T_u^+	69.4	70.0	67.0	70.1	77.5	72.0	67.1	80.0	6.0	13.3
A_u	47.8	----	43.2	44.3	52.4	48.8	42.6	47.8	---	0.9
E_u	54.0	----	50.9	52.0	57.6	53.5	50.4	54.3	---	1.7

I) Self-consistent phonon method. Atom-atom potential. Ref. 129

II) Self-consistent phonon method. Ab-initio potential. Ref. 141

III) Green's functions method. Atom-atom potential. Ref. 90

a) Neutron Scattering data; T = 15 K. Ref. 89

b) Raman data; T = 8 K. Ref. 142; Infrared data; T = 15 K. Ref. 143

For instance, calculations I and II both made by use of the self-consistent phonon method lead to anharmonic shifts that are all positive in the first and all negative in the second case. This is probably due to the fact that the "effective" potential 4.5.2, averaged on the molecular motions, is harder in the first case and softer than the bare harmonic potential in the second case.

The self-consistent phonon method neglects the contribution of the cubic terms of the potential. These are instead correctly taken into account by the more general and complete Green's function approach utilized by Kobashi (Calc. III in the Tables). This author has found that the contribution of the cubic terms to the anharmonic shifts is important and is negative, in contrast to that of the quartic term which is positive. The total anharmonic shifts calculated by Kobashi are positive but they differ considerably from those, also positive, of Raich,Gillis and Anderson shown under heading I in Tables 4.1 and 4.2.

Table 4.2 Anharmonic lattice frequencies of crystalline N_2. γ-phase

	Exp.		I		II		III		
	a	b	har.	anhar.	har.	anhar.	har.	anhar.	2Γ
E_g	55.0	58.4	50.5	58.8	57.9	56.5	53.1	52.8	5.2
B_{1g}	98.1	103.6	74.8	99.8	86.5	85.2	87.6	90.0	10.8
A_{2g}	----	----	----	----	109.7	111.2	100.2	106.5	12.6
E_u	65.0[c]	----	----	----	72.0	69.3	63.4	67.0	3.1
B_{1u}	----	----	----	----	110.3	107.4	105.9	119.6	16.1

I) Self-consistent phonon method. Atom-atom potential. Ref. 129

II) Self-consistent phonon method. Ab initio potential. Ref. 141

III) Green's function method. Atom-atom potential. Ref. 95

a) Ref. 144

b) Ref. 142

c) Ref. 145

The anharmonic shifts of lattice phonons cannot be compared directly to physical quantities, owing to the fact that overtones and combination bands of lattice modes are normally not observed in the spectra.

The data of Tables 4.1 and 4.2 show clearly that the anharmonic shifts are very sensitive to the type of intermolecular potential used. It is, however, difficult to judge from these anharmonic shifts the validity of the approximations used by the authors and the quality of the intermolecular potentials utilized.

More significant, from this point of view, are the calculations of bandwidths or band shapes, since in this case there is a direct comparison with experiments.

The band shapes of both α- and γ- crystalline nitrogen have been calculated by Kobashi [90] and by Kobashi and Chandrasekharan [95] at different temperatures. The bandwidths calculated at 10 K for α-N_2 are compared to the experimental bandwidths in Table 4.1. The agreement is in general quite good except for the triply degenerate translational mode at 69.4 cm^{-1}. In this case, the experimental value is about three times smaller than the calculated one. In principle, this is a surprising result since one expects all calculated values to be lower than the experimental ones. The damping mechanism utilized in the Green's function treatment of Section 4.6 is in fact correct to the second order and at this level involves only three-phonon processes (diagrams d and e of Fig.4.2). Higher order processes, not considered in the derivation of the self-energy 4.6.111, also contribute to the band broadening by means of multi-phonon processes. The bandwidths obtained using 4.6.115 are thus approximated by defect.

The main source of errors in the calculation of bandwidths is the number of resonances that the Dirac δ-functions in 4.6.113 select from the two-phonon density of states in the band region. In a given spectral region the latter consists of the sum of all pairs of lower frequency modes and of differencies involving higher frequencies. Therefore, errors in the low-frequency harmonic phonons affect the two-phonon density of states and may thus lead to incorrect results. The error is further amplified for degenerate modes.

The calculated bandwidths for the γ-phase of solid N_2 are given in Table 4.2 for a temperature of 10 K. No comparison with experiments is available in this case.

Anharmonic calculations of bandwidths have recently been made in our laboratory for the more complex case of crystalline ammonia. Since the crystal structure of NH_3 is known only at 68 K, calculations were made at this temperature (see Section 3.6 for the harmonic calculations). The calculated bandwidths are compared with the experimental results in Table 4.3. The general agreement is again very satisfactory, especially if the indeterminacy of the experimental bandwidths is taken into account. The spectra were in fact measured on polycrystalline samples which always give rise to appreciable scattering and no correction for the finite resolution of the instruments was made, since the measurements were made just to localize the crystal frequencies. Despite of this, the calculated bandwidths follow correctly the pattern of the experimental values. Broad bands have large calculated widths whereas thin bands normally have small calculated widths. The only exception is the band of F species at 360 cm^{-1} which has an experimental width of about 20 cm^{-1} whereas the calculated value is 4.3 cm^{-1}. This band possesses, however, a very peculiar doublet structure which is unexplained for the moment and which probably involves multiphonon processes that are not yet understood. For this reason the discrepancy between calculations and experiments cannot be taken as significant.

The calculated bandwidths are, as expected, smaller than the exper-

Table 4.3 Calculated and observed bandwidths of crystalline NH_3

	Freq.	observed		calculated[c]	
	80 K	80 K[a]	8 K[b]	68 K	
A	313	---	---	1.7	
	138	6	---	4.4	
E	298	8	2.4	1.8	
	107	4.5	1.1	2.7	
F	533	14	---	22	LO
	431	27	18	15.3	LO
	366.5 } 360.5	19	17	4.3	TO
	260	5	2	1.2	LO + TO
	184	10	---	17.6	LO + TO
	141	8	2.3	1.5	LO + TO

a) Ref.96

b) Unpublished data from V.Schettino and S.Califano

c) New calculations from the authors of Ref.146

imental values, except for the two bands at 533 and 184 cm^{-1} . In these two cases the calculated bandwidths are larger than the experimental bandwidths. As discussed in the case of crystalline N_2, the main source of errors in bandwidth calculations is the two-phonon density of states from which the Dirac delta functions select the resonances producing the phonon decay. When a band falls in a flat region of the two-phonon density of states, its bandwidth is rather insensitive to errors in the harmonic frequencies. When however, a band falls in a region where the two-phonon density of states changes rapidly, small errors in the harmonic frequencies may produce large variations in the bandwidth . This is clearly seen in Fig.4.5 where the calculated phonon frequencies are represented by lines overimposed on the two-phonon density of states. It would be sufficient to shift the overall density of only few wavenumbers toward higher frequencies to reduce sensibly the width of the two bands at 184 and 533 cm^{-1}. This would at the same time increase the width of the group of bands in the region of 300 cm^{-1}, in better agreement with the experimental results.

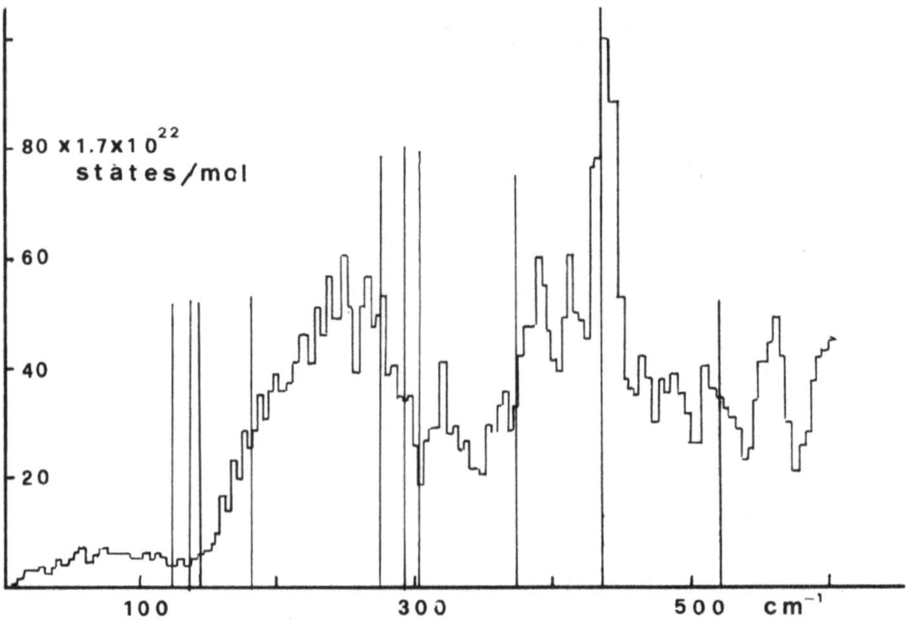

Fig.4.5 Two-phonon density of states of crystalline NH_3. The position of the calculated harmonic frequencies is shown by straight lines.

C H A P T E R 5

TWO-PHONON SPECTRA OF MOLECULAR CRYSTALS

5.1 INTRODUCTION

The object of this chapter is to discuss processes where the absorption (or scattering) of a single photon results in the excitation of two vibrational quanta on different molecules.These processes differ from combination and overtone transitions of isolated molecules in several respects and mainly because they are affected by the intermolecular interactions.The very occurrence of double (or multiple) transitions of the type we want to discuss is a direct consequence of the existence of interaction forces between the molecules.As such double excitations are not specific of the solid state and,in fact, they have been observed in the infrared absorption spectra of compressed gases and of liquid mixtures[147] .However,in ordered solids,because of the translational symmetry,vibrational excitations can be precisely characterized by their frequency ω and wave vector $\mathbf{k}$ and the selection rules for multiple excitations,that we call "multiphonon processes",are of particular significance.For these reasons,the structure of multiphonon bands in crystals can be studied in more detail and,in turn,the information that can be obtained on intermolecular forces and phonon densities of states are particularly valuable.

In this chapter we will give a review of the methods available for the calculation of two-phonon band shapes and intensities in molecular crystals using different approaches.By comparing the theoretical results with the experimental observations,it will be shown that it is possible to identify the mechanism of two-phonon excitation and to establish a correlation between the experimental band shape in the infrared and Raman spectra and the frequency distribution in molecular crystals.In the theoretical treatment of the shape and intensity of two-phonon transitions in crystals,use will be made of the general concepts of anharmonic interactions introduced in the previous Chapter.

5.2 GENERAL CONSIDERATIONS

5.2a Conservation Relations

The origin of two-phonon excitations in crystals and the source of structure in two-phonon infrared and Raman bands can be easily understood qualitatively by considering the basic conservation relations. For simplicity,we shall refer to an absorption process but the overall conclusions apply to Raman scattering as well.We consider a process where a photon of frequency ω_o and wave vector $\mathbf{q}$ is absorbed by the crystal and two phonons i and j are produced (or phonon i is produced and phonon j is destroyed).We characterize the two phonons by their frequencies ω_i and ω_j and wave vectors $\mathbf{k}_i$ and $\mathbf{k}_j$.The energy conservation then requires

$$\omega_o = \omega_i + \omega_j \qquad 5.2.1$$

and the momentum conservation

$$\mathbf{q} = \mathbf{k}_i + \mathbf{k}_j + \mathrm{K} \qquad 5.2.2a$$

where $\mathbf{K}$ is a reciprocal lattice vector. Since in the infrared or visible region the photon wave vector is very small ($\mathbf{q} \simeq 0$) relation 5.2.2a can be written

$$\mathbf{k}_i = - \mathbf{k}_j \qquad 5.2.2b$$

Due to energy conservation 5.2.1,two-phonon absorption occurs at frequencies that are sums and differences of phonon frequencies and corresponds to the classical mixing of normal modes due to non linearities. The momentum conservation has,however,wider implications.In fact,5.2.2b shows that there is no restriction to the wave vector of the individual phonons involved in the second-order process.Therefore,a two-phonon band will contain contributions from all the phonons in the first Brillouin zone. Important possibilities thus arise of using two-phonon absorption profiles as a guide to the phonon frequency distribution and as a test of the intermolecular coupling responsible for the phonon dispersion. The relevance of these possibilities arises from the fact that equivalent information can otherwise be obtained from the much more difficult neutron scattering experiments.

5.2b Mechanisms of Two-Phonon Excitations

In the previous section it has been shown from energy and momentum conservation requirements that two-phonon band profiles may bear a relation to the phonon densities of states. The distinguishing characteristics of this relation depend on the mechanism allowing the processes under consideration to actually occur. In general terms, the phonons must produce an electric moment parallel to the electric vector of the photon. In a two-phonon process two types of mechanisms leading to this kind of coupling should be taken into consideration[148,149].

Mechanism a. The photon couples directly with the two phonons ω_i and ω_j. For dipole transitions the crystal dipole moment must depend quadratically on the two phonon coordinates. This coupling mechanism, associated with the second-order electric moment, is of a very general type. In fact, it may also occur in the infrared spectrum of crystals without polar phonons. It can be qualitatively depicted in the following way. A phonon causes a local distortion of the lattice charge distributions. A second phonon interacts with these induced charges to produce a second-order moment that can couple with the photon.

Mechanism b. The photon couples directly with a transverse optical phonon ω_{TO} through the linear dipole moment. This intermediate "virtual" phonon rapidly decays into two phonons ω_i and ω_j by the third-order mechanical potential. This anharmonic coupling mechanism can thus only occur in crystals with polar phonons. This limitation, however, is of no actual significance in molecular crystals. While the energy conservation should only be obeyed between the initial and final states, the momentum must be conserved also in the intermediate stages of the absorption process.

The two coupling mechanisms are schematically depicted in Fig. 5.1.

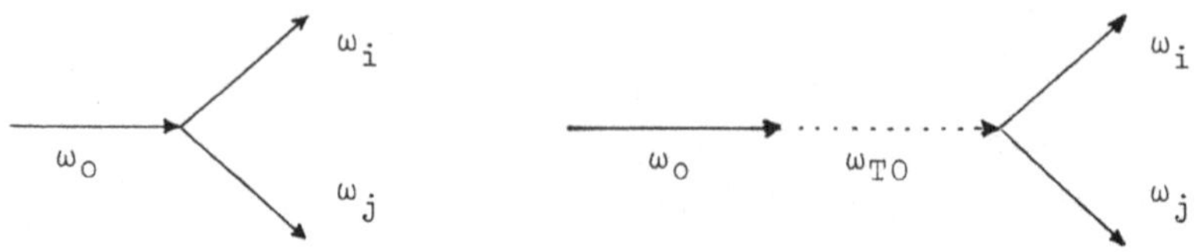

a) direct mechanism b) indirect mechanism

Fig. 5.1. Schematic representation of two-phonon excitation mechanisms

The resulting profile and intensity of a two-phonon band depend on the magnitude of the anharmonic coupling coefficients and of the matrix elements of the first-and second-order moments, and therefore on the relative importance of the two coupling mechanisms. When both processes are possible, the problem of distinguishing between the two mechanisms and of evaluating their relative contribution arises. This problem will be considered in detail in the following discussion.

5.2c Types of Two-Phonon Bands in Molecular Crystals

As it has been discussed in previous Chapters, in a molecular crystal a clear distinction between internal vibrations and low-frequency librational and translational phonons can be made in most cases. Within the validity of this approximation it may turn out to be convenient to distinguish between different types of two-phonon bands although there is not a conceptual difference between them. We may classify two-phonon bands involving:

a) two internal vibrations,
b) one internal and one lattice vibration,
c) two lattice vibrations.

Two-phonon bands of type a) are the simplest. If the internal vibrations have a negligible dispersion, the two-phonon transition will be observed as a sharp peak. When, however, they have an appreciable dispersion, broad bands whose structures are related to the combined density of states will be observed.The simplifying feature of this type of two-phonon bands is that in most cases each internal mode in molecular crystals can be considered independently of all other crystal vibrations.In addition,it may happen that only one of the combining modes has a significant dispersion.

In type b) transitions the internal mode has generally a much lower dispersion than the external phonons.These transitions will then result in broad absorption features appearing on both sides of narrow fundamental bands and are known as "phonon side bands".If also the internal mode is largely dispersed,a single broad absorption band can be observed with side structure.

Transitions of type c) are more difficult to be observed experimentally since they give rise to very broad absorption bands that can only be observed in thick samples where the overcrowding of the spectra becomes severe.

In the following discussion we shall concentrate only on type a) and type b) two-phonon transitions in molecular crystals since these have been reported repeatedly and investigated theoretically in these last few years.

5.2d Review of Available Experimental Data

Although two-phonon bands of ionic and homopolar crystals have been known for a long time,only recently has experimental evidence for their occurrence in molecular crystals been accumulated.Two-phonon absorption in molecular crystals was clearly detected for the first time in the hydrogen chloride crystal where transitions of the type a) and b) were observed[150] ;these were interpreted in terms of single or double molecule transitions.Later,phonon side bands were identified in the infrared spectra of various molecular crystals composed of small molecular units like CO[151,152] ,CO_2[153,154,81] ,N_2O[153,155,156] ,OCS[157,158] ,SiF_4[159] , SF_6[160] and in the Raman spectra of N_2[161] and of some metal hexafluorides[162] .Several attempts of establishing a qualitative correlation between the peaks in the phonon side bands and those in the phonon densities of states or with the critical points in the Brillouin zone or with the fundamental infrared and Raman lattice frequencies have been made. More recently,two-phonon absorption bands of type a) have been observed in the infrared spectra of a series of crystals of simple molecules like CO_2[81,163], N_2O[156], OCS[164], SF_6[160] and in the Raman spectrum of crystalline CO_2[165]. In a few cases,two-phonon bands have been reported for more complicated systems like crystalline benzene[166], naphthalene[167] and others[168].

5.3 TWO-PHONON DENSITY OF STATES

In previous Chapters it has been described in detail how to calculate the vibrational frequencies of molecular crystals as a function of the wave vector and hence how to obtain the one-phonon density of states.The basic quantities required to interpret the two-phonon band shapes are however the combined two-phonon densities of states.Their calculation will be described in this Section.Calculated combined densities of states will then be compared with the experimental infrared and Raman band shapes with the purpose of sorting out the main factors that enter in a two-phonon process.These ideas will then be considered quantitatively in the next Section.At the present stage the approximation of harmonic intra- and intermolecular potentials will be used since this is sufficient to obtain most of the details of two-phonon densities of states.Corrections for the phonon densities of states due to anharmonic interactions will be considered later.

5.3a Internal Vibrations

We want to evaluate the two-phonon density of states for the particular case where the combining modes are two internal vibrations. A simplifying assumption in such cases is that each internal vibration behaves independently of all other crystal vibrations and therefore there is no mixing in the crystal with all other internal or external modes Within this approximation,calculations of the crystal energies can be easily performed using the exciton model[169,170].

Consider a degenerate normal mode f of the isolated molecule. If the unit cells and sites in the crystal are denoted by n (or m) and ν (or μ), respectively, the local normal coordinate for the ith component of the normal mode will be denoted by $q_{fi}^{n\nu}$ and the harmonic oscillator wave function by $\varphi_{fi}^{n\nu}$. The vibrational crystal ground state is

$$|0\rangle = \Pi\, \varphi_{o}^{n\nu} \tag{5.3.1}$$

If we define the one-site harmonic oscillator quotient function[171]

$$|n\nu,fi\rangle = \varphi_{fi}^{n\nu} / \varphi_{o}^{n\nu} \tag{5.3.2}$$

the excitation through the crystal of one quantum of the ith component of the normal mode f can be described by the symmetrized wave functions

$$|fi\nu,\mathbf{k}\rangle = L^{-\frac{1}{2}} |0\rangle \sum_{n} |n\nu,fi\rangle\, e^{i\mathbf{k}\cdot(\mathbf{n}+\boldsymbol{\nu})} \tag{5.3.3}$$

where, as usually, L is the number of unit cells in the crystal, $\mathbf{k}$ a wave vector in the first Brillouin zone and $\mathbf{n}$ and $\boldsymbol{\nu}$ are position vectors for the nth unit cell in the crystal and for the νth site in the unit cell, respectively. The frequency shifts in the crystal can be obtained by diagonalization of the perturbation matrix $\mathbb{H}$

$$\Omega = \mathbb{B}\mathbb{H}\mathbb{B}^{\dagger} \tag{5.3.4}$$

$\mathbb{H}$ has dimensions $d\times Z$, where d is the degeneracy of the mode and Z the number of molecules in the cell. The crystal eigenstates are given by

$$|\alpha f,\mathbf{k}\rangle = \sum_{\nu i} B_{\nu i,\alpha}(\mathbf{k})\, |fi\nu,\mathbf{k}\rangle \tag{5.3.5}$$

and the phonon density of states by

$$n_{f}(\omega) = \sum_{\mathbf{k}\alpha} \delta(\omega - \omega_{\mathbf{k}\alpha}^{f}) \tag{5.3.6}$$

The elements of the perturbation matrix are

$$H^{f}_{\nu i,\mu j}(\mathbf{k}) = \langle fi\nu,\mathbf{k}|V|fj\mu,\mathbf{k}\rangle \qquad 5.3.7$$

where V is an appropriate intermolecular potential.As it has been discussed in previous Chapters the various contributions to the intermolecular potential in a molecular crystal can be conveniently expressed by atom-atom or electrostatic interactions .Without much loss of generality,in the following discussion of the internal vibrations we shall assume that the intermolecular potential can be simply described by multipole interactions.Although inadequacies in some of the cases considered will be noted,in general this assumption will prove very convenient since the discussion will be mostly confined to two-phonon infrared absorption and it has been found that this is generally associated with strong infrared active modes.For these internal vibrations,dipole-dipole interactions play a dominant role.

For the evaluation of the matrix elements 5.3.7 the potential V must be expressed as a quadratic form in the local internal normal coordinates

$$2V = \sum_{n\nu m\mu}\sum_{fij} F^{fi,fj}_{n\nu,m\mu}\, q^{n\nu}_{fi} q^{m\mu}_{fj} \qquad 5.3.8$$

For the dipole-dipole interaction we expand the dipole moment in the normal coordinates and retain only first-order terms

$$\mu_{n\nu} = \mu^{o}_{n\nu} + \mu^{n\nu}_{fi}\, q^{n\nu}_{fi} \qquad 5.3.9$$

where the coefficients $\mu^{n\nu}_{fi} = (\partial\mu_{n\nu} / \partial q^{n\nu}_{fi})_{o}$ can be obtained from infrared intensities and are obviously independent of n.The intermolecular force constants appearing in 5.3.8 are given by[172,173]

$$F^{fi,fj}_{n\nu,m\mu} = \sum_{n'\nu'} \mu^{n\nu}_{fi}\, T^{n\nu,n'\nu'}\left[\, \mathbb{E} + \boldsymbol{\alpha}\,\mathbf{T}\,\right]^{-1}_{n'\nu';m\mu}\, \mu^{m\mu}_{fj} \qquad 5.3.10$$

where $\mathbb{E}$ is the unit matrix, $\boldsymbol{\alpha}$ a block diagonal tensor of the molecular

3, are given by

$$(T^{n\nu,m\mu})_{\alpha\beta} = - \frac{\partial^2}{\partial R_\alpha \, \partial R_\beta} \left(\frac{1}{\mathbf{R}_{m\mu} - \mathbf{R}_{n\nu}} \right) \qquad 5.3.11$$

For the present purposes, the dipole-dipole potential can be more conveniently expressed in a contracted form

$$2V = \tilde{\boldsymbol{\mu}} \, \mathbb{S} \, (\mathbb{E} + \boldsymbol{\alpha} \, \mathbb{S})^{-1} \, \boldsymbol{\mu} \qquad 5.3.12$$

where the quantities $\boldsymbol{\mu}$, $\boldsymbol{\alpha}$ and $\mathbb{S}$ refer to the unit cell and for the evaluation of the frequency shifts of a d-fold degenerate mode in a unit cell with Z molecules have dimensionality 3dZ. The blocks of the lattice sums matrices are

$$S_{\nu i,\mu j} = \sum_m T^{n\nu,m\mu} \, e^{i\mathbf{k}\cdot[(m-n) + (\mu-\nu)]} \qquad 5.3\ 13$$

The evaluation of the lattice sums 5.3.13 in relation to the long-range character of the dipole-dipole interaction has been described in many places and is discussed in Section 3.5 of this book.

In recent years, the calculation of the frequencies of the internal vibrations in a series of simple molecular crystals using the dipole-dipole interaction has been reported. In several cases, it has been shown that a good agreement with the experimental TO-LO and Davydov splittings can be obtained[174]. With this model the one-phonon densities of states have been obtained. In several case, however, it has been found that the simple dipole-dipole potential cannot account for all the effects observed in the infrared and Raman spectra. For instance, the gas-to-crystal frequency shifts are not completely reproduced by the dipole-dipole potential. This clearly shows that there can be an appreciable contribution from shorter-range interactions. For instance, in the case of the ν_2 mode in crystalline CO_2 a Davydov splitting of 5 cm^{-1} is observed, while the dipole-dipole model gives a contribution of 10 cm^{-1} [163,174]. This discrepancy can likewise be ascribed to short-range interactions. On the other hand in this same case, the TO-LO splitting, that specifically depends on the dipole interaction, is reproduced correctly. In the HCl crystal, a large Davydov splitting of 43 cm^{-1} is observed. It has been found that the dipole-dipole potential provides a contribution of only 18 cm^{-1} while the short-range atom-atom potential gives a negligible contribution[79]. For the internal mode of the HCl crystal it has been found that a satisfactory agreement with experiments can however be obtained if quadrupole-quadrupole interactions are taken into account[79,175].

Finally, when using dipole-dipole interactions it should be noted that if the mode under consideration is both infrared and Raman active both μ and α depend on the internal coordinates. In these cases, additional terms quadratic in the normal coordinates are obtained when μ and α are derived with respect to the q's. These additional terms have been found to have a significant effect in the case of the hydrogen chloride crystal [79].

For the evaluation of the combined density of states we define localized two-particle states describing the excitation of two vibrational quanta at different lattice sites

$$|n\nu,fi\rangle|m\mu,gj\rangle = \varphi_{fi}^{n\nu}\,\varphi_{gj}^{m\mu} \,/\, \varphi_{o}^{n\nu}\varphi_{o}^{m\mu} \qquad 5.3.14$$

Symmetrized two-phonon states are then defined in the form

$$|f\nu i,g\mu j;\mathbf{k}_f,\mathbf{k}_g\rangle = L^{-1}|0\rangle\sum_{nm}|n\nu,fi\rangle|m\mu,gj\rangle e^{i[\mathbf{k}_f\cdot(\mathbf{n}+\nu)+\mathbf{k}_g\cdot(\mathbf{m}+\mu)]} \qquad 5.3.15$$

The two-phonon states 5.3.15 can be alternatively expressed in terms of sum and difference wave vectors

$$\mathbf{k}_s = \mathbf{k}_f + \mathbf{k}_g \quad ; \quad \mathbf{k}_d = \tfrac{1}{2}(\mathbf{k}_f - \mathbf{k}_g) \qquad 5.3.16$$

In the manifold of two-phonon states $|f\nu i,g\mu j;\mathbf{k}_f,\mathbf{k}_g\rangle$ we are however only interested in those states for which the total wave vector is zero

$$\mathbf{k}_f + \mathbf{k}_g = 0 \quad ; \quad \mathbf{k}_d = \tfrac{1}{2}(\mathbf{k}_f - \mathbf{k}_g) = \mathbf{k} \qquad 5.3.17$$

since, as stated in the conservation relations 5.2.2a and 5.2.2b, the momentum must be conserved in a two-phonon process.Two-phonon states of interest in optical processes are thus of the type

$$|f\nu i,g\mu j;\mathbf{k},-\mathbf{k}\rangle = L^{-1}|0\rangle\sum_{nm}|n\nu,fi\rangle|m\mu,gj\rangle e^{i\mathbf{k}\cdot[(\mathbf{n}-\mathbf{m})+(\boldsymbol{\nu}-\boldsymbol{\mu})]} \qquad 5.3.18$$

and can be conveniently expressed in terms of the one-phonon states 5.3.3

$$|f\nu i,g\mu j;\mathbf{k},-\mathbf{k}\rangle = |f\boldsymbol{\nu}i,\mathbf{k}\rangle|g\mu j,-\mathbf{k}\rangle \qquad 5.3.19$$

The energies of the two-phonon states 5.3.18 can be obtained by evaluating the two-phonon perturbation matrix

$$H^{fg}_{\nu i,\mu j;\nu' i',\mu' j'}(\mathbf{k}) = \langle f\nu i,g\mu j;\mathbf{k},-\mathbf{k}\,|V|f\nu' i',g\mu' j';\mathbf{k},-\mathbf{k}\rangle \qquad 5.3.20$$

If the interaction potential is assumed to be harmonic as given in 5.3.8 the matrix elements 5.3.20 can be simply expressed in terms of the one--phonon matrix elements 5.3.6

$$H^{fg}_{\nu i,\mu j;\nu' i',\mu' j'}(\mathbf{k}) = H^{f}_{\nu i,\nu' i'}(\mathbf{k}) + H^{g}_{\mu j,\mu' j'}(-\mathbf{k}) \qquad 5.3.21$$

such that the energy of the two-phonon states is given by

$$\omega^{fg}_{\alpha\beta}(\mathbf{k}) = \omega^{f}_{\alpha}(\mathbf{k}) + \omega^{g}_{\beta}(-\mathbf{k}) \qquad 5.3.22$$

and the two-phonon crystal wave functions are

$$|\alpha f,\beta g;\mathbf{k},-\mathbf{k}\rangle = \sum_{\mu j}\sum_{\nu i} B^{f}_{\nu i,\alpha}(\mathbf{k})B^{g}_{\mu j,\beta}(-\mathbf{k})\,|f\nu i,g\mu j;\mathbf{k},-\mathbf{k}\rangle \qquad 5.3.23$$

Having obtained the two-phonon energies as a function of the wave vector,the two-phonon density of states

$$n_{fg}(\omega) = \sum_{\mathbf{k}\alpha\beta}\delta(\omega - \omega_{\alpha\mathbf{k}} - \omega_{\beta\mathbf{k}}) \qquad 5.3.24$$

can be calculated.In the calculation of the combined density of states various simplifying features may occur.

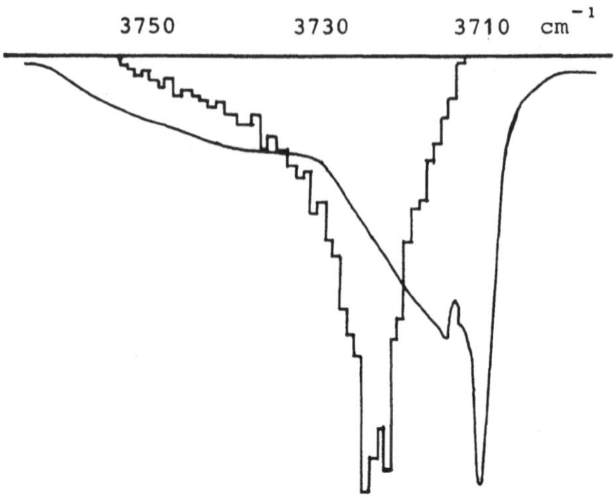

Fig. 5.2a Comparison of calculated harmonic density of states (histogram) in the $\nu_1 + \nu_3$ region of crystalline CO_2 with the observed infrared two-phonon band shape.

a) The simplest situation is encountered when one of the modes involved in the combination is not dispersed. In these cases, the combined density of states coincides with the one-phonon density of states of the mode that is dispersed. This case occurs, for instance, in the CO_2 crystal if we consider the $\nu_1 + \nu_3$ combination [81,174]. In fact the stron infrared active ν_3 mode has a large dispersion ($\sim$ 60 cm^{-1}) while the totally symmetric ν_1 mode is basically non-dispersed. This has been foun both experimentally and on the basis of model calculations[81]. In Fig. 5.2a the calculated density of states is compared with the infrared absorption spectrum in the $\nu_1 + \nu_3$ region. It can be seen that the width of the observed band compares reasonably with the density of states while the same cannot be said of the overall shape. In fact, on the low-frequency side, a sharp peak is observed that is absent in the density of states. Roughly, this peak can be described as being due to the one-molecule combination (corresponding to the states with $n\nu = m\mu$ in the definition 3.5.14) that, because of the intramolecular anharmonicity, occurs outside the two-phonon continuum. Qualitatively, one could say that the interaction between the single molecule and the two-phonon sta

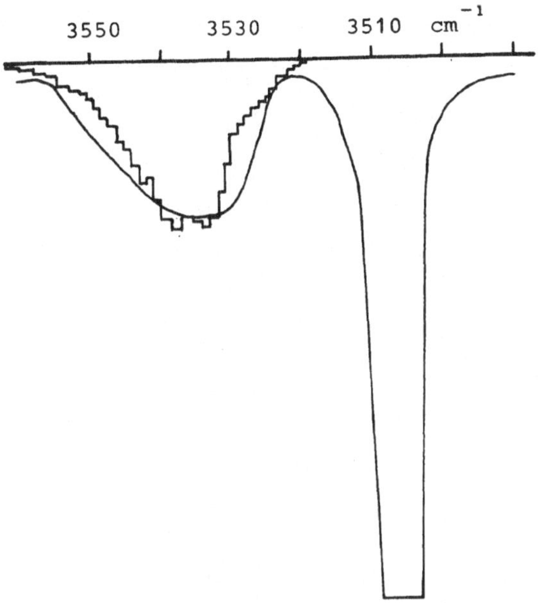

Fig. 5.2b Comparison of calculated harmonic density of states (histogram) in the $\nu_1 + \nu_3$ region of crystalline N_2O with the observed infrared two-phonon band shape.

tes gives rise to a perturbation of the overall band shape such that the comparison with the unweighted density of states cannot be direct. It is obvious that the strenght of the intramolecular anharmonic interaction and hence the position of the molecular combination will greatly affect the resulting band shape in the two-phonon region.

b) If the combining modes both have an appreciable dispersion,the one- and two-phonon densities of states may largely differ.If the one--phonon densities of the two combining modes are structureless with a single peak,the two-phonon density will also exhibit a single peak with an overall width that is the sum of those of the individual phonon bands. This occurs,for instance,in the N_2O crystal[156] .The ν_1 and ν_3 modes have phonon band widths of 12 and 43 cm^{-1} ,respectively.The combined density of states has a width of 55 cm^{-1} and is shown in Fig. 5.2b together with the observed infrared spectrum.By comparison with the $\nu_1 + \nu_3$ absorption in the CO_2 crystal,that has been discussed above,it can be seen that in N_2O the sharp peak,tentatively assigned to the one-molecule com-

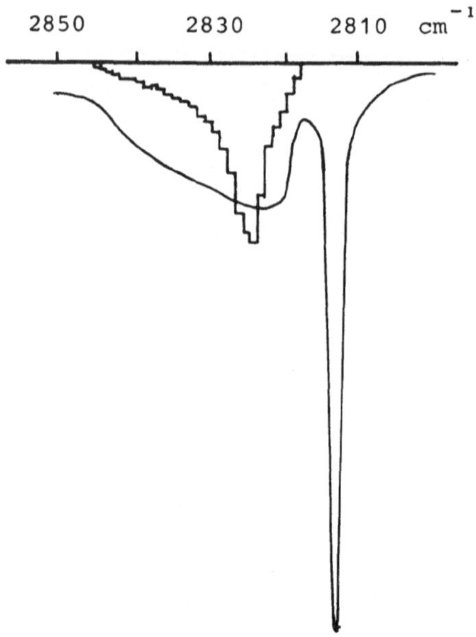

Fig. 5.2c Comparison of calculated harmonic density of states (histogram) in the $\nu_1 + \nu_2$ region of crystalline N_2O with the observed infrared two-phonon band shape.

bination,is farther away from the two-phonon continuum.This is due to the larger intramolecular anharmonicity in the N_2O case.We may thus imagine that for the $\nu_1 + \nu_3$ band of N_2O the perturbation of the observed band shape with respect to the unweighted density of states is much smaller.An intermediate case is found in the case of the $\nu_1 + \nu_2$ band of crystalline N_2O,whose density of states and infrared band shape are shown in Fig. 5.2c.

These general considerations apply to other crystals as well.In carbonyl sulfide[164] the very strong infrared active ν_3 mode has a dispersion extending over 120 cm^{-1} while the other internal modes are not appreciably dispersed.Therefore,the $\nu_1 + \nu_3$ combined density of states coincides with the ν_3 one-phonon density of states.The width of the phonon band compares nicely with the observed width of the infrared combination band.In contrast the density of states of the $2\nu_3$ overtone band is twice as large as that of the fundamental and this is in agreement with the observed width of the infrared combination band.

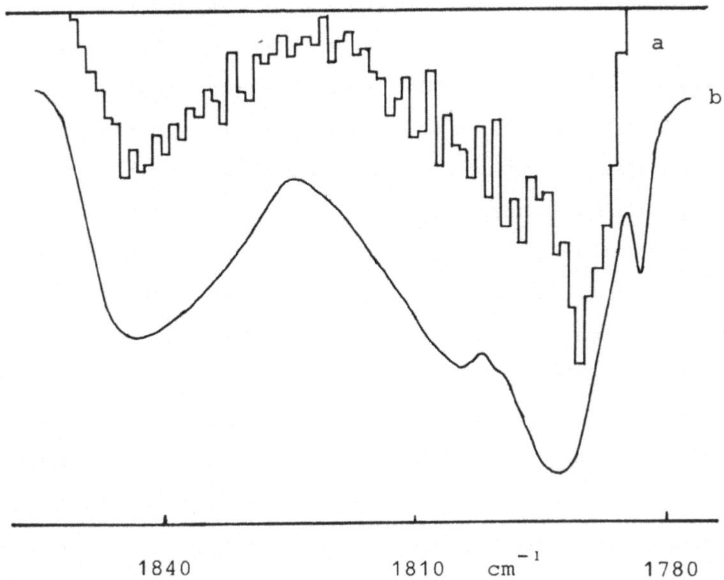

Fig. 5.3 Calculated harmonic density of states (a) and observed infrared band shape (b) in the $\nu_1 + \nu_3$ region of crystalline SiF_4.

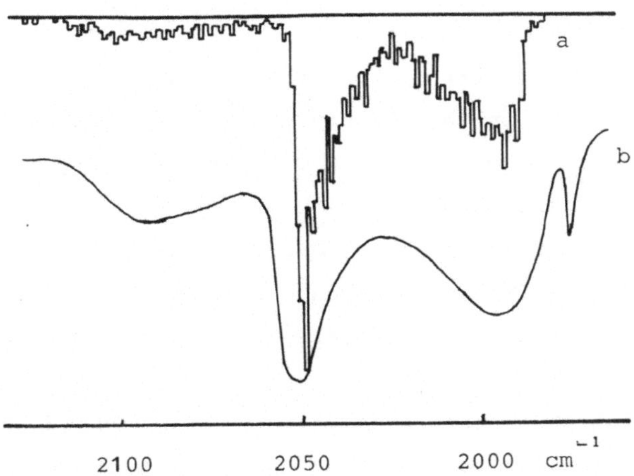

Fig. 5.4 Calculated harmonic density of states (a) and observed infrared band shape (b) in the $2\nu_3$ region of crystalline SiF_4

In the case of the SiF_4 crystal,more complicated situations arise since the infrared-active ν_3 and ν_4 modes have densities of states with two distinct peaks[176,180]. In the combinations with the ν_1 and ν_2 Raman modes,that can be assumed to have a negligible dispersion,the one-phonon and two-phonon densities of states coincide.A comparison with observed infrared absorption band shape in the $\nu_1 + \nu_3$ region is made in Fig.5.3. The fit of the calculated densities of states to the observed band shape is,on the whole,satisfactory but the occurrence of sharp peaks superimposed on the two-phonon continuum can be noticed.

In evaluating the combined density of states in the overtone region,some caution is necessary.If the mode is not degenerate and if there is a single occupancy of the unit cell as,for instance,is the case with the ν_3 mode of crystalline OCS[164], then no additional observations are necessary.If,on the contrary,the mode is degenerate or if there is a multiple occupancy of the unit cell,novel problems arise that have been described by Craig and Schettino[177]. A careful consideration of the symmetry properties of the wavefunctions of the basis set is necessary.In the case of the $2\nu_3$ overtone in SiF_4,the combined density of states greatly differs from the one-phonon density of states both in width and shape.The combined density of states is compared with the observed infrared absorption in Fig. 5.4 where it can be seen that the fit is striking.To a certain extent,the same considerations can be made for the overtone band in the HCl crystal[79]. The results of correct calculations of the two-phonon density of states greatly differ from previous qualitative inferences.

The various features that may occur in the calculation of two-phonon densities of states have been encountered in the SF_6 crystal that has been recently described in detail by Salvi and Schettino[160].

As a conclusion,in this Section it has been shown how the two-phonon densities of states for internal modes in molecular crystals can be calculated, and comparison with the observed infrared absorption bands has been made.It has been found that in general this comparison is rather satisfactory and that therefore the infrared band profile gives a fairly reasonable idea of the density of states.However,in some cases, as described, relevant distortions are observed and, in particular, sharp peaks appear which should, most likely, be assigned to one-molecule combination bands. In addition to the quantitative explanation of the origin of these additional peaks, that involves the anharmonic interactions, the problem of evaluating the intensity distribution in the two-phonon bands still remains.

5.3b Phonon Side Bands

The procedure for the calculation of the two-phonon density of states when one mode is an internal vibration and the other an external vibration does not differ from that described in the previous section if we confine ourselves to the harmonic approximation. In the majority of cases the internal vibration has a dispersion that is much smaller than the low frequency phonon band width. Under these circumstances, the one-phonon density of states is the appropriate quantity to be compared with the observed shape of the phonon side bands.

In recent years, densities of states have been calculated for several molecular crystals using model potentials as described in Chapter 3. Comparison of these quantities with phonon side bands is, however, not available in many cases. Detailed quantitative studies have been reported

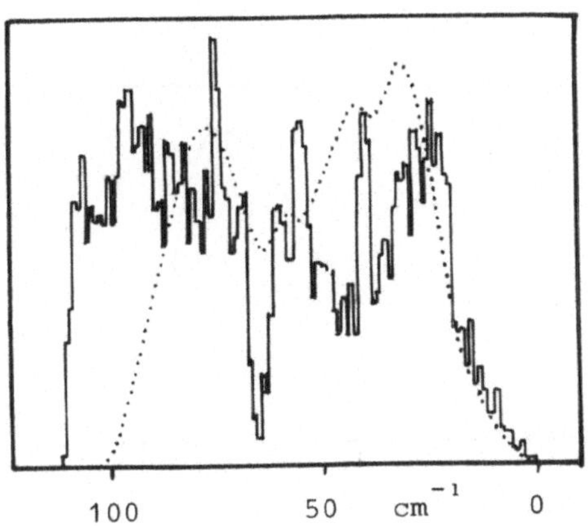

Fig. 5.5 Density of states and infrared phonon side band of the ν_2 mode in crystal OCS

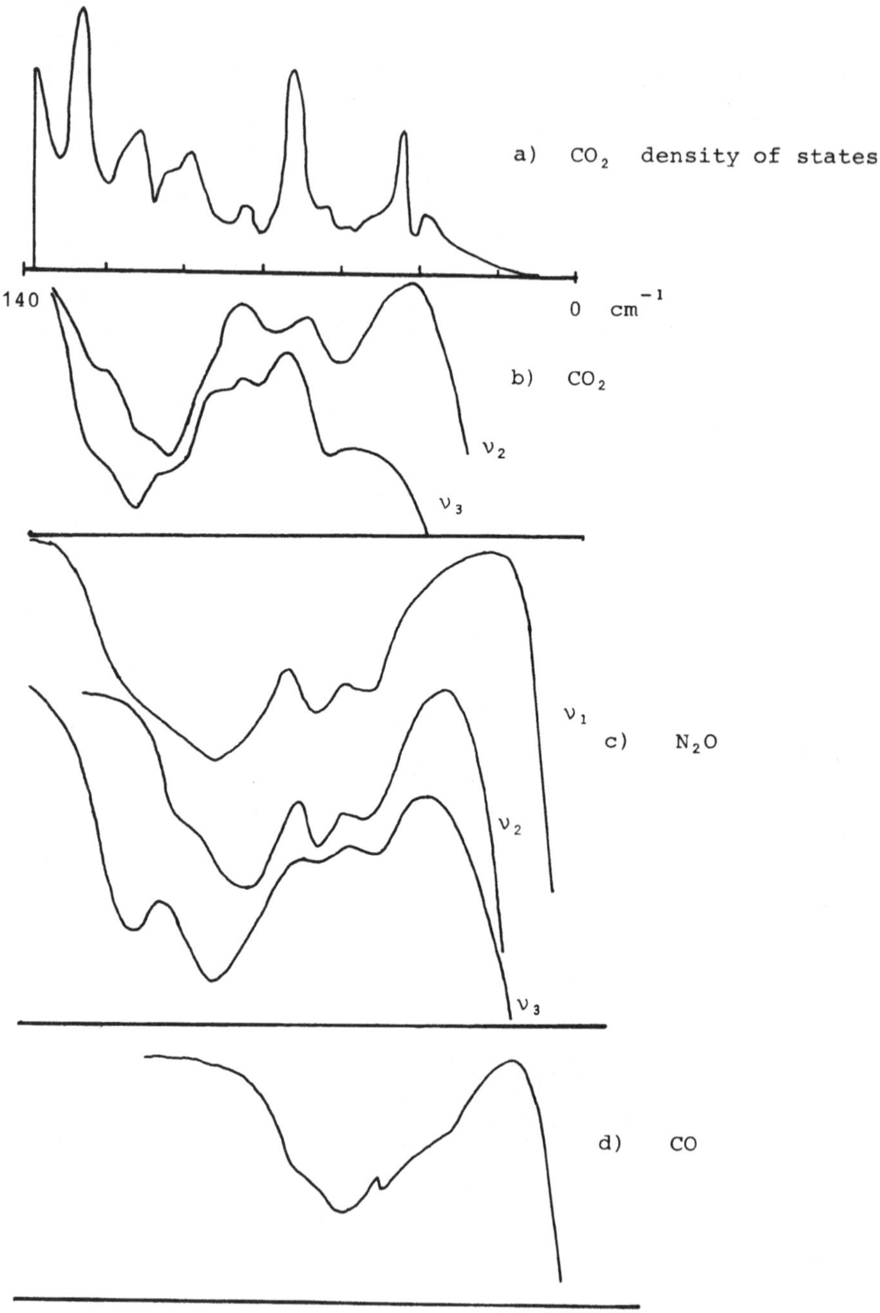

Fig. 5.6 Calculated density of phonon states in crystalline CO_2 (a) and infrared phonon side bands in crystalline CO_2 (b), N_2O (c) and CO (d)

only for the phonon side bands in the infrared spectrum of crystalline OCS[158,178,179] and in the Raman spectrum of α-N_2[161]. In the case of crystalline OCS, a comparison of the phonon side bands with the phonon density of states reveals quite significant differences showing that such kind of comparison is not straightforward. This can be seen in Fig. 5.5 for the ν_2 infrared region. Quantitative calculations for the OCS side bands will be described in the next Section. Qualitatively it is interesting to observe that in the OCS crystal the phonon side bands of the ν_1 and ν_2 modes are quite similar. In the infrared spectra, phonon side bands have also been observed in the case of the CO_2 crystal[81,153,154]. They are particularly well defined in the bending region. In Fig. 5.6 the absorption spectrum of the infrared fundamentals in CO_2 is compared with the density of states calculated by Suzuki and Schnepp[180]. Again quite significant differences are observed. It may be of interest to compare the phonon side bands in crystalline CO_2 and N_2O reported by various authors[81 153-156]. In fact, the crystal and electronic structures of CO_2 and N_2O are quite similar and we may reasonably expect that the densities of states for these molecular crystals should not greatly differ. It can be seen from Fig. 5.6 that there are striking similarities in the phonon side bands of carbon dioxide and nitrous oxide. On the contrary, as it can be seen from Fig. 5.6, in carbon monoxide, which also has a structure similar to that of CO_2 and N_2O, the phonon side band does not show the same structure with several defined peaks.

From the few examples discussed above it could be argued that the structure of the phonon side bands does actually not depend markedly on the particular internal mode involved and, in the case of CO_2 and N_2O, even the change of the molecular system seems to have a slight effect. In contrast, in the more complicated system of crystalline naphthalene there seems to be some dependence of the shape of the phonon side bands on the internal mode involved, as it can be seen from Fig. 5.7. For several internal modes the same number of peaks is observed at the same position, but the relative intensity is rather different. In general, however, also in crystalline naphthalene it can be observed that many details of the phonon density of states do not show up in the infrared phonon side bands.

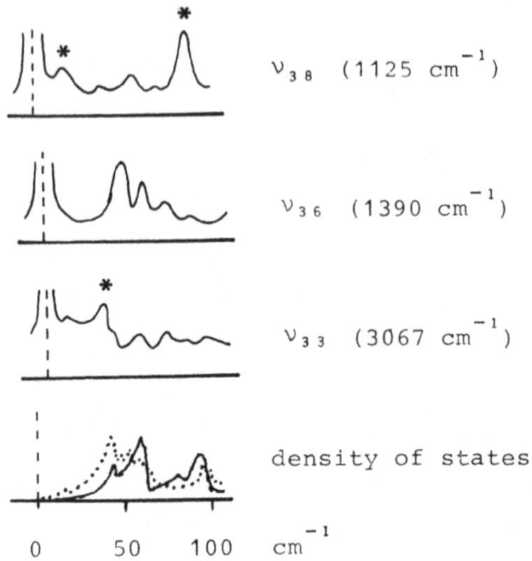

Fig. 5.7 Phonon side bands and translational (dotted line) and rotational (full line) density of phonon states in crystalline naphtalene (* one-phonon feature)

In the Raman spectra of molecular crystals a detailed study of phonon side bands has been reported only in the case of crystalline α-N_2 [161]. In Fig. 5.8 the Raman spectrum in the region of the internal vibration is compared with the phonon density of states. Also in this case, it can be noticed that relevant features of the density of states (and particularly those at higher frequencies) do not contribute appreciably to the observed spectrum. On the contrary, this does not seem to be the case with the phonon side band observed in the Raman spectrum of crystalline benzene in combination with the ν_9 fundamental [165] and shown in Fig. 5.9.

In conclusion, it can be observed that the experimental information on phonon side bands in molecular crystals is not at present very extensive. From the few examples discussed above it can however be said that the relation between the observed band shapes and the phonon densities of states seems to be more complicated in these cases than found in the previous discussion of two-phonon bands involving two internal vibrations.

Finally, it can be of interest to conclude this section by quoting

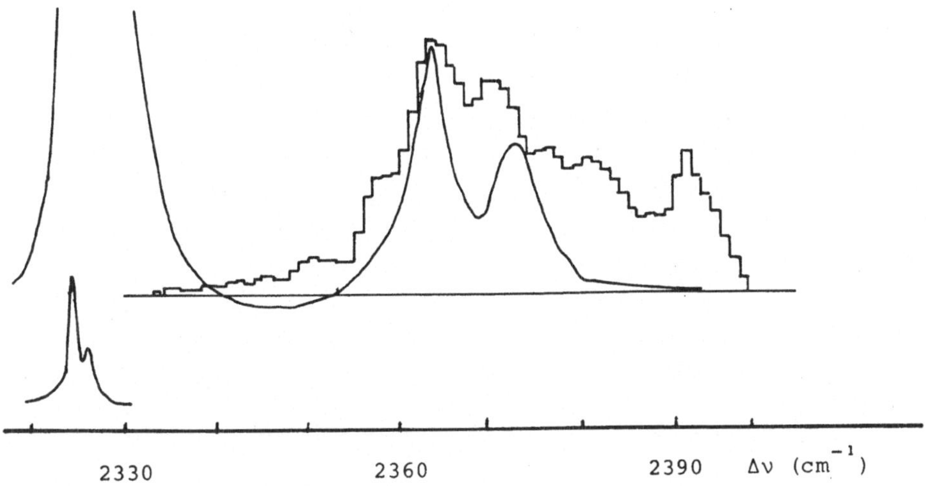

Fig. 5.8 Density of phonon states and phonon side bands in the Raman spectrum of crystalline α-N_2

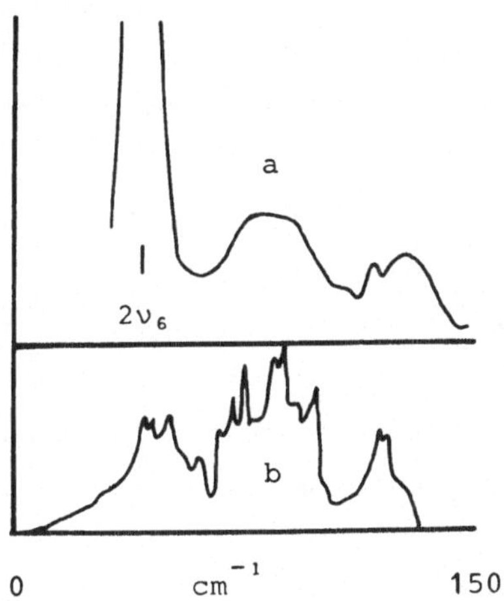

Fig. 5.9 Phonon side band (a) and phonon density of states (b) in crystalline benzene.

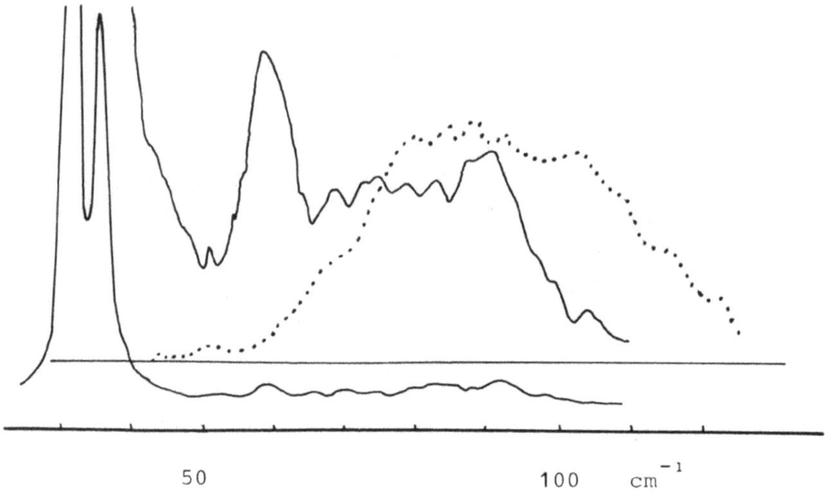

Fig. 5.10 Two-phonon Raman spectrum (full line) and two-phonon density of states (dotted line) in crystalline α-N_2

the results obtained in the study of the two-phonon region involving two external lattice modes in crystalline α-N_2 [161].A definite structure has been observed in the low-frequency Raman spectrum which is compared with the two-phonon density of states in Fig. 5.10.It can be noticed that many features of the density of states are not actually observed in the Raman spectrum.The low-frequency region has also been studied in the Raman spectrum of crystalline benzene but the experimental data do not seem to be sufficiently accurate in the present case [165].Finally, a two-phonon density of states has been calculated for crystalline NH_3 but there are no available experimental data in this case [146].

5.4 TWO-PHONON ABSORPTION COEFFICIENT

In the previous Section the two-phonon densities of states have been discussed and compared with experiments for a series of cases involving both internal and external vibrations in molecular crystals. The importance of this comparison has already been stressed,since it may be very useful to have an optical spectroscopic guide to the phonon density of states.However,the results of the previous Section have revealed that unweighted densities of states may generally differ from two-phonon band profiles and that these differences may be large.Simplifying somehow the notation and indicating by $\langle 0|H_p|2\rangle$ the matrix element connecting the initial (ground) state with the two-phonon states through a perturbation potential H_p ,the intensity distribution in a two-phonon band will behave as

$$I(\omega) \propto \sum_{\mathbf{k}\alpha\beta} |\langle 0|H_p|2\rangle|^2 \, \delta(\omega - \omega_{\alpha\mathbf{k}} - \omega_{\beta\mathbf{k}}) \qquad 5.4.1$$

The two-phonon band would thus be an exact replica of the density of states only if the coupling coefficients $\langle 0|H_p|2\rangle$ were constant.The results discussed in the previous Section clearly show that this is not the case.Various models for calculating the matrix elements appearing in 5.4.1 and hence the two-phonon band intensities have been reported and some of them will be described in this Section.

For combinations of internal vibrations the contribution to the two-phonon absorption,due to the non-linear moments (mechanism a) of section 5.2b),has been estimated by evaluating the dipole moment matrix elements connecting the ground and two-phonon states [81] .It has been found that the contribution of this mechanism is in most cases small, ranging from 1 to 10 % of the experimentally measured intensity.Various simple models have been taken into consideration to account phenomenologically for the anharmonic interactions responsible for mechanism b. For instance,in some cases [164,176] an adaptation of the Fano theory of configuration interaction [181] has been used to account for the mixing of two-phonon states with sharp one-molecule combination states.There is,however,no simple relation between the empirical parameters entering these models and the anharmonic molecular properties.More recently, a Green function approach to the two-phonon absorption coefficient,that considers the perturbation as being due to the intramolecular anharmonic interactions has been reported [182,183] .The results of the application of this approach will be described.

For phonon side bands in molecular crystals we shall briefly describe the calculation of the absorption coefficient that has been recently reported by Della Valle et al.[179] and applied to the carbonyl sulfide crystal.The method uses ordinary perturbation theory and directly evaluates the third-order intermolecular coefficients starting from an atom-atom (or from other model) potential.

5.4.a Internal Vibrations

The interaction with the radiation leading to the two-phonon absorption can occur through linear (mechanism b) or through the second-order (mechanism a) electric moment.In any case,the resulting optical absorption can be related to some appropriate Green function.For the processes of interest in this Section it is necessary to evaluate the one- and the two-phonon Green functions renormalized by the third- and fourth-order anharmonic potentials.A many-body approach to the description of two-phonon states has been previously described by Ruvalds and Zawadowski[184],by Cohen and Ruvalds[185] and by Agranovitch[186].A similar treatment,especially adapted for the application to molecular crystals,has been discussed by Bogani[182,183].

In order to achieve a separation of the various types of two-phonon processes it is convenient to approach the problem in terms of creation ($A^{\dagger}$) and destruction (A^{-}) operators that are usually defined as

$$A_f^{\dagger}\left(\begin{smallmatrix}\mathbf{k}\\ \alpha\end{smallmatrix}\right) = \left(\frac{\omega^f_{\alpha\mathbf{k}}}{2\hbar}\right)^{\frac{1}{2}} \left[Q^f_{\alpha\mathbf{k}} + i\,\frac{\dot{Q}^f_{\alpha\mathbf{k}}}{\omega^f_{\alpha\mathbf{k}}} \right] \qquad 5.4.2$$

$$A_f^{-}\left(\begin{smallmatrix}\mathbf{k}\\ \alpha\end{smallmatrix}\right) = \left(\frac{\omega^f_{\alpha\mathbf{k}}}{2\hbar}\right)^{\frac{1}{2}} \left[Q^f_{\alpha\mathbf{k}} - i\,\frac{\dot{Q}^f_{\alpha\mathbf{k}}}{\omega^f_{\alpha\mathbf{k}}} \right] \qquad 5.4.3$$

where $Q^f_{\alpha\mathbf{k}}$ is the crystal normal coordinate.It is convenient to express the operators 5.4.2 and 5.4.3 in the local representation by means of the transformation

$$Q_f^{\dagger}\binom{n}{\nu_i} = \left(\frac{\omega^f}{2\hbar L}\right)^{\frac{1}{2}} \sum_{\alpha\mathbf{k}} B^f_{\nu_i\alpha}(\mathbf{k})\, e^{i\mathbf{k}\cdot(n+\boldsymbol{\nu})} \frac{A_f^{\dagger}\binom{\mathbf{k}}{\alpha}}{(\omega^f_{\alpha\mathbf{k}})^{\frac{1}{2}}} \qquad 5.4.4$$

and similarly for $Q_f^{-}\binom{n}{\nu_i}$. In 5.4.4 ν_i is a composite index for the site in the unit cell and for the ith component of the degenerate normal mode f. In general, the dispersion of the internal modes is small compared to the molecular frequency. Therefore, although this is not strictly correct, the operators in the local representation can be identified with creation and destruction operators of mode f at a given site in the crystal. Within this approximation, the anharmonic terms contributing to the renormalization of the Green functions will be most important within a single molecule.

<u>One- and two-phonon Green functions</u>.

In the local representation it is necessary to evaluate the Green function

$$G_{fg}\left(\begin{smallmatrix}n\\m\end{smallmatrix}\middle|\begin{smallmatrix}\nu_i\mu_j\\\nu_i'\mu_j'\end{smallmatrix}\right) = -i\Theta(t)\sum_{l_1 l_3} \langle\langle Q_f^{\dagger}\binom{l_1}{\nu_i}|t) Q_g^{\dagger}\binom{l_1+n}{\mu_j}|t); Q_f^{-}\binom{l_3}{\nu_i'}|0) Q_g^{-}\binom{l_3+m}{\mu_j'}|0)\rangle\rangle \qquad 5.4.5$$

where $\langle\langle \cdots\cdots \rangle\rangle$ indicates the statistical average on the system and $\Theta(t)$ is the Heaviside step function. The Green function 5.4.5 describes the probability that modes f and g, after having been destroyed at an initial time t=0 at sites l_3 and l_3+m, are created at a later time t at sites l_1 and l_1+n. In the equation of motion for the Green function 5.4.5

$$i\hbar\frac{d}{dt}G_{fg}\left(\begin{smallmatrix}n'n\\m'm\end{smallmatrix}\middle|\begin{smallmatrix}\nu_i,\mu_j\\\nu_i',\mu_j'\end{smallmatrix}\right) = \hbar\delta(t)\langle[Q_f^{\dagger}\binom{n'}{\nu_i}|t) Q_g^{\dagger}\binom{n}{\mu_j}|t); Q_f^{-}\binom{m'}{\nu_i'}|0) Q_g^{-}\binom{m}{\mu_j'}|0)]\rangle +$$

$$+ \langle\langle[Q_f^{\dagger}\binom{n'}{\nu_i}|t) Q_g^{\dagger}\binom{n}{\mu_j}|t), H]; Q_f\binom{m'}{\nu_i'}|0) Q_g\binom{m}{\mu_j'}|0)\rangle\rangle \qquad 5.4.6$$

it is convenient to separate the contributions deriving from the harmonic and anharmonic parts of the Hamiltonian

$$H = H_o + H_A \qquad 5.4.7$$

The harmonic Hamiltonian H_o includes the harmonic Hamiltonian of the isolated molecule and intermolecular potential 5.3.8.The anharmonic part H_A includes only the intramolecular potential that for two modes f and g can be written in the local representation as

$$H_A = \sum_{n\nu} H_A(n\nu) = \sum_{n\nu}\sum_{g\geq f} X_{fg}\, n_f(n\nu)\, n_g(n\nu) \qquad 5.4.8$$

where the n's are the number operators

$$n_f = Q_f^{\dagger}\binom{n}{\nu} Q_f^{-}\binom{n}{\nu} \qquad 5.4.9$$

and X_{fg} is the anharmonicity constant which is known from gas phase studies.Expression 5.4.8 is incomplete since various terms that only contribute to the renormalization of the frequency of the isolated molecule have been omitted.These terms can be taken into account by using an effective frequency for the isolated molecule.The terms left in 5.4.8 represent the effective perturbation that enters in the evaluation of the anharmonic two-phonon Green function.It is thus possible to separate the harmonic and anharmonic contributions to the Green function.

Fig. 5.11 First-order diagrams contributing to the two-phonon Green function

In terms of diagrams we want to evaluate the contributions that correspond to the summation of the first-order diagrams of Fig. 5.11. The model adopted has the following simplifying features:

a) The system can be essentially considered at T=0K.In fact the internal vibrations have relatively high energies and even at moderate temperatures can their occupation numbers 5.4.9 be considered as zero to a good approximation.

b) The anharmonic potential 5.4.8 does not couple molecules on different sites.

c) The intermolecular harmonic potential 5.3.8 depends only on differences between molecular position coordinates.

Under these conditions, it can be shown[182] that the harmonic two-phonon Green function is given by

$$g_{fg}\begin{pmatrix} n \\ m \end{pmatrix}\begin{vmatrix} \nu_i \mu_j \\ \nu_i' \mu_j' \end{vmatrix} = \sum_{\alpha\beta\mathbf{k}} \frac{B^{f*}_{\nu_i\alpha} B^{f}_{\nu_i'\alpha} B^{g*}_{\mu_j\beta} B^{g}_{\mu_j'\beta}}{\omega - \omega^{f}_{\alpha\mathbf{k}} - \omega^{g}_{\beta\mathbf{k}}} \, e^{i\mathbf{k}\cdot(m+\mu_j'-n-\mu_j)} \, e^{i\mathbf{k}\cdot(\nu_i-\nu_i')} \qquad 5.4.10$$

A further simplifying feature occurs in many cases where one of the combining modes (for instance mode g) has a negligible dispersion. The harmonic two-phonon Green function then reduces to

$$g_{fg}\begin{pmatrix} n \\ m \end{pmatrix}\begin{vmatrix} \nu_i \mu_j \\ \nu_i' \mu_j' \end{vmatrix} = \delta_{\mu_j\mu_j'} \sum_{\alpha\mathbf{k}} \frac{B^{f*}_{\nu_i\alpha} B^{f}_{\nu_i'\alpha}}{\omega - \omega^{g} - \omega^{f}_{\alpha\mathbf{k}}} \, e^{i\mathbf{k}\cdot(\mathbf{m}+\nu_i'-\mathbf{n}-\nu_i)} \qquad 5.4.11$$

Within the approximations discussed above we may write

$$G_{fg}\begin{pmatrix} n \\ m \end{pmatrix}\begin{vmatrix} \nu_i \mu_j \\ \nu_i' \mu_j' \end{vmatrix} = \delta_{\mu_j\mu_j'} \, G_{fg}\begin{pmatrix} n \\ m \end{pmatrix}\begin{vmatrix} \nu_i \mu_j \\ \nu_i' \mu_j \end{vmatrix} \qquad 5.4.12$$

and the equation for the anharmonic Green function becomes

$$G_{fg}\begin{pmatrix} n \\ m \end{pmatrix}\begin{vmatrix} \nu_i \mu_j \\ \nu_i' \mu_j \end{vmatrix} = g_{fg}\begin{pmatrix} n \\ m \end{pmatrix}\begin{vmatrix} \nu_i \mu_j \\ \nu_i' \mu_j \end{vmatrix} + \frac{X_{fg}}{L} \, g_{fg}\begin{pmatrix} n \\ 0 \end{pmatrix}\begin{vmatrix} \nu_i \mu_j \\ \mu_j \mu_j \end{vmatrix} G_{fg}\begin{pmatrix} 0 \\ m \end{pmatrix}\begin{vmatrix} \mu_j \mu_j \\ \nu_i' \mu_j \end{vmatrix} \qquad 5.4.13$$

This equation is formally identical with the equation for a lattice with a point defect at site 0ν[187]. In the present case the role of the impurity is played by the anharmonicity of the single molecule.

If we define

$$g_{fg}(0,\nu_i) = g_{fg}\begin{pmatrix} 0 \\ 0 \end{pmatrix}\begin{vmatrix} \nu_i \nu_i \\ \nu_i \nu_i \end{vmatrix} = \frac{1}{d_f Z} \sum_{\nu_i} g_{fg}\begin{pmatrix} 0 \\ 0 \end{pmatrix}\begin{vmatrix} \nu_i \nu_i \\ \nu_i \nu_i \end{vmatrix} \qquad 5.4.14$$

and

$$f(\omega) = 1 - \frac{X_{fg}}{d_f ZL} g(0,\nu_i) \qquad 5.4.15$$

the two-phonon Green function can be written in a more convenient way as

$$G_{fg}\begin{pmatrix} n & \nu_i\mu_j \\ 0 & \mu_j\mu_j \end{pmatrix} = g_{fg}\begin{pmatrix} n & \nu_i\mu_j \\ 0 & \mu_j\mu_j \end{pmatrix} \Big/ f(\omega) \qquad 5.4.16$$

and the result is the same as in the non-interacting case multiplied by a resonance denominator that is given explicitely by

$$f(\omega) = 1 - \frac{X_{fg}}{d_f ZL} \sum_{\alpha\mathbf{k}} \frac{1}{\omega - \omega^g - \omega^f_{\alpha\mathbf{k}} - i\varepsilon} =$$

$$= \left[1 - \frac{X_{fg}}{d_f ZL} \int_p \frac{n(z)dz}{\omega-\omega^g-z} \right] + i \left[\pi X_{fg} n(\omega-\omega^g) \right] \qquad 5.4.17$$

where $\int_p$ denotes the principal part.

The behavior of the real part of f(ω) for the $\nu_1+\nu_3$ band of crys-

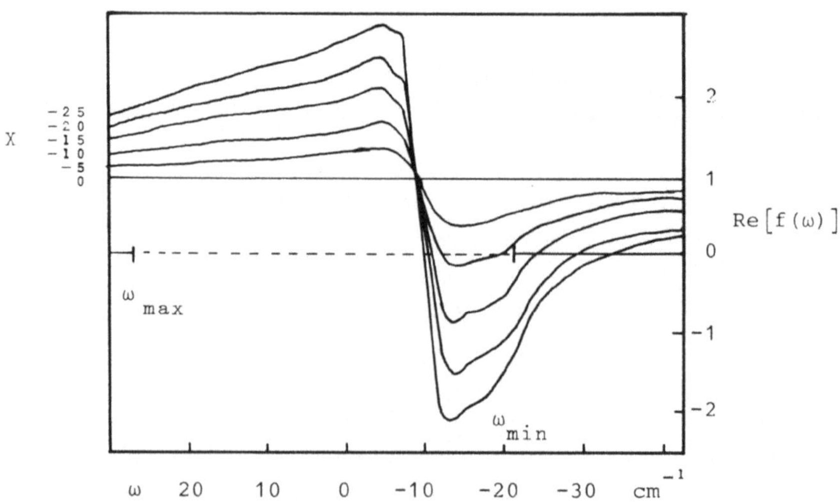

Fig. 5.12 Behavior of Re[f(ω)] as a function of the anharmonicity constant for the $\nu_1+\nu_3$ combination in crystalline CO_2.

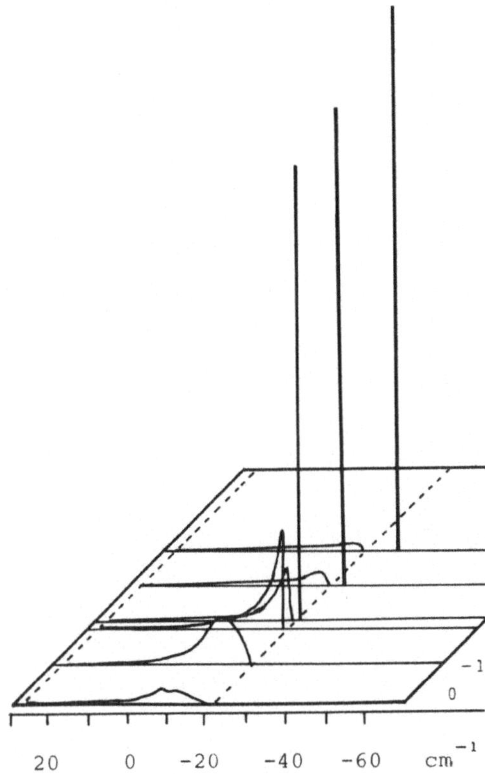

Fig. 5.13 Renormalized density of states of the $\nu_1+\nu_3$ combination of crystalline CO_2 as a function of the anharmonicity constant.

talline CO_2 is shown in Fig. 5.12 as a function of the value of the anharmonicity constant X_{fg}. It is seen that when the anharmonicity is sufficiently large the function Re$[f(\omega)]$ has at least two zeros. If the zeros fall inside the two-phonon continuum they give rise to resonances. If one zero falls outside the continuum, the new pole of the Green function leads to the formation of a bound state that can be interpreted as a bound pair of excitations and may correspond to the occurrence of of sharp peaks in the spectra.

The renormalized two-phonon density of states corresponding to the imaginary part of the Green function is given by

$$\rho(\omega) = \frac{n(\omega-\omega^g)}{[f(\omega)]^2} \qquad 5.4.18$$

within the continuum.If a bound state occurs at frequency ω_B the Green function is given by

$$G^B(\omega\pm i\varepsilon) = L\, X_{fg}^2 \left[\int \frac{n(z)dz}{(\omega_B-\omega^g-z)^2} \right]^{-1} \frac{1}{\omega-\omega_B\pm i\varepsilon} \qquad 5.4.19$$

In Fig. 5.13 the renormalized density of states for the combination $\nu_1+\nu_3$ of CO_2 is shown for various values of the anharmonicity constant. It can be seen that the anharmonic intramolecular interaction can change the harmonic density of states introducing new features in the resulting profile.

It is also necessary to evaluate the one-phonon Green function defined as

$$G_f(\nu_i\nu_i') = -i\Theta(t)\sum_{nm} \langle\langle Q_f^\dagger(\begin{smallmatrix} n \\ \nu_i \end{smallmatrix}|t);Q_f^-(\begin{smallmatrix} n' \\ \nu_i' \end{smallmatrix}|0)\rangle\rangle \qquad 5.4.20$$

The harmonic contribution can be shown to be

$$g_f(\nu_i\nu_i') = L\sum_\alpha \frac{B_{\nu_i\alpha}^{f*}B_{\nu_i'\alpha}^f}{\omega - \omega^f - \omega_{\alpha\mathbf{k}}^f} \qquad 5.4.21$$

The anharmonic contribution of interest derives from the following terms in the potential

$$H_{an} = \sum_{n\nu ij} K_{ffg} q_{n\nu}^{fi} q_{n\nu}^{fi} q_{n\nu}^{gj} \qquad 5.4.22$$

that can be expressed in the operators $Q^\dagger$ and Q^-.The terms of interest contain products of operators of the type $Q_f^\dagger Q_g^\dagger Q_f^-$ that give a non-zero probability that the mode f is destroyed at a lattice site and the modes f and g are later created at different lattice sites.It can be shown that the anharmonic one-phonon Green function is given by

$$G_{fg}(\nu_i\nu_i'') = g_f(\nu_i'\nu_i'') + \frac{K_{ffg}}{L}\sum_{\nu_i} g_f(\nu_i'\nu_i)\; G_{fg}(0|\nu_i\nu_i\nu_i'') \qquad 5.4.23$$

where the new Green function $G_{fg}(0|\nu_i\nu_i\nu_i'')$ is

$$G_{fg}(0|\nu_i\mu_j\nu_i') = \sum_{l_1 l_4} \langle\langle Q_f^{\dagger}\binom{l_1}{\nu_i}\Big|t) Q_g^{\dagger}\binom{l_1+n}{\mu_j}\Big|t); Q_g^{-}\binom{l_4}{\nu_i'}\Big|0)\rangle\rangle \qquad 5.4.24$$

After some manipulation it can be shown that the one-phonon Green function can be expressed in terms of the two-phonon Green function,whose renormalization by the anharmonic potential has already been considered

$$G_{fg}(\nu_i\nu_i') = g_f(\nu_i\nu_i') + \frac{K_{ffg}^2}{L^2}\sum_{\nu_i''} g_f(\nu_i\nu_i'') G_{fg}\left(\begin{matrix}0\\0\end{matrix}\middle|\begin{matrix}\nu_i''\nu_i''\\\nu_i''\nu_i''\end{matrix}\right) g_f(\nu_i''\nu_i') \qquad 5.4.25$$

Using the results obtained before the expression 5.4.25 can be written

$$G_{fg}(\nu_i\nu_i') = g_f(\nu_i\nu_i') + \frac{K_{ffg}^2}{L^2}\sum_{\nu_i''} g_f(\nu_i\nu_i'') g_f(\nu_i''\nu_i') \frac{\sum_{\nu_i} g_f(0|\nu_i)/d_f Z}{f(\omega)} \qquad 5.4.26$$

The equation 5.4.26 thus shows that the one-phonon Green function has poles corresponding to the two-phonon excitations of modes f and g.

Two-phonon absorption coefficient.

The real n and the imaginary k parts of the refractive index obey the relations

$$n^2 - k^2 = 4\pi \mathrm{Re}\chi(\omega) \qquad 5.4.27$$

$$2nk = 4\pi \mathrm{Im}\chi(\omega) \qquad 5.4.28$$

where $\chi(\omega)$ is the dielectric susceptibility that,for a cubic crystal, is given by

$$\chi(\omega) = \frac{1}{3}(\hbar V)^{-1} \int_{-\infty}^{+\infty} dt\, e^{i(\omega - i\varepsilon)t} \langle \mathbf{M}(t)\cdot\mathbf{M}(0)\rangle \qquad 5.4.29$$

In 5.4.29 V is the volume of the unit cell, $\langle\cdots\cdots\rangle$ denotes the statistical average and **M** is the crystal dipole moment.The two-phonon absorption is in general weak and n can be considered as a constant in the region of interest.The two-phonon absorption coefficient can thus be written as

$$\alpha(\omega) = \frac{\omega}{c}2k = \frac{4}{3}\frac{\pi\omega}{nc\hbar V}\mathrm{Im}\int_{-\infty}^{+\infty} dt\ e^{i(\omega-i\varepsilon)t}\langle \mathbf{M}(t)\cdot\mathbf{M}(0)\rangle \qquad 5.4.30$$

Various contributions to the absorption coefficient $\alpha(\omega)$ arise that are clearly recognized when the dipole moment is expanded in the normal coordinates. Using for simplicity n, m, f and g as cumulative indices for $n\nu, m\mu, f_i$ and g_i, respectively, the crystal dipole is

$$\mathbf{M} = \sum_{nf} M_f^n q_f^n + \sum_{nfg} M_{fg}^n q_f^n q_g^n + \sum_{nf}\sum_{mg} M_{fg}^{nm} q_f^n q_g^m \qquad 5.4.31$$

where the linear term has already been defined in 5.3.9 and

$$M_{fg}^n = \left(\frac{\partial^2 M_n}{\partial q_f^n\ \partial q_g^n}\right) \qquad 5.4.32$$

$$M_{fg}^{nm} = \left(\frac{\partial \alpha_n}{\partial q_f^n}\right)_0 T^{nm}\left(\frac{\partial \mu_m}{\partial q_g^m}\right)_0 + \left(\frac{\partial \alpha_m}{\partial q_g^m}\right)_0 T^{nm}\left(\frac{\partial \mu_n}{\partial q_f^n}\right)_0 \qquad 5.4.33$$

The linear term in the dipole operator contributes to the two-phonon absorption only through an indirect mechanism b). By the linear term the infrared-active mode f is excited; the mode f then decays into modes f and g by the third-order potential 5.4.22. It is clearly seen from the form of the first-order dipole operator that this contribution to the absorption coefficient depends on the renormalized one-phonon Green function $G_f(\nu_i\nu_i')$.

The non-linear moment with expansion coefficients given in 5.4.32 is localized on a single molecule and is responsible for the processes where direct interaction with radiation gives rise to two phonons in the crystal. The contribution of this process to the absorption coefficient, $\alpha^2(\omega)$, clearly depends on the two-phonon Green function $G_{fg}\begin{pmatrix}0 & \nu_i\nu_i \\ 0 & \nu_i'\nu_i'\end{pmatrix}$. In addition, we should also consider the interference terms between the first- and second-order localized moment that depend on the Green function $G_{fg}(0|\nu_i\nu_i\nu_j)$.

Using the results obtained before for the various Green functions and after some rearrangements, under the assumption that mode g is not dispersed, these contributions to the absorption coefficient are given by

$$\alpha^{1}(\omega) = C\,\frac{\hbar (M_f^n)^2}{\omega^f}\left[\frac{K_{ffg}}{2\,\omega^g}\right]^2 L\,\rho(\omega) \qquad 5.4.34$$

$$\alpha^{2}(\omega) = C\left[\frac{\hbar\,M_{fg}^n}{2}\right]^2 \frac{1}{\omega^f\omega^g}\,L\rho(\omega) \qquad 5.4.35$$

$$\alpha^{1,2}(\omega) = C\,\frac{\hbar\,M_f^n\,M_{fg}^n\,K_{ffg}}{4\,\omega_f\omega_g}\left[\frac{\hbar}{\omega^g}\right]^{\frac{1}{2}} L\rho(\omega) \qquad 5.4.36$$

where

$$C = \frac{4}{3}\,\frac{\pi\omega}{nc}\,\frac{Zd_f d_g}{\hbar V} \qquad 5.4.37$$

It can be seen from the expressions 5.4.34 to 5.4.36 that the process with a virtual intermediate state gives rise to a renormalization of the second-order moment. If we define a renormalized second-order moment

$$D_{fg} = \frac{1}{2}\,\frac{\hbar}{(\omega^f\omega^g)^{\frac{1}{2}}}M_{fg}^n + \left(\frac{\hbar}{\omega^f}\right)^{\frac{1}{2}}\frac{K_{ffg}}{2\omega^g}\,M_f^n \qquad 5.4.38$$

we may write the contributions 5.4.34 to 5.4.36 simply as

$$\alpha(\omega) = C\,(D_{fg})^2\,L\,\rho(\omega) \qquad 5.4.39$$

The contribution to the absorption coefficient of the non-localized second-order moment 5.4.33, whose significance has been described previously, has been discussed in detail by Dows and Schettino [81]. It contributes to the two-phonon absorption both in the harmonic and anharmonic approximation. It has been found that the harmonic contribution, that is related to the harmonic two-phonon Green function 5.4.10, is the largest and is given by

$$\alpha^{3}(\omega) = \frac{4}{3}\,\frac{\pi\omega}{nc}\,|H(\omega)|^2 \qquad 5.4.40$$

where

$$|H(\omega)|^2 = \sum_{\mathbf{k}\alpha\beta}\left[\sum_{\nu_i\mu_j} B^f_{\nu_i\alpha}(\mathbf{k}) B^g_{\mu_j\beta}(\mathbf{k}) \sum_m M^{0\nu m\mu}_{f_i g_j} e^{i\mathbf{k}(\mathbf{m}+\boldsymbol{\mu}-\boldsymbol{\nu})}\right]^2 \delta(\omega-\omega^f_{\alpha\mathbf{k}}-\omega^g_{\beta\mathbf{k}}) \qquad 5.4.41$$

Expressions 5.4.40 and 5.4.41 are valid also in the case that both the modes f and g are dispersed.It can be seen from 5.4.41 that the contribution from the non-localized second-order moment is essentially given as a weighted two-phonon density of states.As a matter of fact,the results of various calculations [81,156,163,164,79,177] show that in the case of combinations of two internal vibrations the weighting coefficients in 5.4.41 are slowly varying functions of the frequency or of the wave vector.The anharmonic contribution to the non-localized mechanism and the cross terms with the first- and second-order localized moments will not be considered here since in all practical applications described so far they have been found to give a negligible contribution to the absorption coefficient.Expressions for these contributions have been obtained by Bogani[183].

Finally,for sufficiently large anharmonicities it is necessary to express the absorption coefficient for the bound state at frequency ω_B. This can be obtained from the Green function 3.5.19[183].

The theory of two-phonon absorption discussed above has been applied to a number of cases that will now be mentioned separately.

<u>Nitrous oxide.</u> The $\nu_1+\nu_3$, $\nu_2+\nu_3$ and $\nu_1+\nu_2$ bands have been studied. For the first it was reported[156] that the contribution of the non-localized second-order moment according to 5.4.40 can account for most of the observed intensity in the broad feature observed in the spectrum (see Fig.5.2c).In fact,the experimental intensity is 0.5×10^{-3} the intensity of the ν_3 fundamental as compared with a calculated intensity of $0.4 \cdot 10^{-3}$.In contrast,for the $\nu_2+\nu_3$ it is found that the indirect mechanism is responsible for all the observed intensity.Fig. 5.14a shows the harmonic and renormalized two-phonon density of states obtained by using an anharmonicity constant $X_{23} = -13\ cm^{-1}$ which is close to $-14\ cm^{-1}$ found in the gas phase.The calculated and observed transmission spectra are shown in Fig. 5.14b; a good agreement with experiments is obtained with a value of the renormalized second-order moment equal to the value found for the isolated molecule.It can be seen that in the present case a bound state is observed.An example of resonances is in-

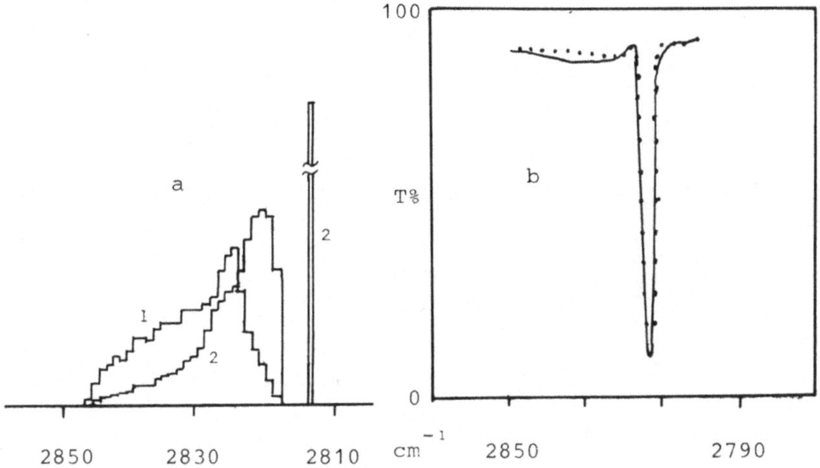

Fig. 5.14 $\nu_2+\nu_3$ combination band in crystalline N_2O. a) Harmonic (1) and renormalized (2) density of states; b) calculated (•••••••••) and observed (————) transmission curves.

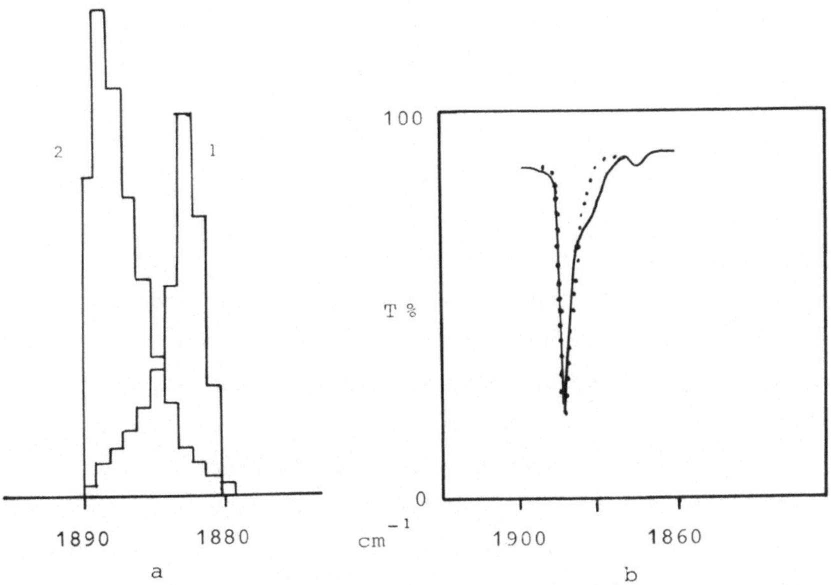

Fig. 5.15 $\nu_1+\nu_2$ combination band of crystalline N_2O. a) Harmonic (1) and renormalized (2) density of states; b) calculated (•••••••••) and observed (————) transmission curves.

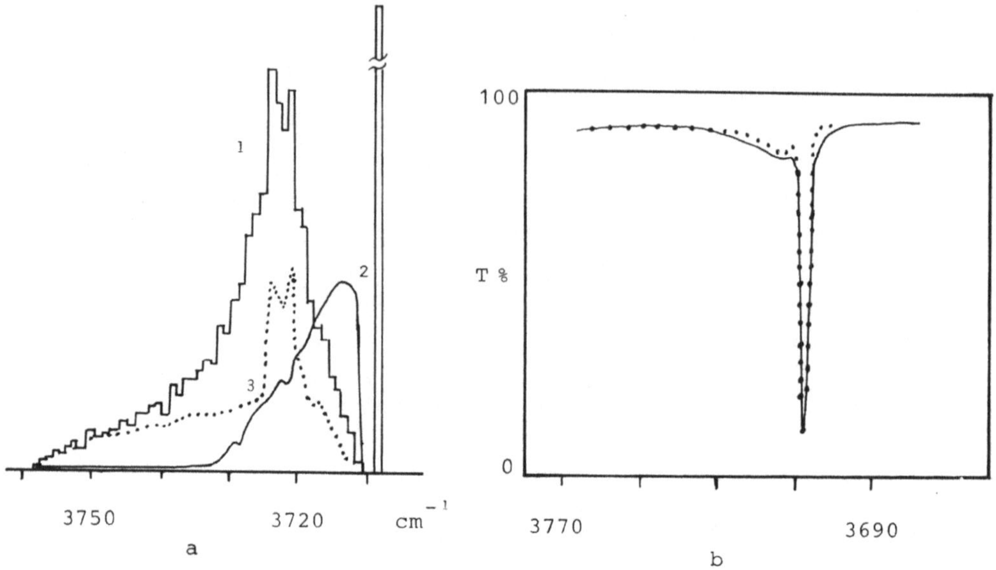

Fig. 5.16 $\nu_1+\nu_3$ combination band in crystalline CO_2. a) Harmonic (1) and renormalized (2) density of states and contribution of the direct mechanism (3); b) calculated (•••••••••) and observed (————) transmission curves.

stead observed in the region of $\nu_1+\nu_2$ .Using an anharmonicity constant X_{12}= - 4 cm^{-1} (against the gas phase value of -5 cm^{-1}),the renormalized density of states of Fig.5.15a is obtained.In Fig. 5.15b the spectrum calculated using the value of the renormalized second-order moment obtained in the gas phase is compared with the observed spectrum.

Carbon dioxide. For the $\nu_1+\nu_3$ combination we report in Fig. 5.16a the harmonic density of states,the contribution of the non-localized direct mechanism,that gives a contribution of 10% of the observed intensity,and the renormalized density of states obtained with X_{13}=-16 cm^{-1} (to be compared with the gas phase value of - 19.66 cm^{-1}).The transmission curve calculated using the gas phase value of the second-order moment is compared with experiments in Fig. 5.16a.An excellent agreement is obtained.

In crystalline CO_2 the $\nu_1+\nu_2$ combination band has also been studied[183] but probably the available density of states is not sufficiently

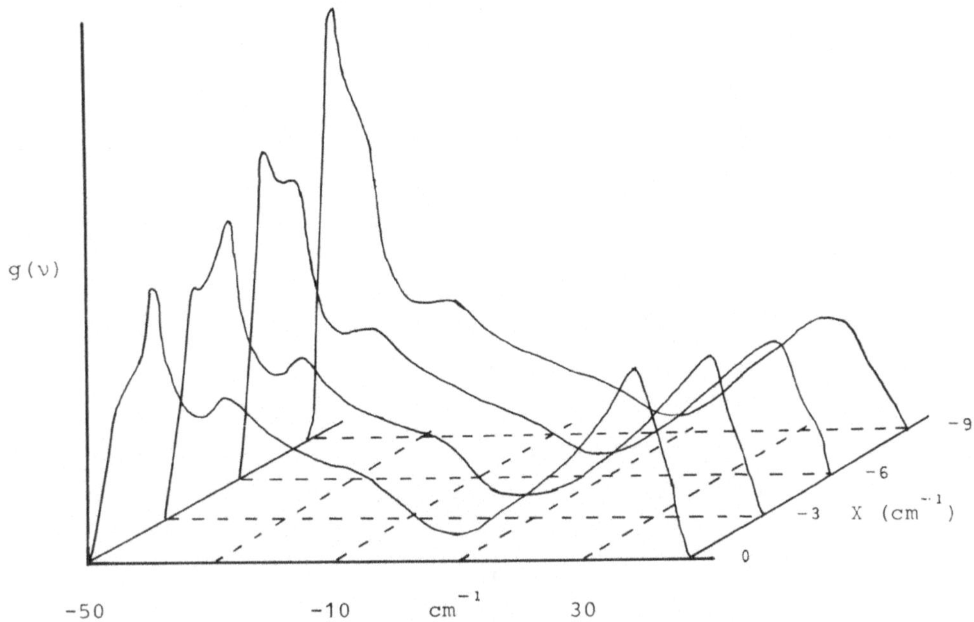

Fig. 5.17 Renormalized density of states for the combination bands in crystal SF_6.

accurate[174] to obtain a quantitative agreement with the observed shape.

Sulfur hexafluoride. The following combination bands have been studied in some detail : $\nu_1+\nu_3$, $\nu_2+\nu_3$, $\nu_1+\nu_4$ and $\nu_5+\nu_4$[160]. It has been assumed

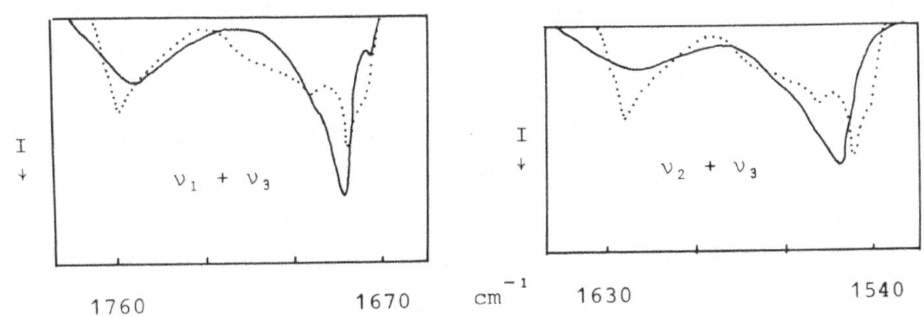

Fig. 5.18 $\nu_1+\nu_3$ and $\nu_2+\nu_3$ combinations bands in crystalline SF_6. ———— observed •••••••• calculated.

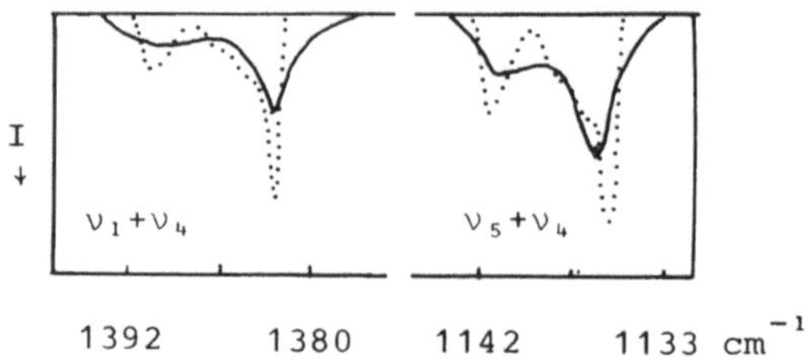

Fig. 5.19 $\nu_1+\nu_4$ and $\nu_5+\nu_4$ combination bands in crystalline SF_6. ———— observed ·········· calculated.

that the ν_1,ν_2 and ν_5 modes are not dispersed in comparison with the ν_3 and ν_4 modes,an approximation that is not completely satisfactory in some cases.The renormalized density of states for various values of the anharmonicity constant is reported in Fig. 5.17 for the ν_3 and ν_4 combinations.It has been found that the contribution of the non-local direct mechanism is small in all the cases considered.In Figs 5.18 and 5.19 the observed spectra are compared with the results of calculation according to the indirect mechanism.A reasonable agreement is obtained. It can be noted,however,that the observed bands extend over a wider region than those calculated.This can be ascribed to the fact that modes ν_1,ν_2 and ν_5 have a proper width that is not negligible.

Silicon tetrafluoride.Silicon tetrafluoride and sulfur hexafluoride have a body-centered crystal structure and,in the dipole approximation, the internal vibrations of both molecules have similar dispersion curves and densities of states.Therefore,apart from a scale factor,the renormalized densities of states of SF_6 (Fig. 5.17) are also appropriate to compare with the combination bands of SiF_4.In fact the $\nu_1+\nu_3,\nu_2+\nu_3$ and $\nu_1+\nu_4$ combinations of SiF_4 [159] can be very satisfactorily explained by the indirect mechanism,using the renormalized density of states obtained for SF_4.

In all the applications discussed above it has been assumed that only one of the modes has a dispersion.Although this is a satisfactory approximation in many cases it cannot be assumed to be generally valid. In particular,in the case of overtone bands of non-centrosymmetric crystals an extension of the theory is necessary.For single occupancy of the unit cell,as is the case with crystalline OCS,the treatment of overtones is straightforward along the lines discussed above[183].In the case of the $2\nu_3$ overtone the anharmonicity constant is small (-11 cm^{-1}) compared to the frequency dispersion of the ν_3 mode and only resonances are found without much distortion of the density of states on renormalization. Using a second-order moment of 0.04 Debye a reasonable fit of the observed band shape is obtained.

An extension of the theory to overtones in the case of double occupancy of the unit cell has been made recently with particular applications to the HCl and HBr crystals[188].It has been shown that it is possible to explain the observed spectrum in the overtone region of HCl and HBr more satisfactorily than in previous attempts of Jortner and Rice[189] and Avrillier et al.[190].

As a conclusion,it has been shown in this Section that the two-phonon absorption involving two internal vibrations in molecular crystals can be satisfactorily accounted for assuming that the crystal dipole moment is expanded to the second-order and that the only higher-order terms in the mechanical potential are due to intramolecular anharmonicity.It is found that in the majority of cases the most important contribution to the absorption coefficient derives from the indirect mechanism where a transverse phonon is excited by the linear electric moment and then decays into two phonons by the intramolecular potential. Using single-molecule gas-phase properties an excellent agreement with the observed band shapes and intensities can be obtained for all the cases that have been studied so far.

5.4b Phonon Side Bands

Also for the two-phonon processes leading to phonon side bands it is possible to conceive a direct mechanism in which the two modes are excited directly through non-linear terms in the electric moment.This mechanism has been considered in detail by Salvi et al.[158] by direct evaluation of the matrix elements of the dipole operator M connecting the ground state to the two-phonon states $|\alpha,\beta;\mathbf{k},-\mathbf{k}\rangle$

$$M_{\alpha\beta}(\mathbf{k}) = \langle 0|\mathbf{M}|\alpha,\beta;\mathbf{k},-\mathbf{k}\rangle \qquad 5.4.42$$

In 5.4.42 α and β are notations for the internal and external vibrations, respectively,whose eigenvectors are known from lattice dynamics calculation.The dipole operator is expanded to second order and only the non-localized terms 5.4.33,due to mutual polarization of the molecular units, are retained in the present case.From 5.4.33 it can be seen that this mechanism can only be effective if the combination involves one infrared- and one Raman-active mode.The intensity due to this direct mechanism is proportional to

$$I(\omega) \propto \sum_{\mathbf{k}\alpha\beta} |M_{\alpha\beta}(\mathbf{k})|^2 \ \delta(\omega - \omega_{\alpha\mathbf{k}} - \omega_{\beta\mathbf{k}}) \qquad 5.4.43$$

For the phonon side bands in carbonyl sulfide it has been found[158] that in the intensity profile calculated according to 5.4.43 many features of the phonon density of states,and particularly those at higher frequencies,do not contribute appreciably.This is in agreement with the experimental observations.However,the calculated integrated intensity is only 3% of the experimental value.Therefore,it is necessary to assume that also in this case the major contribution comes from the indirect mechanism where the internal vibration at $\mathbf{k} = 0$ acts as an intermediate state decaying into the two final phonons.This mechanism involving higher-order intermolecular coupling has been discussed in detail recently by Della Valle et al.[179] and this treatment will be illustrated in the following.

The Hamiltonian of the system composed of the crystal and of the radiation field is written as

$$H = H_o + H_p = H_2 + H_f + H_3 + H_i \qquad 5.4.44$$

where H_o is the unperturbed Hamiltonian including the harmonic part of the crystal Hamiltonian H_2 and the free radiation field Hamiltonian H_f. The perturbation H_p includes the third-order mechanical potential H_3 and the photon-phonon interaction H_i.Defining in the usual way the creation and destruction operators for the phonons $a^\dagger_{j\mathbf{k}}$ and $a_{j\mathbf{k}}$ and for the photons $b^\dagger_{\lambda\mathbf{q}}$ and $b_{\lambda\mathbf{q}}$ the various terms of the Hamiltonian 5.4.44 can be expressed as

$$H_2 = \sum_{j\mathbf{k}} \hbar\omega_{j\mathbf{k}} (a^{\dagger}_{j\mathbf{k}} a_{j\mathbf{k}} + \tfrac{1}{2}) \qquad 5.4.45$$

$$H_f = \sum_{\lambda\mathbf{q}} \hbar\omega_{\mathbf{q}} b^{\dagger}_{\lambda\mathbf{q}} b_{\lambda\mathbf{q}} \qquad 5.4.46$$

$$H_3 = \sum_{\mathbf{k}\mathbf{k}'\mathbf{k}''jj'j''} B\binom{jj'j''}{\mathbf{k}\mathbf{k}'\mathbf{k}''} (a_{j\mathbf{k}} + a^{\dagger}_{j-\mathbf{k}}) (a_{j'\mathbf{k}'} + a^{\dagger}_{j'-\mathbf{k}'}) (a_{j''\mathbf{k}''} + a^{\dagger}_{j''-\mathbf{k}''}) \qquad 5.4.47$$

$$H_i = -\mathbf{M}\cdot\mathbf{E} \qquad 5.4.48$$

In 5.4.45 to 5.4.48 j denotes the phonon branch, $\mathbf{q}$ and λ are the photon wave vector and polarization, $\mathbf{M}$ is the crystal dipole moment and $\mathbf{E}$ the electric field.

The third-order coupling coefficients B appearing in 5.4.47 have been defined in 4.3.12 and the simplifying features introduced by the invariance and symmetry requirements and by the assumption of two-body interactions have been discussed in the preceding Chapter. The practical calculation of the coupling coefficients using an atom-atom and multipole potential has been discussed by Della Valle et al.[179]. Since in the case of phonon side bands the intermediate phonon must be infrared-active and the basic conservation relation 5.2.2a must be obeyed, the only coefficients that are actually needed are of the type $B\binom{jj'j''}{0\,-\mathbf{k}\mathbf{k}}$.

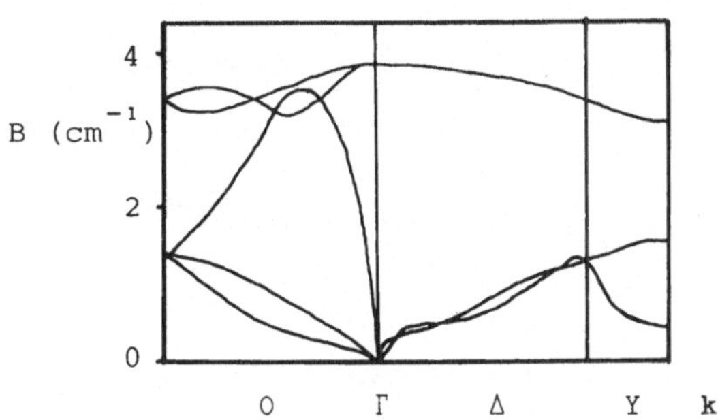

Fig. 5.20 Third order coupling coefficient for the ν_2 side band in crystalline OCS as a function of the wave vector of the low-frequency phonon.

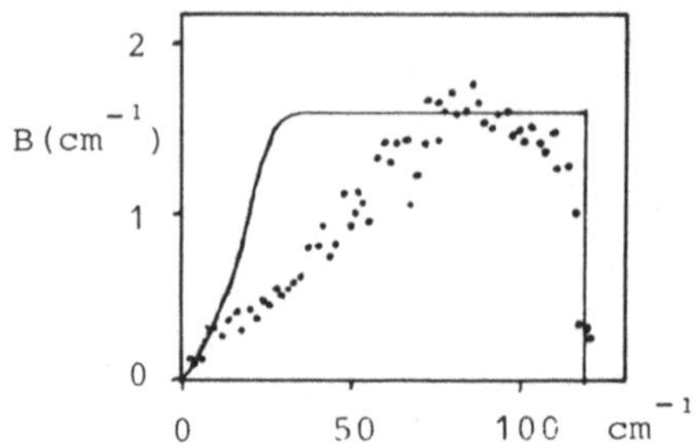

Fig. 5.21 Average third-order coupling coefficient for the ν_2 side band in crystalline OCS as a function of phonon frequency. ••••••••• : calculated from atom-atom potentials; ———— :empirical parameter used in Ref. 158.

The third-order coupling coefficients of interest for the phonon side bands of the ν_2 internal mode of crystalline OCS are shown in Fig. 5.20 for various branches of the lattice mode $\mathbf{k}''j''$. It can be seen that the coefficients strongly depend on the branch and on the wave vector and for acoustic branches approach zero as $\mathbf{k} \to 0$. It is also possible to calculate a third-order coupling coefficient as a function of the frequency of the lattice mode averaged over the Brillouin zone. This average coupling coefficient in the case of the ν_2 mode of crystalline OCS is shown in Fig. 5.21 and is compared with the empirical parameter used in a previous study by Salvi et al.[158] to obtain the best agreement with the experimental band shape.

An explicit expression for the photon-phonon interaction is obtained by expanding the crystal dipole moment in the crystal normal coordinates to the first order. In the second quantization representation the expansion is given by

$$\mathbf{M} = L^{-\frac{1}{2}} \sum_{j\mathbf{k}} \left(\frac{\hbar}{2\omega_{j\mathbf{k}}}\right)^{\frac{1}{2}} (a^{\dagger}_{j-\mathbf{k}} + a_{j\mathbf{k}}) \sum_{n\mu l} M_{l}^{n\mu}\, e(\mu l|j\mathbf{k})\, e^{i\mathbf{k}\cdot\mathbf{n}} \qquad 5.4.49$$

where $M_{l}^{n\mu}$ are expansion coefficients in the local normal coordinates of the type defined in 5.3.9. The photon-phonon interaction is then obtained by introducing 5.4.49 into 5.4.48 and using the expression for the electric field in terms of photon creation and destruction operators

$$H_i = L^{-\frac{1}{2}} i\hbar \sum_{j\mathbf{k}\lambda\mathbf{q}} \sum_{n\mu\lambda} \left(\frac{\pi\omega_q}{V\varepsilon\omega_{j\mathbf{k}}}\right)^{\frac{1}{2}} \upsilon_{\lambda\mathbf{q}}\, e(\mu l|j\mathbf{k})\,(a_{j\mathbf{k}} + a^{\dagger}_{j-\mathbf{k}})\,(-b_{\lambda\mathbf{q}} + b^{\dagger}_{\lambda-\mathbf{q}}) \times$$
$$\times\, M_l^{n\mu} \exp[i(\mathbf{q}+\mathbf{k})\cdot\mathbf{n}] \qquad 5.4.50$$

where $\upsilon_{\lambda q}$ is the photon polarization vector,V the volume of the crystal and ε the dielectric constant.Since $M_l^{n\mu}$ is independent of the unit cell, the second part of 5.4.50 implies that $\mathbf{k} + \mathbf{q} = \mathbf{G}$ where $\mathbf{G}$ is a reciprocal lattice vector.This shows that the momentum conservation must be obeyed also in the intermediate processes.

On the basis of the second-order perturbation theory the rate of transition from an initial state $|i\rangle$ to a final two-phonon state $|f\rangle$ is given by

$$W_{f\leftarrow i} = \frac{2\pi}{\hbar} \sum_m \left| \frac{\langle f|H_p|m\rangle\langle m|H_p|i\rangle}{E_i - E_m} \right|^2 \delta(E_i - E_f) \qquad 5.4.51$$

where $|m\rangle$ is an appropriate intermediate state which,in the present case is the internal vibration at $\mathbf{k} = 0$. Since the perturbation Hamiltonian contains two terms it is possible to separate 5.4.51 into two undistinguishible pathways

$$W_{f\leftarrow i} = \frac{2\pi}{\hbar} \sum_m \left| \frac{\langle f|H_3|m\rangle\langle m|H_i|i\rangle}{E_i - E_m} \quad \frac{\langle f|H_i|m\rangle\langle m|H_3|i\rangle}{E_i - E_m} \right|^2 \delta(E_i - E_f) \qquad 5.4.52$$

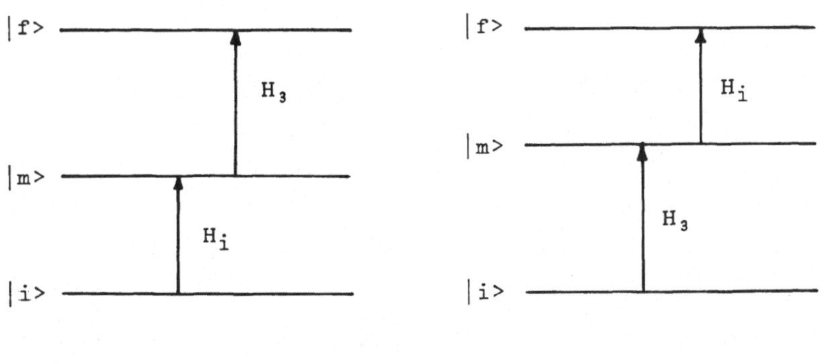

Fig. 5.22 Pathways for two-phonon excitations according to the second-order perturbation theory

The two routes leading to the two-phonon excitation are schematically depicted in Fig. 5.22.

Considering the properties of the creation and destruction operators the matrix elements appearing in 5.4.52 can be easily evaluated. If $\bar{n}_{j\mathbf{k}}$ denotes the occupation number 4.3.24 and we define the quantity

$$P(jk\lambda\mathbf{q}) = -i\hbar\sum_{\mu l}\Delta(\mathbf{k}+\mathbf{q})\left(\frac{\pi\omega_q L}{\varepsilon\omega_{j\mathbf{k}}V}\right)^{\frac{1}{2}}M_l^{n\mu}e(\mu l|j\mathbf{k})\upsilon_{\lambda\mathbf{q}} \qquad 5.4.53$$

The absorption rate for the sum band is given by

$$W_{sum} = \frac{2\pi}{h}\sum_{j'\mathbf{k}'j''\mathbf{k}''}\delta(h\omega-h\omega_{j'\mathbf{k}'}-h\omega_{j''\mathbf{k}''})(1+\bar{n}_{j'\mathbf{k}'}+\bar{n}_{j''\mathbf{k}''})\times$$

$$\times\left|\sum_{j\mathbf{k}}3B\binom{j\,j'\,j''}{\mathbf{k}\,\mathbf{k}'\,\mathbf{k}''}P(jk\lambda\mathbf{q})\frac{1}{\hbar}\left(\frac{1}{(\omega_{j\mathbf{k}}+\omega_{j'\mathbf{k}'}+\omega_{j''\mathbf{k}''})}+\frac{1}{(-\omega_{j\mathbf{k}}+\omega_{j'\mathbf{k}'}+\omega_{j''\mathbf{k}''})}\right)\right|^2 \qquad 5.4.54$$

and for the difference band by

$$W_{dif} = \frac{2\pi}{\hbar}\sum_{j'\mathbf{k}'j''\mathbf{k}''}\delta(h\omega-h\omega_{j'\mathbf{k}'}+h\omega_{j''\mathbf{k}''})(\bar{n}_{j''\mathbf{k}''}-\bar{n}_{j'\mathbf{k}'})\times$$

$$\times\left|\sum_{j\mathbf{k}}3B\binom{j\,j'\,j''}{\mathbf{k}\,\mathbf{k}'\,\mathbf{k}''}P(jk\lambda\mathbf{q})\left\{\frac{1}{\hbar(\omega_{j\mathbf{k}}+\omega_{j'\mathbf{k}'}-\omega_{j''\mathbf{k}''})}+\frac{1}{\hbar(-\omega_{j\mathbf{k}}+\omega_{j'\mathbf{k}'}-\omega_{j''\mathbf{k}''})}\right\}\right|^2 \qquad 5.4.55$$

In 5.4.54 and 5.4.55 the contributions due to stimulated absorption and emission processes have been included.The absorption coefficients are obtained by dividing the transition rate by the velocity of light in the medium.

The calculated and observed transmission curves in the ν_2 region of carbonyl sulfide are compared in Fig. 5.23.It can be seen that although it has not been possible to obtain a complete fit to the experimental spectrum the main features are reproduced correctly.Irregularitie in the calculated band shape can be ascribed to an insufficiently dense sampling in the Brillouin zone.The incomplete agreement with the observed band shape can be due to the various approximations introduced in the treatment or,more likely,to the use of an intermolecular atom-atom potential that is not completely adequate for the description of the car-

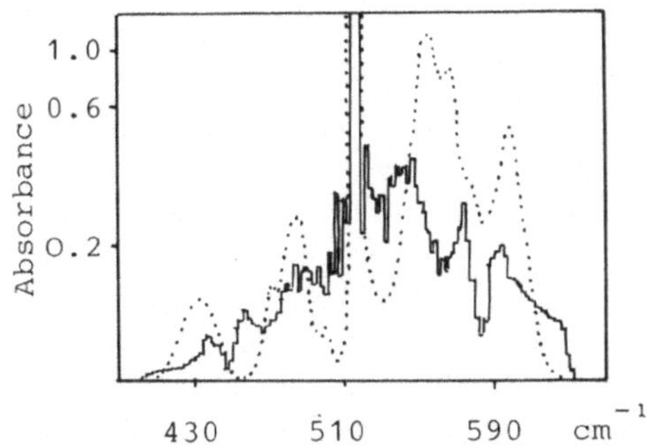

Fig. 5.23 Calculated (full line) and observed (dotted line) infrared spectra in the ν_2 region of crystalline OCS.

bonyl sulfide crystal.From the study of phonon side bands in OCS the following conclusions can be drawn:

a) the atom-atom potential can account for most of the anharmonic effects in a molecular crystal;

b) the structure of the phonon side bands may differ greatly from the unweighted density of states.This is due to the strong dependence of the third order coupling coefficients on the phonon wave vector.

In the particular case of phonon side bands in carbonyl sulfide it has been found that the two-phonon absorption mainly results from the anharmonicity in the mechanical potential.However,this cannot be taken as a general conclusion as is clearly demonstrated by the study of the Raman spectrum of α-N_2[161] indicating that the intensity of the phonon side bands can be accounted for by the non-linear terms in the crystal polarizability.At present it seems that more extensive investigations are necessary before a complete understanding of the phonon side bands is obtained.

C H A P T E R 6

INFRARED AND RAMAN INTENSITIES IN MOLECULAR CRYSTALS

6.1 INTRODUCTION

In the majority of the experimental and theoretical investigations of the vibrational properties of molecular crystals the attention has been concentrated on the frequencies of the vibrational transitions.The principal aim has been to establish a correlation between the frequency shifts and splittings of the internal vibrations or the frequencies of the external vibrations and the intermolecular forces being effective in the crystal under consideration.These attempts have been described at length in most of the previous Chapters of this book.There are,however, several other vibrational properties that can be of interest in connection with the study of intermolecular forces.In this Chapter the problem of vibrational intensities in the infrared and Raman spectra of molecular crystals will be discussed.

At first sight,it may be surprising that the vibrational intensities in molecular crystals have received comparatively little attention up to now.As a matter of fact,limiting for the moment the attention to the internal vibrations,it can be noted that the frequency shifts and the frequency splittings are in general very small when compared with the molecular frequencies.In contrast,the infrared and Raman intensities may change on condensation by a factor of 2 or 3 in many cases,and even by an order of magnitude in some others like,for instance,in hydrogen-bonded systems.As an example,in the weakly hydrogen-bonded hydrogen chloride crystal a gas-to-crystal frequency shift of the order of 5% of the molecular frequency has been observed (a particularly large shift for a molecular crystal)[191] while the infrared intensity in the crystal is ∿10 times larger than in the gas phase[192].There are,however,various reasons why the study of vibrational intensities in molecular crystals has not progressed as rapidly as the study of crystal frequencies.

This is mainly due to the great experimental difficulties encountered in the measurements of precise infrared and Raman intensities. In the infrared region, it is necessary to work with very thin films for transmission experiments. While it is generally impossible to obtain suitable samples for infrared measurements in the case of single crystals more progress has been made with powders and particularly with vapor deposited samples, using the technique for thickness measurements devised by Hollemberg and Dows[193]. The technical problems for infrared intensity measurements are enhanced in the far infrared region where the external vibrations are observed. In addition, for theoretical considerations the intensities for single crystals in various polarizations are actually needed. These data, however, have been obtained only in very few instances. Because of the technical problems mentioned above, the absolute intensities in the middle or far infrared region have only been measured on a limited number of crystals composed of simple molecules. In many other cases, only data on relative intensities are available. Concerning the Raman intensities, considerable progress has been made with the advent of the laser as a Raman source but again absolute intensities on single crystals in various polarizations are not available except in very few cases. Thus, theoretical models must be limited to the consideration of relative intensities.

A second motivation for the slow progress in the study of vibrational intensities in molecular crystals is connected with the great theoretical difficulties in their treatment. In fact, it is apparent from a fairly large number of investigations of vibrational intensities in liquid and solid state that the small changes of the electronic charge distributions that occur upon condensation can affect quite drastically the intensities. At present, there is no simple method to take these effects into account. In addition to the effect of small changes in the electronic charge distribution, there are electrostatic interactions between the molecular units that may also greatly affect the infrared and Raman intensities. Although in the past, these electrostatic effects were occasionally taken into consideration by various authors, only very recently has a more complete theory of the electrostatic interaction effects on the infrared and Raman intensities been discussed in a unified and compact form. It is important to point out that although many vibrational studies of molecular crystals have only been concerned with relative intensities,these qualitative considerations have played an important role in the formulation of a theory of infrared and Raman intensities and,in turn,in the interpretation of infrared and Raman spectra.

This Chapter first gives a brief historical survey of the connection between the interpretation of the vibrational spectra and the vibrational intensities.Then the available electrostatic theory of the infrared and Raman intensities of both the internal and external vibrations will be discussed considering particularly the manifestation of local field effects.A review of the various applications of the theory for the calculation of infrared and Raman intensities of lattice vibrations will illustrate the present degree of comprehension of the available information.

6.2 HISTORICAL SURVEY

Before discussing in detail the available theory of local field effects on the vibrational intensities in molecular crystals,it may be convenient to briefly review the studies made over the past 20 or 30 years on infrared and Raman intensities.This will clarify the present status of the theoretical comprehension of the subject and the aspects that are still waiting for substantial progress.

Simplifying somehow the problem,it can be said that studies on vibrational intensities in molecular crystals have been concerned with three different aspects :

a) study of the relative intensities of the components of the Davydov splittings (dichroic ratios);
b) study of the intensity changes of internal vibrations on condensation
c) study of the intensity of lattice vibrations.

These various aspects will now be discussed separately.

6.2a Dichroic Ratios

In the early vibrational investigations of molecular crystals that followed the pioneering works of Hornig[194] and of Winston and Halford [195,196] the interest was concentrated on the infrared spectra of single crystals and on the discussion of the directional properties of the infrared absorption.The purpose of these studies was twofold.On the one hand,attempts of correlating the polarization and multiplicity of the infrared (and Raman) bands with the crystal structure have been made. This,in principle,amounts to a straightforward application of the group theoretical methods described in Chapter 2.This type of approach has led to many significant applications intended to deduce the crystal structure from infrared and Raman data,particularly in cases of crystals composed of small molecular units that crystallize at low temperature where X-ray investigations are accomplished with some difficulty.It is not possible to review here the large amount of work developed in this field.The reader is referred to several available reviews[84,197-200].We only want to mention as an example the investigation of the crystal structure of ethylene performed by many authors,and particularly by Dows[201].As mentioned in Chapter 3,early X-ray investigations could not exactly locate the position of the hydrogen atoms and thus did not completely elucidate the structure.Therefore,several attempts to deduce the exact crystal structure from infrared and Raman spectra and also from model calculations[72,202-204] have been made.The inferences made have recently been confirmed by X-ray[205] and neutron[206] scattering investigations.

The second aspect of interest has been the connection of the directional properties of the infrared absorption with the symmetry properties of the vibrations of the isolated molecules.This was of great importance for the vibrational assignments of complicated systems.The basis for the interpretation of the infrared dichroism was the oriented gas model developed by Pimentel and others[207,208] following previous suggestions by Fruhling[209] and Kastler and Rousset[210].

The oriented gas model assumes that the crystal is composed of rigidly oriented molecular units that do not interact with each other.In this limit of vanishing intermolecular interactions the intensity of a vibrational transition along a given crystallographic axis depends on the composition of the projections along the axis of the transition dipole moments of the various molecules in the unit cell.The basis of the model can be simply illustrated using the vibrational exciton theory described in Section 5.3.Consider an isolated infrared-active mode f of the free molecule with transition dipole moment $(\partial\mu/\partial q)_0$ along the direction defined by a unit vector **e**.We are interested in the transitions from the crystal ground state 5.3.1 to the various crystal eigenstates 5.3.5,the infrared intensity being controlled by the square of the dipole moment matrix element at $\mathbf{k} = 0$

$$\mathbf{M}_\alpha = \langle 0|\mathbf{M}|\alpha f,\mathbf{k}\rangle \qquad 6.2.1$$

where **M** is the crystal dipole operator and is taken as the sum of the molecular moments expanded according to 5.3.9.Using 5.3.2 and 5.3.3 one obtains

$$\mathbf{M}_\alpha = \langle 0|\mathbf{M}|\alpha f,0\rangle = (\partial\mu/\partial q)_0 \langle 0|q|1\rangle \textstyle\sum_\nu B_{\nu\alpha}(0)\mathbf{e}_\nu \qquad 6.2.2$$

where $\langle 0|q|1\rangle$ is the usual harmonic oscillator matrix element.It can be seen from 6.2.2 that,if the crystal dipole operator is simply the sum of the isolated molecular dipoles,the crystal transition dipole involves only vectorial addition of the molecular moments.By symmetry requirements the transition dipole moment for the α-th component of the factor group splitting is parallel to a crystal axis.The dichroic ratios are thus given by the ratios of the squares of the quantities 6.2.2 and depend on the molecular orientation and on the eigenvectors.

In many cases,the eigenvectors are completely determined by the symmetry of the unit cell.Then,if the crystal structure is known,it is very simple to figure out the oriented gas model prediction of the dichroic ratios for the free molecule active vibrations.Whenever the orientation of the molecular moment is completely fixed for a given symmetry species,the predicted dichroic ratios are the same for all the vibrations of that species.This property has been the basis for the use of polarized infrared spectra in vibrational assignments.The process can, however,be reversed and observed polarization properties in the infrared spectra can be used to draw conclusions on the unknown crystal structure.

In practice the following circumstances may limit the applicability of the oriented gas model.

a) The oriented gas model simply involves the composition of the dipole moments of the free molecule and the changes of these molecular properties due to small variations in the crystal of the electronic charge distribution are neglected.Therefore,the model is strictly applicable only in the limit of vanishingly small intermolecular interactions.

b) There are several instances where the eigenvectors are not completely determined by symmetry and must be obtained from lattice dynamics calculations.In these cases the eigenvectors depend on the intermolecular potential used.As an example,this occurs for the ν_2 bending mode in crystalline carbon dioxide.CO_2 crystallizes in the cubic T_h^6 space group with four molecules in the unit cell located on S_6 sites[211].The molecular Π_u vibration gives rise in the crystal to eight modes classified as $E_u + 2F_u$ in the T_h factor group.There are therefore two components of the Davydov splitting,belonging to the same irreducible representation (F_u) of the factor group,that are active in the infrared spectrum. In such cases,it is not possible to make any prediction of the relative intensities by symmetry arguments.The high and low frequency infrared components of the ν_2 mode in CO_2 are separated by 5 cm^{-1} with a 1:1 intensity ratio [81].Using a resonant dipole-dipole potential,a splitting of 10 cm^{-1} and an oriented gas intensity ratio of 10:1 have been calculated[174].The discrepancy in the intensity ratio can be due to local field effects or inadequacy of the calculated eigenvectors.However,in the present case it is also found that the intensity change upon condensation,which should be connected to local field effects,is small[212]. Therefore,the deviation from the predictions of the oriented gas model is likely due to inadequacy of the eigenvectors.From another point of view,there are similar difficulties in the application of the oriented gas model to molecules and crystals with low symmetry.For instance,in molecules with C_{2h} symmetry,the direction of the transition dipole moment is uniquely fixed for modes belonging to the A_u species and the oriented gas model predictions can be easily figured out.In contrast,for modes of B_u species,the direction of the transition moment is not defined and symmetry only requires that it lies in the plane perpendicular to the symmetry axis.Therefore,different modes of the same symmetry species may exhibit different dichroic ratios.Some aspects of this problem have been discussed recently by Luty and Rohleder[213].

c) In monoclinic and triclinic systems,a frequency dependence of the polarization occurs.This introduces difficulties into the application of the oriented gas model.This has been discussed by several authors and invoked in many cases as a qualitative explanation of observed deviations from the predictions of the oriented gas model.

d) Extension of the oriented gas model to the Raman spectrum has been made following the work of Suzuki et al.[214] on naphthalene and anthracene single crystals.For the Raman spectrum expression 6.2.2 must be replaced by

$$A^{\alpha}_{ij} = C \sum_{\nu} B_{\nu\alpha}(0)\ \alpha^{\nu}_{ij} \qquad 6.2.3$$

where C is a constant and α^{ν}_{ij} a component of the polarizability of molecule at site ν projected on the crystal reference frame.The dichroic ratios for vibrations that induce a change of a single component of the polarizability tensor are simply governed by the symmetry of the unit cell.If the molecular vibration induces changes of more than one component of the polarizability tensor,the dichroic ratios also depend on the relative magnitude of the components and are thus not the same for all the vibrations of the same symmetry species.

e) The oriented gas model allows the prediction of dichroic ratios for modes that are active in the isolated molecule.It is not possible,on the basis of symmetry considerations,to make any prediction of the polarization properties of the inactive modes of the isolated molecule that become active in the crystal spectrum.

Despite these limitations,there have been many studies of the infrared and Raman spectra in polarized light mainly for the purpose of obtaining vibrational assignments or making inferences on the crystal structure.

6.2b Intensity Changes on Condensation

Absolute infrared intensities of internal vibrations in molecular

Table 6.1 Infrared Intensities of Internal Vibrations in the Gas and in the Solid State

		Gas		Crystal		
		ν_g (cm^{-1})	A_g ($cm.mole^{-1}$)	ν_c (cm^{-1})	A_c ($cm.mole^{-1}$)	A_c/A_g
HCl[a]		2868	3900	2730	24000	6.
HBr[a]		2560	1340	2422	17600	12.
CO[b]		2143	5800	2138	5810	1.
CO_2[c]	ν_2	668	5400	658	7700	1.4
	ν_3	2349	63500	2345	45900	0.7
N_2O[d]	ν_1	1285	5900	1293	4400	0.75
	ν_2	589	820	589	1150	1.4
	ν_3	2224	36600	2235	23800	0.7
CS_2[e]	ν_2	397	5700	394	850	0.15
	ν_3	1523	56700	1504	80000	1.4
OCS[f]	ν_1	859	800	856	740	0.9
	ν_2	521	290	518	760	2.5
	ν_3	2062	59000	2040	68000	1.2
SF_6[g]	ν_3	947	106500	906	104000	1.
	ν_4	615	6900	608	12200	1.8
C_2H_4[h]	ν_{10}	810	53	823	47	0.9
	ν_7	949	7976	945	9503	1.2
	ν_4	1027	115	1039	21	0.2
	ν_{12}	1443	976	1438	1645	1.8
	ν_{11}	2989	1351	2974	699	0.6
	ν_9	3105	2490	3091	751	0.3

a-H.B.Friedrich,W.B.Person:. J.Chem.Phys., 39, 811 (1963)

b-C.J.Jiang,W.B.Person , K.G.Brown: J.Chem.Phys., 62, 1201 (1975)

c-H.Yamada,W.B.Person: J.Chem.Phys., 41, 2478 (1964)

d-H.Yamada,W.B.Person: J.Chem.Phys., 41, 2478 (1964)

e-H.Yamada,W.B.Person: J.Chem.Phys., 40, 309, (1964)

f-H.Yamada,W.B.Person: J.Chem.Phys., 43, 2519, (1965)

g-D.A.Dows,G.M.Wieder: Spectrochim.Acta, 18, 1567, (1962)

h-G.M.Wieder,D.A.Dows: J.Chem.Phys., 37, 2990, (1962)

crystals have not been studied very extensively.However,a sufficiently large number of crystals composed of simple molecules have been investigated and significant comparisons of gas phase and crystal intensities can be made.This problem has been reviewed recently[215] and some of the available data are collected in Table 6.1. It can be seen that the intensity ratio between the solid and the gas phase can vary over a wide range of values.The interpretation of the observed intensity changes on condensation is a very difficult problem.In general,we may invoke three different contributions which are responsible for the observed effects:

a) local field effects;

b) changes in the normal coordinates;

c) electronic effects.

Because of intermolecular interactions,the field actually experienced by the molecules in the crystal is different from the external applied field.There is an additional contribution produced by the multipoles of the surrounding molecules.In crystals with weak interactions the local field effects can entirely account for the small changes occurring in the infrared and Raman intensities.In the simplest approach, these effects can be taken into account by extending the analogous treatment of liquids.The absorbing molecule is considered in a cavity that, in an anisotropic medium,can be assumed to be ellipsoidal.The local field effects can be taken into account more precisely since it is well known how to calculate the effective field at a lattice site of an ordered solid.The role of the local field effects on the infrared and Raman intensities of molecular crystals will be discussed in detail in the next Section for both the internal and external vibrations.

The second source of intensity variation on condensation is the change of normal coordinates that may arise in the solid as a consequence of intermolecular interactions.This,for instance,has been invoked by Dows and Wieder to explain the effect of condensation on the infrared intensities of crystalline SF_6[216] where,as it can be seen from Table 6.1,the intensity of the ν_3 mode decreases and that of ν_4 increases in going from the gas to the crystal.According to Dows and Wieder,this is due to the fact that in the crystal there is a larger mixing of the molecular coordinates giving rise to an intensity borrowing from the more intense ν_3 to the weaker ν_4 transition.Such an explanation would imply that the total intensity within given symmetry species does not change on condensation.This intensity sum rule has been found to be obeyed at least approximately in SF_6 and in many other weakly interacting systems. However,the fact that these intensity sum rules are only approximately obeyed in several cases,while in many others they are not obeyed at all shows that in general intermode mixing is only a minor or concomitant cause of intensity variation.It is evident that mixing of the internal modes besides affecting the absolute intensity of each band,can likewise influence the dichroic ratios in the spectra in polarized light.In fact,the reduction of symmetry that generally occurs in the crystal can mix modes of different symmetry in the isolated molecule.Intermode mixing has thus in many cases been considered as the explanation for the observed deviations from the predictions of the oriented gas model.

As stated in the introduction,the effect on the vibrational intensities of small changes in the crystal of the electronic charge distribution can be very large,since the derivatives of the dipole moment and of the polarizability are rather sensitive to the charge distribution. For instance,in hydrogen-bonded systems the change in the electron distribution is likely to be one of the main reasons for the relevant variations that the condensation induces in the infrared spectra.In the absence of quantum mechanical calculations it is not possible to discuss the observed affects even qualitatively or for specific systems.

6.2c Infrared and Raman Intensities of Lattice Vibrations

As it has been discussed in Chapter 1 the appearance in the low frequency regions of bands arising from the translatory or rotatory motion of the molecules is one of the characteristic features of the spectra of molecular crystals.When molecules have an intrinsic electric dipole moment or an anisotropic polarizability,the librational motions may acquire intensity by the same mechanism responsible for the rotational infrared and Raman spectra in the gas phase.The infrared and Raman intensities of lattice vibrations have been measured for a certain number of crystals,mainly composed of simple molecules.These data and the early theoretical treatments have been reviewed recently[84,217].Consideration of the available data leads to the following conclusions.

a) In many cases the relative intensities of librational modes do not conform to the expectations for intensity arising only from reorientation of permanent dipole moments or of anisotropic polarizabilities.

b) Lattice bands are observed also for crystals composed of non-polar molecules or of molecules with isotropic polarizability.

c) Modes of prevailing translational character can in several cases exhibit an intensity comparable with that of the librational modes.

These observations clearly show that the simple model that ascribes the lattice modes intensity to molecular reorientation with respect to the crystal axes is unsatisfactory.It has been suggested that in the infrared spectra the observed effects can be due to modulation of the electric moments induced by the molecular multipoles.In the same way because of the electrostatic interactions the effective polarizability of relevance for the Raman scattering is different from the sum of the free molecule polarizablities.Deviations from the oriented gas model are thus expected. An unified theory of electrostatic effects on the vibrational intensities in crystals has been discussed recently by Schettino and Califano[97] and by Luty and others[218,219].This theory will be described in detail in the next Section together with various applications and early experimental studies.The electrostatic model can only account for the local field effects discussed above but is sufficiently flexible for application to external and internal vibrations.

6.3 ELECTROSTATIC MODEL OF INFRARED AND RAMAN INTENSITIES IN MOLECULAR CRYSTALS

The simplest approach to vibrational intensities in molecular crystals is based on the assumption that the crystal dipole moment $\mathbf{P}$ and the crystal polarizability $\mathbf{A}$ are the sum of the corresponding properties of the molecules in the crystal

$$\mathbf{P} = \sum_{n\nu} \mathbf{m}^{n\nu} \qquad 6.3.1$$

and

$$\mathbf{A} = \sum_{n\nu} \alpha^{n\nu} \qquad 6.3.2$$

As shown in 3.3.28 the electric moment and polarizability of molecule $n\nu$ in the crystal frame are expressed in terms of the quantities in the molecular frame through a direction cosine matrix $\Gamma^{n\nu}$

$$\mathbf{m}^{n\nu} = \Gamma^{n\nu}\mathbf{m} \qquad 6.3.3$$

and

$$\alpha^{n\nu} = \Gamma^{n\nu}\ \alpha\ \tilde{\Gamma}^{n\nu} \qquad 6.3.4$$

The infrared and Raman intensities associated with the crystal normal coordinate Q_ℓ depend,respectively,on

$$I_{ir} \propto (\partial P/\partial Q_\ell)^2 \qquad 6.3.5$$

and

$$I_{raman} \propto (\partial A/\partial Q_\ell)^2 \qquad 6.3.6$$

In turn,as shown in Chapter 1,the normal coordinate Q_ℓ can be expressed in terms of some local coordinates.Therefore,in order to obtain the infrared and Raman intensities it is necessary to evaluate the derivatives of the operators 6.3.1 and 6.3.2 with respect to the local coordinates. Concerning the external vibrations,these operators are independent of the translational coordinates,as it is seen from 6.3.3 and 6.3.4,and depend only on the rotational coordinates through the direction cosines matrices.For the internal vibrations,only the molecular moment $\mathbf{m}$ and polarizability α depend on the local coordinates.This approach is the basis for the already discussed oriented gas model that completely neglects the interactions between the molecules in the crystal.Since the oriented gas model is a particular case of the more general model,it will not be further considered as such.

Let us consider a crystal embedded in an externally applied field Ξ.The instantaneous dipole moment at site $n\nu$ is given by

$$\mathbf{P}^{n\nu} = \mathbf{m}^{n\nu} + \alpha^{n\nu}\mathbf{F}^{n\nu} \qquad 6.3.7$$

where $\mathbf{F}^{n\nu}$ is the local field at site $n\nu$.The local field includes the external field Ξ and the field produced by the other dipoles and multipoles in the crystal.Taking into account only the contributions due to the molecular dipoles and quadrupoles we have,using B and B' as cumulative indices for $n\nu$ and $n'\nu'$,

$$\mathbf{F}^{n\nu} = \Xi + \sum_{B'} T_2^{BB'} \mathbf{P}^{B'} + \frac{1}{3} \sum_{B'} T_3^{BB'} \Theta^{B'} \qquad 6.3.8$$

The tensors T_2 and T_3 have already been defined in 3.3.25.Introducing 6.3.8 into 6.3.7 we obtain

$$\mathbf{P}^{n\nu} = \mathbf{m}^{n\nu} + \sum_{B'} \alpha_B T_2^{BB'} \mathbf{P}^{B'} + \frac{1}{3} \sum_{B'} \alpha_B \, T_3^{BB'} \Theta^{B'} + \alpha^{n\nu}\Xi \qquad 6.3.9$$

and the crystal dipole operator is given by

$$\mathbf{P} = \sum_{n\nu} \mathbf{P}^{n\nu} \qquad 6.3.10$$

It is convenient to express 6.3.8 and 6.3.9 in a more compact form by performing the lattice sums at $\mathbf{k} = 0$,as described in Section 3.5,and defining the dipole operator for the unit cell as a vector with 3Z components.Defining the unit cell quantities

$$\mathrm{m} = \begin{pmatrix} m^1 \\ m^2 \\ \cdot \\ \cdot \\ \cdot \end{pmatrix} \qquad \alpha = \begin{pmatrix} \alpha^1 & & & 0 \\ & \alpha^2 & & \\ & & \cdot & \\ 0 & & & \cdot \end{pmatrix} \qquad S = \begin{pmatrix} S^{11} & S^{12} & \cdot & \cdot \\ S^{21} & S^{22} & \cdot & \cdot \\ \cdot & \cdot & \cdot & \cdot \\ \cdot & \cdot & \cdot & \cdot \\ \cdot & \cdot & \cdot & \cdot \end{pmatrix} \qquad 6.3.11$$

the expression 6.3.9 can be rewritten for the unit cell as

$$\mathbf{P} = \mathrm{m} + \alpha \mathbf{S}_2 \mathbf{P} + \frac{1}{3} \alpha \mathbf{S}_3 \Theta + \alpha \Xi \qquad 6.3.12$$

and rearranged as

$$P = (E - \alpha S_2)^{-1}(m + \frac{1}{3}\alpha S_3\Theta) + (E - \alpha S_2)^{-1}\alpha\, \mathcal{E} \qquad 6.3.13$$

where E is the unit matrix.The first part of the right-hand side of 6.3.16 is the operator for the infrared intensity;the second term allows an effective polarizability of the crystal to be defined

$$A = (E - \alpha S_2)^{-1}\,\alpha \qquad 6.3.14$$

which is the operator responsible for the Raman intensity.If we assume that the intermolecular interactions are vanishingly small (i.e. the matrices S_2 and S_3 are zero) the oriented gas model approximation is recovered.

In Ref. 97 expression 6.3.13 has been used to calculate the infrared and Raman intensities of lattice vibrations but the expansion

$$(E - \alpha S_2)^{-1} = E + \alpha S_2 + \alpha S_2\alpha S_2 + \cdots\cdots \qquad 6.3.15$$

was used and only the first two terms were retained in the computations. In Refs.207 and 208 the full expression was used.The limits for a significant expansion of matrices of the type 6.3.15 are that each individual element of the matrix αS_2 is less than unity[220].In the present context this is the case with not too large molecular polarizabilities;however, for very polarizable molecules the expansion 6.3.15 is not applicable. The differences when using the full expression 6.3.13 or only the first two terms in the expansion 6.3.15 will be discussed for a particular example in the next Section.

In the operator 6.3.13 the unit cell quantities m,α and Θ depend on the internal coordinates and also,through the direction cosines,on the rotational coordinates while the lattice sums S_2 and S_3 are a function of the translational coordinates.Therefore,on the basis of the expression 6.3.13,it is seen that:

a) the intensity of the internal vibrations in the crystal can change as a consequence of the mutual polarization of the molecules.This,in turn,can lead to deviations of the dichroic ratios from the expectations of the oriented gas model.

b) The translational modes can have an infrared and Raman intensity.

c) The intensity of librational modes may not conform to the predictions of the oriented gas model.

d) Non-polar molecules may exhibit an infrared intensity in the lattice region if they have a non-zero quadrupole (or higher order multipole) moment.

e) Spherical molecules can have a Raman lattice spectrum due to the translational modes.

Let us first consider the case of lattice vibrations. In order to obtain the infrared and Raman intensities it is necessary to evaluate the derivatives of the relevant operator with respect to the local rotational and translational coordinates. Using the mass-weighted rotational and translational coordinates defined in 1.6.9 and recalling the definitions given in Chapter 3 we have

$$\frac{\partial S^{\mu\nu}_{\alpha\beta}}{\partial t^{n\nu}_{\lambda}} = - M^{-\frac{1}{2}} \sum_{\gamma} S^{\mu\nu}_{\alpha\beta\gamma} \Lambda^{\mu}_{\gamma\lambda} \qquad 6.3.16$$

$$\frac{\partial S^{\mu\nu}_{\alpha\beta\gamma}}{\partial t^{n\nu}_{\lambda}} = - M^{-\frac{1}{2}} \sum_{\delta} S^{\mu\nu}_{\alpha\beta\gamma\delta} \Lambda^{\mu}_{\delta\lambda} \qquad 6.3.17$$

$$\frac{\partial \Gamma^{n\nu}_{\alpha\rho}}{\partial r^{n\nu}_{\lambda}} = I_{\lambda}^{-\frac{1}{2}} \sum_{\tau} \Lambda^{\nu}_{\alpha\tau} \delta_{\tau\rho\lambda} \qquad 6.3.18$$

where $\delta_{\tau\rho\lambda}$ is the Levi-Civita symbol. On the basis of these properties the derivatives of the crystal polarizability with respect to a translational coordinate at site ν assumes the form

$$\frac{\partial A}{\partial t^{\nu}} = (E - \alpha S_2)^{-1} \alpha S_2' \alpha \widetilde{(E - \alpha S_2)}^{-1} \qquad 6.3.19$$

and with respect to a rotational coordinate

$$\frac{\partial A}{\partial r^{\nu}} = (E - \alpha S_2)^{-1} \alpha' - (E - \alpha S_2)^{-1} \alpha' S_2 \alpha \widetilde{(E - \alpha S_2)}^{-1} \qquad 6.3.20$$

In 6.3.19 and 6.3.20 S_2' is defined in 6.3.16 and α' is a block diagonal matrix containing a single non-zero 3×3 block with elements

$$\frac{\partial \alpha^{\nu}}{\partial r_{\lambda}^{\nu}} = I_{\lambda}^{-\frac{1}{2}} \sum_{\tau\rho} \alpha_{\tau\tau} (\Lambda_{\alpha\rho}^{\nu} \Lambda_{\beta\tau}^{\nu} + \Lambda_{\alpha\tau}^{\nu} \Lambda_{\beta\rho}^{\nu}) \delta_{\rho\tau\lambda} \qquad 6.3.21$$

For clarity, we may consider the particular case of a crystal with two molecules per unit cell and the translation along the λ molecular axis of the molecule at site 1. By indicating by V the matrix $(E - \alpha S_2)^{-1}$ the expression 6.3.19 is explicitly given by

$$\frac{\partial A}{\partial t_{\lambda}^{1}} = \begin{pmatrix} V^{11} & V^{12} \\ V^{21} & V^{22} \end{pmatrix} \begin{pmatrix} \alpha^{1} & \\ & \alpha^{2} \end{pmatrix} \begin{pmatrix} S'^{11} & S'^{11} \\ S'^{21} & S'^{22} \end{pmatrix} \begin{pmatrix} \alpha^{1} & \\ & \alpha^{2} \end{pmatrix} \begin{pmatrix} \tilde{V}^{11} & \tilde{V}^{21} \\ \tilde{V}^{12} & \tilde{V}^{22} \end{pmatrix} \qquad 6.3.22$$

Similarly for a rotational coordinate we obtain

$$\frac{\partial A}{\partial r_{\lambda}^{1}} = \begin{pmatrix} V^{11} & V^{12} \\ V^{21} & V^{22} \end{pmatrix} \begin{pmatrix} \alpha'^{1} & \\ & 0 \end{pmatrix} + \begin{pmatrix} V^{11} & V^{12} \\ V^{21} & V^{22} \end{pmatrix} \begin{pmatrix} \alpha'^{1} & \\ & 0 \end{pmatrix} \begin{pmatrix} S^{11} & S^{12} \\ S^{21} & S^{22} \end{pmatrix} \begin{pmatrix} \alpha^{1} & \\ & \alpha^{2} \end{pmatrix} \begin{pmatrix} \tilde{V}^{11} & \tilde{V}^{21} \\ \tilde{V}^{12} & \tilde{V}^{22} \end{pmatrix} \qquad 6.3.23$$

where the matrix α'^{1} is defined in 6.3.21.

If the matrix V is expanded and only the leading terms are retained, expressions 2-19 and A-12 and A-13 of Ref. 97 are obtained. Considering the properties

$$(E - \alpha S_2)^{-1} \alpha = \alpha (E - S_2 \alpha)^{-1} \qquad 6.3.24$$

and

$$\widetilde{(E - \alpha S_2)}^{-1} = (E - S_2 \alpha)^{-1} \qquad 6.3.25$$

which are easily obtained from expansion 6.3.15, expression 6.3.20 can be rewritten as

$$(\partial A/\partial r_{\lambda}^{\mu}) = (E - \alpha S_2)^{-1} \alpha \ \widetilde{(E - \alpha S_2)}^{-1} \qquad 6.3.26$$

which is equivalent to Eq. 18 of Ref.218.Similarly expression 6.3.19 is equivalent to Eq. 23 of Ref.218.

With a completely analogous procedure it is possible to express the derivatives of the crystal dipole operator with respect to the translational coordinates

$$\frac{\partial P}{\partial t^{\mu}_{\lambda}} = (E - \alpha S_2)^{-1}(\alpha S_2^{!} P + \frac{1}{3}\alpha S_2^{!}\Theta) \qquad 6.3.27$$

and to the rotational coordinates

$$\frac{\partial P}{\partial r^{\mu}_{\lambda}} = (E - \alpha S_2)^{-1}\left[m' + \frac{1}{3}(\alpha' S_3\Theta + \alpha S_3\Theta')\right] + (E - \alpha S_2)^{-1}\alpha' S_2 P \qquad 6.3.28$$

where m' and Θ' are obtained with the same procedure leading to 5.3.22. Again, in Ref. 97 only the leading terms of 6.3.27 and 6.3.28 were considered.The explicit form of 6.3.27 and 6.3.28 in the case of multiple occupancy of the unit cell can be found by analogy with 6.3.22 and 6.3.23

Concerning the <u>internal vibrations</u> it is necessary to consider the change in the molecular parameters m,α and Θ with respect to the internal coordinates .These quantities can be obtained for the isolated molecule from the measurement of the infrared and Raman intensities or can be estimated by theoretical calculations.The expression of the dipole moment derivative with respect to a normal coordinate on molecule ν is then identical with expression 6.3.28 once the different nature of the quantities m',α' and Θ' is taken into account.Similarly,the derivative of the crystal polarizability with respect to intramolecular normal coordinates assumes the form 6.3.26.

6.3a Applications

In the infrared region the important experimental quantity is the integrated absorption coefficient

$$\varkappa = (1/dl)\int_{band} \ln(I_0/I)d\nu \qquad 6.3.29$$

where d and l are the sample density and thickness, respectively. The absorption coefficient is connected to the theoretical quantities obtained above by the relation

$$\varkappa = (N\pi/3c^2)\ (\partial P/\partial Q_\ell)^2 \qquad 6.3.30$$

where N is Avogadro's number and c the velocity of light. Relation 6.3.30 is appropriate for isotropic irradiation and contains an averaging over molecular orientations. In the case of spectra of single crystals in polarized light is to be substituted by

$$\varkappa = (N\pi/c^2)\ |(\partial P/\partial Q_\ell)_\alpha|^2 \qquad 6.3.31$$

where α denotes the component of the transition moment of relevance for the experimental arrangement. As already noted in the introduction the experimental difficulties in the measurements of absolute infrared intensities allow in many cases only a comparison with observed relative intensities to be made.

For the comparison of theoretical and experimental Raman intensities, we refer to the usual right angle scattering geometry in which the crystal is irradiated along the crystallographic α direction with a laser beam polarized along the β direction. If we observe the radiation scattered along γ and polarized along α, the scattered intensity I is related to the exciting intensity I_0 by

$$I = \frac{d\sigma}{d\Omega} I_0 \qquad 6.3.32$$

where the differential scattering cross section per molecule is given by

$$\frac{d\sigma}{d\Omega} = \frac{2\pi^2}{Z\,L}\ \frac{h(\nu_0 - \nu_m)^4}{\nu_m\left[1 - \exp(-h\nu_m/kT)\right]}\ A'^{\,2}_{\alpha\beta} \qquad 6.3.33$$

where ν_0 is the frequency of the exciting radiation, ν_m the Raman shift, T the absolute temperature and $A'_{\alpha\beta} = (\partial A/\partial Q_\ell)_{\alpha\beta}$ a component of the polarizability derivative discussed above.

In many cases, Raman measurements in molecular crystals are not performed on single crystals but on polycrystalline samples obtained, for instance, by vapor deposition. In such cases, assuming a random orientation of the crystallites, it is necessary to average over the crystal orientations. Proceeding as for a liquid sample[11], we define the average value of the scattering tensor

$$A' = (1/3)\sum_\alpha A'_{\alpha\alpha} \qquad 6.3.34$$

and its anisotropic part

$$\Gamma'^2 = \frac{1}{2}\left[\sum_{\alpha\neq\beta} (A'_{\alpha\alpha} - A'_{\beta\beta})^2 + 6\sum_{\alpha<\beta} A'^2_{\alpha\beta}\right] \qquad 6.3.35$$

Using these quantities, the total scattering cross section for perpendicular observation[11] is given by

$$\left(\frac{d\sigma}{d\Omega}\right)_\perp = \frac{2\pi^2}{ZL}\,\frac{h(\nu_0 - \nu_m)^4}{\nu_m[1 - \exp(-h\nu_m/kT)]}\;\frac{45A'^2 + 7\Gamma'^2}{45} \qquad 6.3.36$$

and for parallel observation

$$\left(\frac{d\sigma}{d\Omega}\right)_\parallel = \frac{2\pi^2}{ZL}\,\frac{h(\nu_0 - \nu_m)^4}{\nu_m[1 - \exp(-h\nu_m/kT)]}\;\frac{6\Gamma'^2}{45} \qquad 6.3.37$$

Expressions 6.3.36 and 6.3.37 do not take into account the effect of light propagation and depolarization within the crystalline sample and at the surfaces. To account for this it has been suggested to average between $(d\sigma/d\Omega)_\perp$ and $(d\sigma/d\Omega)_\perp + (d\sigma/d\Omega)_\parallel$. However, it has been found that calculated intensities do not change much when using expression 6.3.36 or averages of the type discussed above[97]. The problem of crystal optics in relation to Raman intensities has recently been considered by Munn et al.[221] who have likewise found by model calculations that the effect of crystal optics is small.

<u>Infrared intensities of lattice modes.</u> The various theoretical studies of infrared intensities of lattice modes of molecular crystals that have been reported will now be reviewed in order to illustrate the efficacy of the various degrees of approximation in the electrostatic

theory discussed in the previous Section.

Carbon dioxide. There are two infrared active lattice modes in the CO_2 crystal, both of translational character, observed at 114 and 68 cm^{-1}. The far infrared intensities have been calculated by Ron and Schnepp[222] on the basis of the theory developed by Schnepp[223]. The intensity is ascribed entirely to moments induced by the quadrupolar field with neglect of the local field effects. Therefore, by comparison with 6.3.27 the transition dipole moment is calculated as

$$(\partial P/\partial t_\lambda^\mu) = (1/3)\alpha S'_\mu \Theta \qquad 6.3.38$$

Comparison with experiments is difficult in the present case since various experimental measurements are in gross disagreement with each other both as to the absolute and relative intensities[222,224-226]. It can be said, however, that the model proposed gives the correct order of magnitude for the absolute intensity. It is rather of more interest to note the pronounced temperature dependence of the infrared intensities reported by Sataty and Ron[226]. In the electrostatic model, intensity changes with temperature are connected with the lattice expansion and consequently with changes in the S matrices and in the eigenvectors. The effect of these changes on calculated intensities has never been tested.

α-Nitrogen. The far infrared intensities of the α-N_2 crystal have been calculated by Ron and Schnepp[222] using the same model as for the CO_2 crystal. The comparison with the experimental measurements of St. Louis and Schnepp[143] shows a considerable disagreement, the calculated intensity being ~5 times larger than that observed. Further complications arise from the temperature dependence of the intensity which decreases rapidly with increasing temperature. The effect is quite large and cannot be ascribed simply to the lattice expansion. Rather, it has been suggested that it could be due to an increasing orientational disorder that reduces the effective molecular quadrupole moment. In conclusion, the simple electrostatic model is not able to account for the measured far infrared intensities in the α-N_2 crystal. It would be of interest to test if the contribution of the dipole moments induced by the molecular octupoles or the local field effects could improve the agreement. While the absolute intensities are not in agreement with experiments,the observed intensity ratio of the two infrared active bands is well reproduced by the calculation[222].

Halogen crystals.The far infrared intensities of the two active translational modes of the halogen crystals Cl_2,Br_2 and I_2 have been measured by David and Person[227].The authors did not attempt to calculate the intensity of the lattice modes but found that,in agreement with the quadrupole induced model already used for CO_2 and α-N_2 crystals,the intensity pattern in the three halogen crystals followed in the ratio of $\alpha^2\Theta^2$ where α is the mean polarizability and Θ the quadrupole moment. Recently,the absolute intensities of the iodine crystal have been calculated by Luty and Munn[219].This calculation is independent of model of lattice dynamics of the crystal,since the eigenvectors are uniquely fixed by symmetry.The authors have only considered the moments induced by the molecular quadrupoles and found that the neglect of the local field effects gives completely unsatisfactory results with regard to both the absolute and relative intensities. The inclusion of the local field effects considerably improves the integrated intensity while the relative intensities do not change appreciably.The authors ascribe this discrepancy to the strong interactions between the iodine molecules, which could produce a significant distortion of the charge distribution around the molecules,or to the change of the molecular parameters within the crystal.According to the electrostatic model,the inclusion of the octupolar field could likewise provide some significant contribution.

Cyanogen halides.The far infrared absorption of ClCN and BrCN has been studied by Friedrich[228].There are two molecules in the orthorhombic unit cell and two librational modes are active in the infrared spectrum. Friedrich measured the absolute intensities in disoriented thin films and evaluated the theoretical intensities due to the reorientation of the permanent dipole moments and of the moments induced by the dipolar field.The local field effects are thus included in the calculation but in an incomplete form,since the dependence of the polarizability on the librational coordinates was neglected.In addition,the moments induced by the quadrupolar field were not considered,on the assumption that thei effect is negligible in view of the large value of the permanent dipole moment.The calculations clearly demonstrate the importance of the local field effects whose inclusion increases the intensity by 100 %.The author also investigated the effect on the calculated intensities of taking different locations for the point dipole in the molecule and of small changes in the molecular polarizability.Calculated and observed total intensities (i.e. sum of the intensities of the two modes) are in good agreement.In the absence of lattice dynamics calculations and of crystal eigenvectors,the author was unable to discuss the relative intensities of the two modes.

Hydrogen halides.The infrared intensity of the lattice modes in the orthorhombic HCl and HBr crystals has been calculated by Carlson and Friedrich[229] within the same approximation discussed above for the XCN crystals[228] (i.e. with neglect of the polarizability dependence on the librational coordinates) but with the inclusion of the quadrupole induced contributions.The authors show that the local field effects are substantial and,in the absence of lattice dynamics calculations,they used the calculated intensities to elucidate the band assignment.On the basis of experimental observations and comparison with calculated intensities,they come to the conclusion that the effective molecular dipole moment in the crystal is lower than in the gas phase.Later work has revealed that the assignment of the lattice modes of the hydrogen halides [230-232] is not yet completely clarified.Therefore,also the problem of the vibrational intensities should be reconsidered in connection with a reasonable model of the lattice dynamics.

Carbonyl sulfide. The crystal structure of carbonyl sulfide is very simple with one molecule in the trigonal cell and therefore there is only one infrared active librational normal mode,whose form is completely determined by symmetry.Friedrich[233] has compared the experimental absolute intensity with the intensity predicted by the simple oriented gas model,with the inclusion of an empirical local field correction deduced from the intensity change on condensation of the internal modes. The agreement between experiments and calculations is excellent.

Benzene. Since the benzene crystal is centrosymmetric the six infrared active lattice modes have a translational character.The far infrared intensities have been calculated by Schettino and Califano[97] assuming

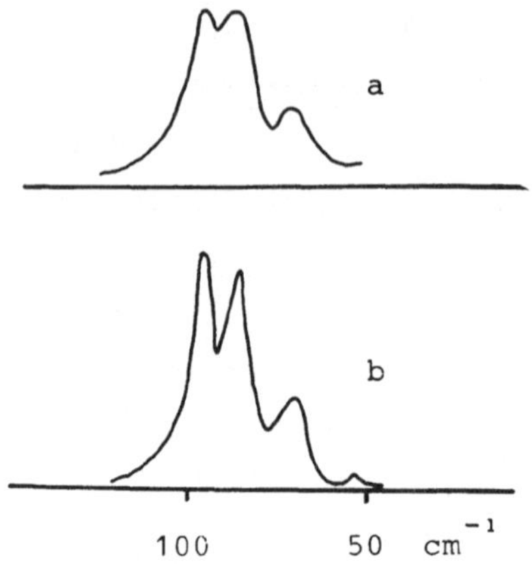

Fig. 6.1 Observed (a) and calculated (b) far infrared spectrum of benzene. A 5 cm^{-1} half width was assumed in the calculated spectrum.

that the intensity exclusively arises from the quadrupole-induced moments,neglecting the local field effects.As a matter of fact,in the calculation of Raman intensities to be described later[97] it has been found that the deviations from the oriented gas model predictions,and thus the local field corrections,are small in this case.The calculated spectrum is in excellent agreement with that of vapor deposited samples reported by Harada and Shimanouchi[234,235] as shown in Fig. 6.1.However, the comparison with experiments is not obvious since there are significant discrepancies among the relative intensities reported by different authors.Harada and Shimanouchi[234-236] compared the spectra of samples deposited from the vapor or crystallized from the liquid and found that orientational effects can produce marked changes in the relative intensities.In measurements at 20 K and by crystallization from the liquid Sataty et al.[237] obtained a quite different set of relative intensities. In particular,the highest frequency mode at 108 cm^{-1},which was not observed in previous studies,is quite prominent in their spectra.Unfortunately,Sataty et al.[237] did not report the intensities at higher temperatures and therefore it is not possible at present to know if the observed differences can,to some extent,be ascribed to the different temperatures of the experiments.A still different set of relative intensities can be estimated from the spectra in polarized light reported by Wincke and Hadni[238].Due to the present uncertainty in the experimental data it is difficult to draw conclusions on the efficacy of the theory for crystalline benzene.A different type of intensity calculation has been reported by Sanquer and Contreras[239],using a model that associates a characteristic dipole moment and polarizability with each bond in the molecule.By a comparison with the single crystal spectra[238] it can be seen,however,that Sanquer and Contreras use an incorrect assignment of the lattice modes and therefore it is not possible to draw conclusions from their results.

Ammonia.The far infrared intensities of crystalline ammonia have been considered by Schettino and Califano[97].The crystal structure is non centrosymmetric and in principle a coupling of librational and translational modes can occur.However,it has been found by lattice dynamics[17] calculations that the lattice modes have essentially a pure librational or translational character.In this situation,it has been found that the far infrared intensity mainly arises from the reorientation of the permanent molecular dipole moments.The inclusion of the effect of the dipolar and quadrupolar fields gives a rather small contribution except for the two weak lowest frequency modes that are almost pure translational modes.The computed and experimental spectra are compared in Fig. 6.2 and are in excellent agreement.

Naphthalene. The far infrared spectra of crystalline naphthalene

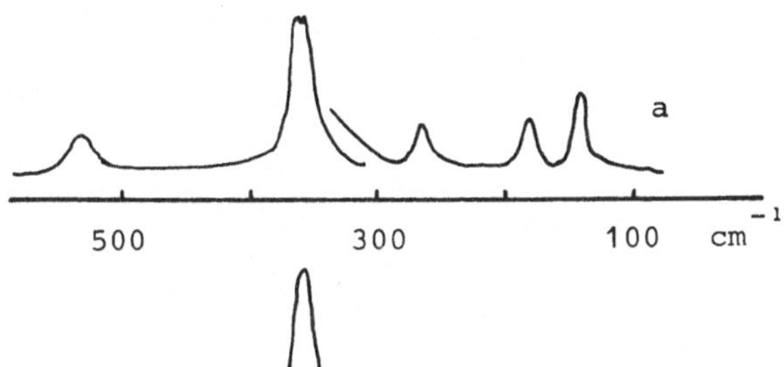

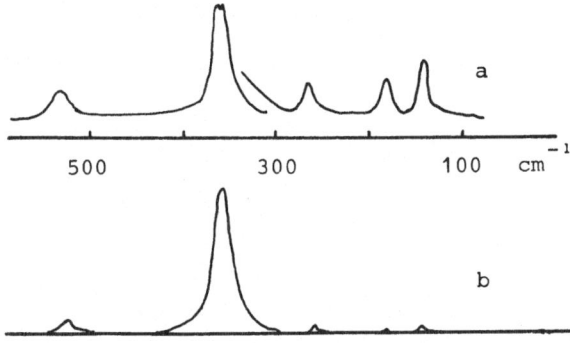

Fig. 6.2 Observed (a) and calculated (b) far infrared spectrum of ammonia.In the calculated spectrum a 15 cm^{-1} half width was assumed.

have been discussed by Hadni et al.[240] and by Sanquer and Contreras[239]. There are three infrared active translational modes,one of symmetry A_u with polarization in the ac plane of the monoclinic C_{2h}^3 unit cell,and two of symmetry B_u with polarization parallel to the b axis.In the spectra of the single crystal in polarized light only one mode of symmetry B_u has been detected,the second having a vanishingly small intensity. The observed modes of symmetry A_u and B_u have a comparable intensity, and the A_u mode is found to be completely polarized parallel to the a axis although there are no symmetry requirements in this sense.An attempt to calculate the infrared intensities has been made by Hadni et al.[240] using a simple model of point charges on the carbon and hydrogen atoms and considering only the interactions with nearest neighboring molecules.This simplified model correctly predicts a polarization along the a axis for the A_u mode.The infrared intensities have also been calculated by Sanquer and Contreras[239] by means of a model of bond dipoles. They found that one of the B_u modes has a much smaller intensity than the others and that the A_u mode has an intensity comparable to that of the second B_u mode.

Thiourea.The far infrared intensities of thiourea and thiourea-d_4 crystals have been studied by Takahashi et al.[241].The intensities have been calculated considering only the reorientation of the permanent dipole moments and neglecting the local field effects.The authors have found a qualitative agreement with experiments.However,on the basis of preliminary lattice dynamics calculations[242],it is likely that the mode assignment in thiourea should be subjected to some revision.This will give a basis for reconsidering the intensity calculation with a more satisfactory set of eigenvectors.

Cyanogen.In crystalline cyanogen there are only translational modes with infrared activity.Being the molecule centrosymmetric,their intensities are exclusively due to induced moments.The infrared intensities have been calculated by Richardson and Nixon[243] considering only the contribution deriving from the molecular quadrupole field and neglecting the local field correction.The fit of calculated relative intensities to experiment is only within a ratio of 2 or 3 except for the lowest frequency mode which shows a larger discrepancy.This disagreement could be due,at least in part,to the inadequacy of the intermolecular potential used in the frequency calculation and thus of the calculated eigenvectors.

Raman intensities of lattice modes.The molecular crystals mentioned above,as well as others,have also been studied in relation to their Raman intensities in the lattice region.At least in one sense,the Raman studies are more detailed,whenever polarization properties have been determined experimentally.Therefore,the role of the various approximations in the theory can,in some cases,be discussed more precisely.The various investigations of the Raman intensities will now be reviewed.

Carbon dioxide.The Raman intensities of the lattice modes in crystalline CO_2 have been measured by Cahill and Leroi[244].The authors have also calculated the Raman intensities in the oriented gas approximation. In the absence of lattice dynamics calculations and of crystal eigenvectors they could only calculate the total intensity for symmetry species (E_g and T_g).The obtained intensities were then used for the vibrational assignment of the lattice modes and the derivation of eigenevctors for the two T_g modes.As a whole,the oriented gas model is capable of giving a satisfactory fit the Raman intensities in the present case.

α-Nitrogen.The Raman spectra of α-nitrogen were first studied by Cahill and Leroi[244] and by Brith et al.[245].Three Raman-active lattice modes are predicted for the α-nitrogen crystal belonging to the E_g+ $2T_g$ species of the T_h factor group.Both groups of authors,however,were able to observe only two bands in the Raman spectrum and on the basis of the intensities calculated according to the oriented gas model concluded that the E_g and one of the T_g modes coincide accidentally.With this assumption the eigenvectors of the two T_g modes were adjusted to give the best agreement between observed and calculated relative intensities. This interpretation was confirmed by a lattice dynamics calculation by Kuan et al.[246].Later studies,however,using neutron scattering[89] and Raman[247] spectroscopy identified the third librational band as a separate peak.The experimental intensity ratio of the three Raman bands is, in the order of increasing frequency, 2.8 : 1 : 0.1 while the prediction of the oriented gas model yield the ratio 2.82 : 1 : 0.12[244].

The agreement is complete but could be partly fortuitous. It is seen, however, that in α-N_2, like in CO_2, the oriented gas model gives quite satisfactory results.

<u>α- and β-Oxygen</u>. The oriented gas model has also been applied to the calculation of the Raman intensities of the librational modes in α- and β-oxygen crystals[248]. α- and β-oxygen both crystallize with one molecule per unit cell in the monoclinic system (space group C_{2h}^3) and in the rhombohedral system (space group D_{3d}^5), respectively. Therefore, there are two and one Raman-active modes in α- and β-oxygen, respectively. The oriented gas model gives the correct order of magnitude for the Raman intensity in both phases but underestimates the intensity of the low frequency mode in the α phase by a factor of 2.

<u>Nitrous Oxyde</u>. The Raman spectrum of crystalline N_2O is very similar with those of the isomorphous CO_2 and α-N_2 crystals[244]. The Raman intensities have been calculated using the oriented gas model and the agreement with experiments is excellent.

<u>Benzene</u>. The Raman spectrum of crystalline benzene in the lattice region has been studied by several authors. In particular, the relative Raman intensities in powdered samples have been measured by Elliott and Leroi[249] and the Raman spectra of single crystals in polarized light reported by Bonadeo et al.[250]. Benzene crystallizes in the orthorhombic system, space group D_{2h}^{15}, with four molecules per cell on inversion centers. The Raman lattice modes are librational modes around the molecular

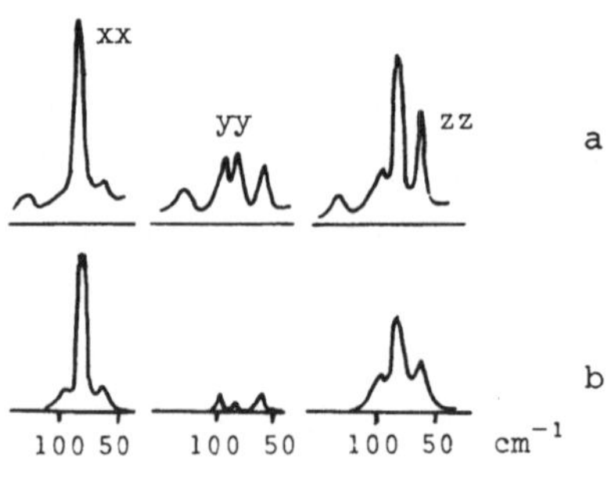

Fig. 6.3 Observed (a) and calculated (b) Raman spectra of the A_g modes of crystalline benzene. Half width of 5 cm^{-1} were assumed in the calculated spectra.

Table 6.2 Relative Raman Intensities of Lattice Vibrations in Polycrystalline Benzene

Symmetry	Raman shift (cm^{-1})		Intensity		
	calc	exp	calc I	calc II	exp
B_{1g}	129	128	18	18	} 74
B_{3g}	129	128	22	20	
B_{2g}	97	-	0.6	1.3	} 3
B_{1g}	91	100	5	3.7	
A_g	90	92	16	14	} 33
B_{2g}	87	90	3	1.4	
B_{2g}	82	73	2	2.2	--
B_{3g}	81	-	0	0.08	--
A_g	70	79	100	100	100
B_{3g}	58	61	3	2	} 3
B_{1g}	53	57	3	3	
A_g	47	57	26	29	36

I - Oriented gas model.
II- Inclusion of local field.

axes.The most recent calculation of the Raman intensities has been reported by Schettino and Califano[97] who tried to evaluate the local field effects by retaining the first two terms in the expansion of the local field factor $(E - \alpha S_2)^{-1}$.It has been found that the local field correction is actually small in the benzene crystal,as it was already argued in the early studies of the Raman scattering in this system[209].The agreement with experiments is reasonable in the present case both with respect to the relative intensities in polycrystalline samples and to the intensities in single crystals in polarized light.This can be seen from Table 6.2 and from Fig. 6.3.For the latter,account should be taken of the fact that in the experimental polarized spectra some spurious depolarization of the laser beam is evident,due to crystal imperfections

Table 6.3 Relative Raman Intensities of Lattice Vibrations in Solid Ammonia

Symmetry	Raman shift (cm^{-1})			Intensity		
		calc	exp	calc I	calc II	exp
A		304	315	0	0.008	1.3
		145	---	0	0.25	-
E		298	301	100	100	100
		125	109	1.8	62	69
F	LO	531	---	1.9	1.5	-
	TO	523	528	1.6	1	0.5
	LO	426	429	9.2	9.3	4
	TO	359	363	24.5	23	15
		271	260	1	0.5	2.1
		184	---	0.03	14.4	-
		132	141	1.2	64	27

I - Oriented gas model
II - Inclusion of local field

caused by cooling the sample at 78 K.The results of the calculations by Schettino and Califano [97] differ appreciably from the oriented gas computations by Burgos et al.[251] and,to a minor extent,from the calculations of Elliott and Leroi[249].In the latter case,the difference can be ascribed to the different eigenvectors used in the two calculations.In fact,Elliott and Leroi employed eigenvectors based on the lattice dynamics calculation of Bernstein[252] while Schettino and Califano [97] utilized the eigenvectors obtained by Righini et al.[101].

Ammonia. Crystalline ammonia offers an interesting example of stron deviations from the oriented gas model.Ammonia crystallizes in the cubic system,space group T^4,with four molecules per unit cell.The 21 lattice

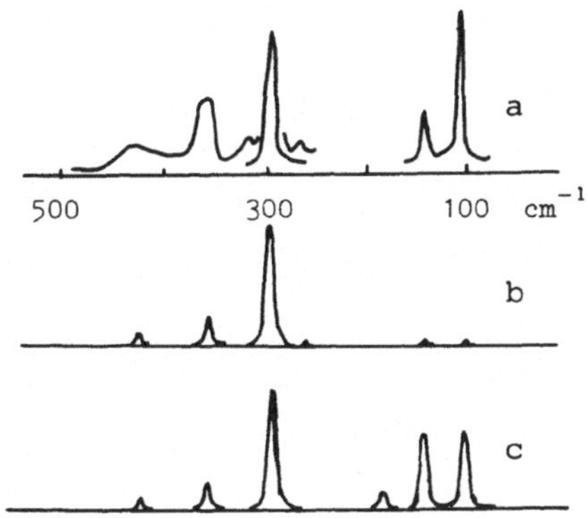

Fig. 6.4 Raman spectrum of crystalline ammonia. a : experimental; b : calculated with the oriented gas model; c : calculated with inclusion of local field effects. In the calculated spectra band widths of 5 cm^{-1} were assumed.

modes are all Raman active and are classified in the factor group as 2A + 2E + 5F.The Raman spectra of polycrystalline samples have been reported by Binbrek and Anderson [96] and the relative intensities estimated by Schettino and Califano[97].The lattice dynamics calculation by Righini et al. described in Section 3.6 have given a reasonable set of eigenvectors.Using these data,Schettino and Califano[97] have calculated the Raman intensities both in the oriented gas approximation and with the inclusion of the first two terms in the expansion of the local field factor.The results are shown in Table 6.3.It can be seen that the oriented gas model gives a very poor fit to the observed spectrum.In fact the two lowest frequency Raman modes,that are essentially translational modes,show up in the spectrum as strong Raman bands while in the oriented gas approximation they are expected to have a very low intensity arising from mixing with rotational modes.The local field has a marked effect on these modes that are among the strongest in the spectrum.The calculated and observed Raman spectra are compared in Fig. 6.4.Part of the residual disagreement with experiments (for instance the relatively high calculated intensity of the band at 180 cm^{-1}) is thought to arise from the inadequacy of the eigenvectors,as it can be argued from the study of the effect of small changes in the eigenvectors on the calculated intensities.

Naphthalene.The Raman intensities of the lattice modes in crystalline naphthalene have been discussed by many authors starting from the early work by Kastler and Rousset[210].The Raman spectra in polarized light have been measured recently by Suzuki et al.[214] at various temperatures and the oriented gas approximation has been used to calculate the Raman intensities by Burgos et al.[251] and by D'Orazio[253].More recently the local field effect has been investigated by Luty et al.[218] and the obtained results can be used as a starting point for the discussion of the naphthalene spectrum.Luty et al.[218] have first shown that the calculated intensities strongly depend on the eigenvectors employed. This is found by using the eigenvectors obtained by Pawley[254] and by Burgos et al.[251].However,it should be noted that the two sets of eigenvectors differ considerably,particularly for the A_g symmetry species where in Pawley's calculation the average error between calculated and observed frequencies is 25 % while in Burgos' calculation it is only 2.5 %.Because of the randomness of the calculated intensities,due to changes in the eigenvectors or to other causes that will be discussed in the following,Luty et al. concluded that a comparison of calculated and observed spectra would not be significant.However,using Burgos' eigenvectors the oriented gas model gives a reasonable agreement with experiments.Independently of the set of eigenvectors used it has been found that the inclusion of the local field effects greatly decreases the intensities in the c'c' polarization and,to a minor extent,in the ac' polarization.In contrast,the intensity in the aa and bb spectra increases rather uniformily by a factor of 2 or 3.This seems on the whole in agreement with experimental observations.The most important point, however,is that it has been found experimentally that the most intense band is the highest frequency A_g mode at 125 cm^{-1} in the bb polarization: it is not possible to reproduce this feature either with the oriented gas model or with the inclusion of the local field factor.

In all the calculations of lattice modes intensities reported so far,multipole moments and polarizabilities of the free molecule have been used.However,we have already discussed that these molecular properties may change in the crystal,due to small overlap of the electron clouds of neighboring molecules.Luty et al. have considered these effects in the case of naphthalene suggesting that in calculations of the Raman intensities one should not use the polarizability of the free molecule but rather the effective polarizability measured in the crystal.The determination of the effective polarizabilities from measured dielectric tensors has been discussed by several authors[255-261].The procedure is not straightforward for crystals with multiple occupancy of the unit cell where it is not possible to obtain a unique result and various types of approximations have been suggested. Luty et al. have calculated the Raman intensities in naphthalene,using either the free molecule polarizability (a) or two different sets of calculated effective polarizabilities[251,252](b,c) which are shown below.

$$\overset{(a)}{\begin{bmatrix} 20.2 & & \\ & 18.8 & \\ & & 10.7 \end{bmatrix}} \quad \overset{(b)}{\begin{bmatrix} 30.7 & -3.4 & 0.5 \\ & 15.1 & 1.2 \\ & & 12.3 \end{bmatrix}} \quad \overset{(c)}{\begin{bmatrix} 43.4 & -1.6 & 8.5 \\ & 22.6 & -5.0 \\ & & 7.7 \end{bmatrix}} \qquad 6.3.39$$

The authors have found that the calculated intensities greatly change with the polarizability tensor used.This is not surprising considering that the Raman intensity essentially depends on the anisotropy of the molecular polarizability which is greatly different for the three tensors 6.3.39. Unfortunately,the changes in calculated intensities obtained by using the effective polarizabilities of naphthalene are not in the desired direction as compared with experiments.The computation of naphthalene intensities using effective polarizabilities have been repeated by Schettino[262] using the eigenvectors of Burgos et al. which,as discussed above,appear to be more reliable than those of Pawley used by Luty et al. but the general conclusions have been the same.Schettino also performed small changes in the free molecule polarizability,either increasing or decreasing the anisotropy of the tensor.However,it was found impossible to find a molecular polarizability that consistently improves the agreement with experiments.It is conceivable that small changes of the molecular polarizability around the free molecule value coupled with minor adjustements of the crystal eigenvectors could give more satisfactory results.The results of the calculations described above suggest,however, that the effective polarizabilities are not more suited to represent the real polarizability of the molecules in the crystal.

Schettino[262] has also compared the Raman intensities of naphthalene obtained using the full expression 6.3.16 or expanding the local field factor and retaining only the first two terms in the expansion. Although naphthalene is a highly polarizable molecule,the differences found are not particularly significant.

Finally,the calculation of Raman intensities in naphthalene by D'Orazio[253] should be mentioned. D'Orazio measured the absolute Raman intensities and found that calculated intensities in the oriented gas approximation are lower by a factor of 4 or 5 than those observed. The calculations by Luty et al.[218] and by Schettino[262] clearly show that inclusion of the local field factor greatly improves the absolute intensities.

Miscellaneous.There is an additional number of molecular crystals for which the Raman intensities have been calculated,mainly in the oriented gas approximation.It is not worthwhile to discuss these studies in detail.They are only briefly mentioned in the following.

Burgos et al.[251] have reported oriented gas model calculations for a series of chlorinated benzenes including hexachlorobenzene,α- and β-paradichlorobenzene, 1,2,4,6-tetrachlorobenzene and 1,3,5-trichlorobenzene.In all cases,significant deviations are found between observed and calculated intensity patterns.These deviations can be ascribed both to the inadequacy of the eigenvectors,since for these systems the intermolecular potential is not well established,and to the local field effects.As a matter af fact,in the case of hexachlorobenzene,the local field effect has been taken into account by Luty et al.[218] who used a different set of eigenvectors.A marked variation of the relative intensities with the inclusion of these effects is found.However,the agreement with experiments is not improved substantially.

Among the aromatic hydrocarbons,Raman intensity calculations have been reported by Burgos et al.[251] for anthracene and biphenyl in the oriented gas approximation.For the latter,the fit with experiments is reasonable,but in the case of anthracene the deviations are remarkable, particularly for certain polarizations.

Similar calculations have been reported for thiourea[241] and cyanogen[243] crystals.For both systems the lattice dynamics and intensity models are rather crude.Therefore,from the calculations it can only be stated that there is an order of magnitude agreement with experiments.

Finally,the work of D'Orazio,which was already mentioned in connection with the absolute intensities in naphthalene,must be reconsidered. D'Orazio discusses the absolute intensities in melanine,thiourea and hexamethylbenzene.In all cases,it has been found that calculated absolute intensities according to the oriented gas model are much smaller than those observed,the differences exceeding in some cases one order of magnitude.However,as it has been discussed above,the inclusion of the local field effects ,besides changing the relative intensities,can considerably increase the absolute intensities.

REFERENCES

1 O. Bastiansen: Acta Chem. Scand. 3, 408 (1949); A. Gamba, G. F. Tandardini, M. Simonetta: Spectrochim. Acta A28, 10 (1972); J. Trotter: Acta Cryst. 14, 1135 (1961); A. Hargreaves, H. Rizvi: Acta Cryst. 5, 365 (1962)

2 S. Jamet-Delcroix: Acta Cryst. B29, 977 (1973); A. Yokozeni, K. Kuchitsu: Bull. Chem. Soc. Jap. 43, 2644 (1970)

3 M. Born, K. Huang: Dynamical Theory of Crystal Lattices. Clarendon Press, Oxford 1954

4 A. A. Maradudin, E. W. Montroll, G. H. Weiss, P. Ipatava: Theory of Lattice Dynamics in the Harmonic Approximation. Academic Press, New York 1971

5 M. Born, T. von Karman: Phys. Z. 13, 297 (1912); 14, 15 (1913)

6 D. C. Wallace: Thermodynamics of Crystals. John Wiley, New York 1972

7 N. Neto, O. Oehler, R. M. Hexter: J. Chem. Phys. 58, 5661 (1973)

8 B. M. Powell: Sol. State Commun. 8, 2157 (1970)

9 R. Martin: Sol. State Commun. 9, 2269 (1971)

10 C. Scheringer: Acta Cryst. A30, 295 (1974); A30 359 (1974)

11 A. B. Wilson, J. C. Decius, P. C. Cross: Molecular Vibrations. McGraw Hill, New York 1955

12 A. R. Hoy, I. M. Mills, G. Strey: Mol. Phys. 24, 1265 (1972)

13 S. Califano: Vibrational States. John Wiley, London 1976

14 C. Eckart: Phys. Rev. 47, 552 (1935); A. Sayvetz: J. Chem. Phys. 6, 383 (1937)

15 D. A. Oliver, S. H. Walmsley: Mol. Phys. 15, 41 (1968)

16 G. Taddei, H. Bonadeo, M. P. Marzocchi, S. Califano: J. Chem. Phys. 58, 1966 (1973)

17 R. Righini, N. Neto, S. Califano, S. H. Walmsley: Chem. Phys. 33, 345 (1978)

18 N. Neto, D. Kirin: Chem. Phys. 44, 245 (1979)

19 T. Shimanouchi, M. Tsuboi, T. Miyazawa: J. Chem. Phys. 35, 1597(1961)

20 G. S. Pawley: Phys. Stat. Sol. b39, 475 (1972); G. Dolling, G. S. Pawley, B. M. Powell: Proc. Roy. Soc. London A333, 363 (1973)

21 F. Seitz: Z. Krist. 88, 433 (1934); 90, 289 (1935); 91, 336 (1935); 94, 100 (1936)

22 G. F. Koster, J. O. Dimmock, R. G. Wheeler, H. Statz: Properties of the Thirty-Two Point Groups. Cambridge 1963

23 E. P. Wigner: Group Theory (translated by J.J.Griffin). Academic Press, New York 1959

24 S. L. Altmann, P. Cracknell: Rev. Mod. Phys. 37, 19 (1965)

25 H. W. Streitwolf: Group Theory in Solid State Physics. MacDonald, London 1971

26 G. Ya. Lyubarskii: The Application of Group Theory in Physics. Pergamon Press, Oxford, 1960

27 C. J. Bradley, A. P. Cracknell: The Mathematical Theory of Symmetry in Solids. Pergamon Press, Oxford 1972

28 W. Doring: Zeits. Naturf. 14A, 343 (1959)

29 A. C. Hurley: Phil. Trans. Roy. Soc. 260A, 1108 (1966)

30 O. V. Kovalev: Irreducible Representations of Space Groups (translated Gordon and Breach). Kiev, Academy of Sciences USSR 1965

31 S. C. Miller, W. F. Love: Tables of the Irreducible Representations of Space Groups and Corepresentations of Magnetic Groups. Boulder, Colorado, Pruett 1967

32 J. Zak, A. Casher, M. Glück, Y. Gur: The Irreducible Representations of Space Groups. Benjamin, New York 1969

33 T. G. Worlton, J. L. Warren: Comput. Phys. Commun. 3, 88 (1969)

34 N. Neto: Acta Cryst. A29, 464 (1973)

35 N. Neto: Comput. Phys. Commun. 9, 231 (1975)

36 C. Herring: Phys. Rev. 52, 361 (1937)

37 A. A. Maradudin, S. H. Vosko: Rev. Mod. Phys. 40, 1 (1968)

38 G. Venkataraman, V. C. Sahni: Rev. Mod. Phys. 42, 409 (1970)

39 J. Zak: Lattice Dynamics and Intermolecular Forces. Academic Press, New York (1975)

40 J. O. Hirschfelder, C. F. Curtiss, R. B. Bird: Molecular Theory of Gases and Liquids. John Wiley, New York 1967

41 H. Margenau, N. R. Kestner: Theory of Intermolecular Forces. Pergamon Press, Oxford 1969

42 T. Kihara: Intermolecular Forces. John Wiley, New York 1977

43 A. D. Buckingham: Vibrational Spectroscopy of Molecular Liquids and Solids. Edited by S. Bratos and R. Pick. Plenum Press, New York 1980

44 W. G. Stirling, W. Press, H.H.Stiller: J.Phys. C : Solid State Phys. 10,3959(1977)

45 B.Schrader, W.Maier,K.Gottlieb,H.Agatha,H.Barentzen, P.Bleckmann: Ber. Bunsenges. Physik.Chem. 75,1263(1971) and references therein

46 H.C.Longuet-Higgins, L. Salem : Proc.Roy.Soc. A259,433 (1961)

47 D.E.Williams :J.Chem.Phys.45,3770(1966)

48 D.E.Williams :J.Chem.Phys.47,4680(1967)

49 A.I.Kitaigorodsky : "Molecular Crystals and Molecules". Academic Press(New York) 1973

50 A. I. Kitaigorodsky: Chem. Soc. Revs.7, 133 (1978)

51 S. Lifson, A. T. Hagler, P. Dauber: J. Am. Chem. Soc. 101, 5131 (1979); 101, 5122 (1979) and references therein

52 A. Koide, T. Kihara: Chem. Phys. 5, 34 (1974)

53 B. J. Berne, P. Pechukas: J. Chem. Phys. 56, 4213 (1972)

54 D. A. Dows: Lattice Dynamics and Intermolecular Forces. LV Varenna Course, edited by S. Califano, Academic Press, London, 1975

55 G. Taddei, R. Righini, P. Manzelli: Acta Cryst. A33, 626 (1977)

56 G. Filippini, C. M. Gramaccioli, M. Simonetta, G. B. Suffritti: Mol. Phys. 35, 1659 (1978)

57 J. C. Raich, N. S. Gillis: J. Chem. Phys. 66, 846 (1977)

58 R. Righini, M. L. Klein: J. Chem. Phys. 68, 5553 (1978)

59 P. A. Reynolds: J. Chem. Phys. 59, 2777 (1973)

60 A. I. M. Rae: J. Chem. Phys. C5, 3309 (1972)

61 L. Jansen: Physica 23, 599 (1957)

62 A. D. Buckingham: Discussion Faraday Soc. 40, 232 (1966)

63 N. Neto, R. Righini, S. Califano, S. H. Walmsley: Chem. Phys. 29, 167 (1978)

64 A. D. Buckingham: Physical Chemistry. An Advanced Treatise. Vol. 4 edited by H. Eyring, W. Jost, D. Henderson, Academic Press, New York, 1970

65 R. Giua, V. Schettino: to be published

66 The Hydrogen Bond. Recent Developments in Theory and Experiments. Vol. I, II and II, edited by P. Schuster, G. Zundel and C. Sandorfy North Holland, Amsterdam, 1976

67 E. R. Lippincott, R. Shroeder: J. Chem. Phys. 23, 1131 (1955)

68 Sheng Hsien Lin: Physical Chemistry. An advanced Treatise. Vol. 5 edited by H. Eyring, W. Jost, D. Henderson, Academic Press, New York, 1970

69 W. C. Hamilton, J. A. Ibers: Hydrogen Bonding in Solids. Benjamin, New York, 1968

70 N. Neto, G. Taddei, S. Califano, S. H. Walmsley: Mol. Phys. 31, 457 (1976)

71 E. Burgos, H. Bonadeo: Mol. Phys. in press

72 G. Taddei, E. Giglio : J. Chem. Phys. 53, 2768 (1970)

73 J. C. Decius, R. M. Hexter: Molecular Vibrations in Crystals. Mc Graw Hill, New York, 1977

74 C. Kittel: Introduction to Solid State Physics. John Wiley, New York, 1953

75 B. R. A. Nijboer, F. W. de Wette : Physica, 23, 309 (1957)

76 F. W. de Wette, G. E. Schacher: Phys. Rev. 137A, 78 (1965)

77 P. P. Ewald: Ann. Phys. Lpz. 64, 253 (1921)

78 H. Kornfeld: Z. Phys. 22, 27 (1924)

79 V. Schettino, P. R. Salvi: Chem. Phys. 41, 439 (1979)

80 V. Schettino: Lattice Dynamics and Intermolecular Forces. LV Varenna Course, edited by S. Califano, Academic Press, London, 1975

81 D. A. Dows, V. Schettino: J. Chem. Phys. 58, 5009 (1973)

82 H. Poulet, J. P. Mathieu: Spectres de Vibration et Symmétrie des Cristaux. Gordon and Breach, Paris, 1970

83 G. Turrell: Infrared and Raman Spectra of Crystals. Academic Press London, 1972

84 O. Schnepp, N. Jacobi: Adv. Chem. Phys. 22, 205 (1972)

85 B. Schrader: Chem. Soc. Specialist Reps. Mol. Spectr. Vol. 5 (1977)

86 S. Califano:Vibrational Spectroscopy of Molecular Liquids and Solids. Edited by S. Bratos, R. Pick, Plenum Press, New York, 1980

87 L. Colombo, J. P. Mathieu: Molecular Interactions. Edited by H. Ratajczak, W. J. Orville-Thomas, John Wiley, New York, 1980

88 T. A. Scott: Phys. Reports 27C, 89 (1976)

89 J. K. Kjems, G. Dolling: Phys. Rev. B11, 1639 (1975)

90 K. Kobashi: Mol. Phys. 36, 225 (1978)

91 T. Luty, G. S. Pawley: Chem. Phys. Letters 28, 593 (1974)

92 M. M. Thiery, V. Chandrasekharan: J. Chem. Phys. 67, 3659 (1977)

93 T. Luty: Mol. Phys. 35, 501 (1978)

94 K. Kobashi, T. Kihara: J. Chem. Phys. 72, 378 (1980)

95 K. Kobashi, V. Chandrasekharan: Mol. Phys. 36, 1645 (1978)

96 O. S. Brinbrek, A. Anderson: Chem. Phys. Letters 15, 421 (1972)

97 V. Schettino, S. Califano: J. Chim. Phys. 76, 197 (1979)

98 M. L. Klein, I. R. Mc Donald, R. Righini: J. Chem. Phys. 71, 3673 (1979)

99 A. Hinchliffe, D. G. Bounds, M. Klein, I. R. Mc Donald, R. Righini: J. Chem. Phys. 74, 1211 (1981)

100 I. Natkaniec, E. L. Bokhenkov, B. Dorner, J. Kalus, G. A. Mackenzie, G. S. Pawley, U. Schmelzer, E. F. Sheka: J. Phys. C: Solid State Phys. 13, 4265 (1980)

101 S. Califano, R. Righini, S. H. Walmsley: Chem. Phys. Letters, 64, 491 (1979)

102 R. Righini, S. Califano, S. H. Walmsley: Chem. Phys. 50, 113(1980)

103 T. Wasiutynski, A. van der Avoird, R. M. Berns: J. Chem. Phys. 69, 5288 (1978)

104 G. Filippini, C. M. Gramaccioli, M. Simonetta: J. Chem. Phys. 73, 1376 (1980)

105 M. L. Cangeloni, V. Schettino: Mol. Cryst. Liq. Cryst. 31, 219(1975)

106 H. Takeuchi, G. Allen, S. Suzuki, A. J. Dianoux: Chem. Phys. 51, 197 (1980)

107 C. G. Windsor, D. H. Saunderson, J. N. Sherwood, D. Taylor, G. S. Pawley: J. Phys. C; Solid State Phys. 11, 1741 (1978)

108 P. G. Grout, J. W. Leech: Can. J. of Phys. 56, 851 (1979)

109 R. Righini, S. Califano: Chem. Phys. 17, 45 (1976)

110 M. Kobayashi: J. Chem. Phys. 66, 32 (1977)

111 D. Bougeard, R. Righini, S. Califano: Chem. Phys. 40, 19 (1979)

112 B. M. Powell, G. Dolling, H. Bonadeo: J. Chem. Phys.69, 2428 (1978)

113 E. B. Halac, E. M. Burgos, H. Bonadeo, E. A. D'Alessio: Acta Cryst. A33, 86 (1977)

114 H. Bonadeo, E. A. D'Alessio, E. Halac, E. Burgos: J. Chem. Phys. 68, 4714 (1978)

115 E. Burgos, H. Bonadeo: Chem. Phys.Letters 49, 475 (1977); 57, 125 (1978)

116 C. Faerman, H. Bonadeo:Chem. Phys. Letters 69, 91 (1980)

117 Z. Gamba, H. Bonadeo: Chem. Phys. Letters 69, 525 (1980)

118 J. M. Clugston, R. G. Gordon: J. Chem. Phys. 66, 239 (1977); 66, 244 (1977)

119 A. A. Abrikosov, L. P. Gorkov, I. E. Dzyaloshinsky: Methods of Quantum Field Theory in Statistical Physics. Dover, New York, 1975

120 A. L. Fetter, J. D. Walecka: Quantum Theory of Many Particles Systems. Mc Graw Hill, New York, 1971

121 D. N. Zubarev: Soviet Phys. Uspekhi 3, 320 (1960); E. N. Economou: Green's Funcyions in Quantum Physics. Springer Series in Solid State Physics 7, Springer, Berlin, 1979

122 D. Wallace: Phys. Rev. 152, 247 (1966)

123 J. A. Reissland: The Physics of Phonons. John Wiley, London, 1973

124 N. R. Werthamer: Phys. Rev. B, 1, 572 (1970)

125 M. L. Klein, G. K. Horton: J. of Low Temperature Phys.9, 151 (1972)

126 W. Götze, K. H. Michel: Dynamical Properties of Solids. Edited by J. K. Horton, A. A. Maradudin, Vol. 1, North Holland, Amsterdam 1974

127 M. L. Klein, V. V. Goldman, G. K. Horton: J. Phys. Chem. Solids 31, 244 (1970)

128 J. C. Raich, R. D. Etters: J. Chem. Phys. 55, 3901 (1971)

129 J. C. Raich, N. S. Gillis, A. B. Anderson: J. Chem. Phys.61, 1399 (1974)

130 T. Wasiutynski: Phys. Stat. Sol. 76, 175 (1976)

131 N. M. Plakida, T. Siklôs: Phys. Stat. Sol. 33, 103 (1969)

132 V. Samathiyakanit, H. R. Glyde: J. Phys. C, 6, 1166 (1973)

133 T. R. Koehler: Phys. Letter 33A, 359 (1970); see also ref. 6

134 A. A. Maradudin, A. E. Fein: Phys. Rev. 128, 2589 (1962)

135 R. A. Cowley: Phonons in Perfect Lattices. Edited by R. W. H. Stevenson, Oliver and Boyd, Edinburg, 1966

136 W. Cochran, R. A. Cowley: Enciclopedia of Physics.Vol. XXV:2a, Springer Verlag, Berlin, 1967

137 T. H. K. Barron, M. L. Klein: Dynamical Properties of Solids. Vol.1 Edited by G. K. Horton, A. A. Maradudin, North Holland, Amsterdam, 1974

138 G. Leibfried, W. Ludwig: Solid State Physics. Vol.12 edited by F. Seitz, D. Turbull, Academic Press, New York, 1961

139 P. F. Choquard: The Anharmonic Crystal. Benjamin, New York, 1967

140 S. S. Mitra: Optical Properties of Solids. Edited by S. Nudelman, S. S. Mitra, Plenum Press, New York, 1969

141 T. Luty, A. Van der Avoird, R. M. Berns: J. Chem. Phys. 73, 5305 (1980)

142 F. D. Medina, W. B. Daniels: J. Chem. Phys. 64, 150 (1976)

143 R. V. St. Louis, O. Schnepp: J. Chem. Phys. 50, 5177 (1969)

144 M. M. Thiery, D. Fabre: Mol. Phys. 32, 257 (1976)

145 J. Obriot, F. Fondere, P. Marteau, H. Vu: Proceedings of the 6-th AIRAPT Conference, edited by K. Timmerhaus

146 R. G. Della Valle, P. F. Fracassi, R. Righini, S. Califano, S. H. Walmsley: Chem. Phys. 44, 189 (1979)

147 G. Birnbaum: Vibrational Spectroscopy of Molecular Liquids and Solids. Edited by S. Bratos, R. M. Pick, Plenum Press, New York, 1980

148 M. Lax, E. Burstein: Phys. Rev. 97, 39 (1955)

149 B. Szigeti: Proc. Roy. Soc. (London) 258A, 377 (1960)

150 A. Ron, D. F. Hornig: J. Chem. Phys. 39, 1129 (1963)

151 G. E. Ewing, G. C. Pimentel: J. Chem. Phys. 35, 925 (1961)

152 G. Zumofen: J. Chem. Phys. 68, 3747 (1978)

153 D. A. Dows: Spectrochim. Acta 13, 308 (1959)

154 M. E. Jacox, D. E. Milligan: Spectrochim. Acta 17, 1196 (1961)

155 D. A. Dows: J. Chem. Phys. 26, 745 (1957)

156 V. Schettino, P. R. Salvi: Spectrochim. Acta 31A, 399 (1975)

157 J. E. Cahill, G. E. Leroi: J. Chem. Phys. 51, 97 (1969)

158 P. R. Salvi, R. Righini, V. Schettino: J. Phys. C10, 11 (1977)

159 V. Schettino: Chem. Phys. Letters 18, 535 (1973)

160 P. R. Salvi, V. Schettino: Chem. Phys. 40, 413 (1979)

161 G. Zumofen, K. Dressler: J. Chem. Phys. 64, 5198 (1976)

162 E. R. Bernstein, G. R. Meredith: Chem. Phys. 24, 301 (1977)

163 D. A. Dows, V. Schettino: Spectrochim. Acta 30A, 1451 (1974)

164 V. Schettino, P. R. Salvi: Spectrochim. Acta 31A, 411 (1975)

165 S. D. Colson, P. B. Klein: Chem. Phys. Letters 34, 17 (1975)

166 R. Kopelman, F. K. Ochs, P. N. Prasad: J. Chem. Phys. 57, 5409 (1972)

167 V. L. Broude, L. M. Umarov, E. F. Sheka: Phys. Stat. Sol.(b) 78, 325 (1976)

168 R. Kopelman, F. W. Ochs, P. N. Prasad: Chem. Phys. Letters 29, 134 (1974)

169 D. A. Davydov: Theory of Molecular Excitons. Mc Graw-Hill, New York, 1962

170 R. M. Hexter: J. Chem. Phys. 37, 1347 (1962)

171 D. P. Craig: Advances in Chemical Physics. Edited by I. Prigogine, Vol.VIII, Interscience, New York, 1965

172 M. Mandel, P. Mazur: Physica 24, 116 (1958)

173 J. C. Decius: J. Chem. Phys. 49, 1387 (1968)

174 F. Bogani, V. Schettino: J. Phys. C 11, 1275 (1978)

175 P. N. Ghosh: J. Phys. c 9, 2673 (1976)

176 R. Righini, P. R. Salvi, V. Schettino: Mol. Cryst. Liq. Cryst. 43, 223 (1977)

177 D. P. Craig, V. Schettino: Chem. Phys. Letters 23, 315 (1973)

178 P. R. Salvi, R. Righini, V. Schettino: Molecular Spectroscopy of Dense Phases. Edited by M. Grossman, S. G. Elkomoss, J. Ringeissen, Elsevier, Amsterdam, 1976

179 R. G. Della Valle, P. F. Fracassi, V. Schettino, S. Califano: Chem. Phys. 43, 385 (1979)

180 M. Suzuki, O. Schnepp: J. Chem. Phys. 55, 5349 (1971)

181 U. Fano: Phys. Rev. 124, 1866 (1961)

182 F. Bogani: J. Phys. C 11, 1283 (1978)

183 F. Bogani: J. Phys. C 11, 1297 (1978)

184 M. H. Cohen, J. Ruvalds: Phys. Rev. Letters 23, 1378 (1969)

185 J. Ruvalds, A. Zawadowski: Phys. Rev. B 2, 1172 (1970)

186 W. M. Agranovich: Sov. Phys. Solid State 12, 430 (1970)

187 R. J. Elliott: Lattice Dynamics and Intermolecular Forces. Edited by S. Califano, Academic Press, London 1975

188 F. Bogani, R. Giua, V. Schettino: to be published

189 J. Jortner, S. A. Rice: J. Chem. PHys. 44, 3364 (1966)

190 S. Avrillier, S. S. Mitra, H. Vu: J. Chem. Phys. 64, 2202 (1976)

191 D. F. Hornig, W. Osberg: J. Chem. Phys. 23, 662 (1955)

192 H. B. Friedrich, W. B. Person: J. Chem. PHys. 39, 811 (1963)

193 J. L. Hollenberg, D. A. Dows: J. Chem. Phys. 34, 1061 (1961)

194 D. F. Hornig: J. Chem. Phys. 11, 1063 (1948)

195 R. S. Halford: J. Chem; Phys. 14, 8 (1946)

196 H. Winston, R. S. Halford: J. Chem. Phys. 17, 607 (1949)

197 D. F. Hornig: Disc. Faraday Soc. 9, 115 (1950)

198 W. Vedder, D. F. Hornig: Advances in Spectroscopy. Edited by W. H. Thompson, Interscience, London, 1961

199 D. A. Dows: Physics and Chemistry of the Organic Solid State. Edited by D. Fox, M. M. Labes, A. Weissberger, Interscience, New York, 1963

200 O. Schnepp: Advances in At. and Mol. Phys. 5, 155 (1969)

201 D. A. Dows: J. Chem. Phys. 36, 2833 (1962)

202 C. Brecher, R. S. Halford: J. Chem. Phys. 35, 1109 (1961)

203 G. R. Elliott, G. E. Leroi: J. Chem. Phys. 59, 1217 (1973)

204 F. Vovelle, G. G. Dumas: C. R. Acad. Sc. Paris B281, 239 (1975)

205 G. J. H. van Nes, A. Vos: Acta CRyst. B33, 1653 (1977)

206 W. Press, J. J. Eckert: J. Chem. Phys. 65, 4362 (1976)

207 G. C. Pimentel, A. L. McClellan: J. Chem. Phys. 20, 270 (1952)

208 G. C. Pimentel, A. L. McClellan, W. B. Person, O. Schnepp: J. Chem. Phys. 23, 234 (1955)

209 A. Fruhling: Ann. Phys. (Paris) 6, 40 (1951); J. Chem. Phys. 18, 1119 (1950)

210 A. Kastler, A. Rousset: J. Phys. Radium 2, 49 (1941)

211 J. De Smedt. W. H. Keesom: Z. Kristallogr. 62, 312 (1920)

212 H. Yamada, W. B. Person: J. Chem. Phys. 41, 2478 (1964)

213 T. Luty, J. W. Rohleder: Mol. Cryst. LiQ. Cryst. 19, 87 (1972)

214 M. Suzuki, T. Yokoyama, M. Ito: Spectrochim. Acta 24A, 1091 (1968)

215 W. B. Person, D. Steele: Molecular Spectroscopy. The Chemical Society Burlington House, London, 1974

216 D. A. Dows, G. M. Wieder: Spectrochim. Acta 18, 1567 (1962)

217 O. Schnepp: Lattice Dynamics and Intermolecular Forces. Edited by S. Califano, Academic Press, London 1975

218 T. Luty, A. Mierzejewski, R. W. Munn: Chem. Phys. 29, 353 (1978)

219 T. Luty, R. W. Munn: Chem. Phys. 43, 295 (1979)

220 R. Courant, D. Hilbert: Methods of Mathematical Physics, vol.1. John Wiley (Interscience), New York, 1953

221 R. W. Munn, T. Luty, A. Mierzejewski: Chem. Phys. 34, 1 (1978)

222 A. Ron, O. Schnepp: J. Chem. Phys. 46, 3991 (1967)

223 O. Schnepp: J. Chem. Phys. 46, 3983 (1967)

224 K. G. Brown, W. T. King: J. Chem. Phys. 52, 4437 (1970)

225 T. S. Kuan: Thesis, University of California, 1969

226 Y. A. Sataty, A. Ron: J. Chem. Phys. 61, 5471 (1974)

227 J. C. David, W. B. Person: J. Chem. Phys. 48, 510 (1968)

228 H. B. Friedrich: J. Chem. Phys. 52, 3005 (1970)

229 R. E. Carlson, H. B. Friedrich: J. Chem. Phys. 54, 2794 (1971)

230 T. S. Sun, A. Anderson: Chem. Phys. Letters 17, 104 (1972)

231 J. E. Evesel, B. H. Torrie: Can. J. Phys. 55, 975 (1977)

232 R. Righini: private communication

233 H. B. Friedrich: J. Chem. Phys. 47, 4269 (1967)

234 I. Harada, T. Shimanouchi: J. Chem. Phys. 46, 2708 (1967)

235 I. Harada, T. Shimanouchi: J. Chem. Phys. 55, 3605 (1971)

236 G. W. Chantry, H. A. Gebbie, B. Lessier, G. Willie: Nature 214, 163 (1967)

237 Y. A. Sataty, A. Ron, M. Brith: Chem. Phys. Letters 23, 500 (1973)

238 B. Wincke, A. Hadni: C. R. Acad. Sc. Paris B275, 825 (1972)

239 M. Sanquer, O. Contreras: Mol. Cryst. Liq. Cryst. 39, 7 (1977)

240 A. Hadni, B. Wincke, G. Morlot, X. Gerbaux: J. Chem. Phys. 51, 3514 (1969)

241 H. Takahashi, B. Schrader, W. Meier, K. Gottlieb: J. Chem. Phys. 47, 3842 (1967)

242 G. Klausberger, V. Schettino: unpublished

243 P. M. Richardson, E. R. Nixon: J. Chem. Phys. 49, 4276 (1968)

244 J. E. Cahill, G. E. Leroi: J. Chem. Phys. 51, 1324 (1969)

245 M. Brith, A. Ron, O. Schnepp: J. Chem. Phys. 51, 1318 (1969)

246 T. Kuan, A. Warshel, O. Schnepp: J. Chem. Phys. 52, 3012 (1970)

247 A. Anderson, T. S. Sun, M. C. A. Donkersloot: Can. J. Phys. 48, 2265, (1970)

248 J. E. Cahill, G. E. Leroi: J. Chem. Phys. 51, 97 (1969)

249 G. R. Elliott, G. E. Leroi: J. Chem. Phys. 58, 1253 (1973)

250 H. Bonadeo, M. P. Marzocchi, E. Castellucci, S. Califano: J. Chem.

Phys. 57, 4299 (1972)
251 E. Burgos, H. Bonadeo, E. D'Alessio: J. Chem. Phys. 63, 38 (1975)
252 E. R. Bernstein: J. Chem. Phys. 52, 4701 (1970)
253 M. D'Orazio: J. Raman Spectrosc. 6, 135 (1977)
254 G. S. Pawley: Phys. Stat. Sol. 20, 347 (1967)
255 D. A. Dunmur: Mol. Phys. 23, 109 (1972)
256 P. G. Cummings, D. A. Dunmur: Chem. Phys. Letters 22, 519 (1973)
257 P. G. Cummings, D. A. Dunmur: Chem. Phys. Letters 36, 199 (1975)
258 A. H. Price, J. O. Williams, R. W. Munn: Chem. Phys. 14, 413 (1976)
259 P. J. Bounds, R. W. Munn: Chem. Phys. 24, 343 (1977)
260 T. Luty: Chem. Phys. Letters 44, 335 (1976)
261 F. P. Chen, D. M. Hanson, D. Fox: J. Chem. Phys. 63, 3878 (1975)
262 V. Schettino: unpublished

SUBJECT INDEX

Lecture Notes in Chemistry

Vol. 1: G. H. Wagniere, Introduction to Elementary Molecular Orbital Theory and to Semiempirical Methods. V, 109 pages. 1976.

Vol. 2: E. Clementi, Determination of Liquid Water Structure. VI, 107 pages. 1976.

Vol. 3: S. R. Niketic and K. Rasmussen, The Consistent Force Field. IX, 212 pages. 1977.

Vol. 4: A. Graovac, I. Gutman and N. Trinajstić, Topological Approach to the Chemistry of Conjugated Molecules. IX, 123 pages. 1977.

Vol. 5: R. Carbo and J. M. Riera, A General SCF Theory. XII, 210 pages. 1978.

Vol. 6: I. Hargittai, Sulphone Molecular Structures. VIII, 175 pages. 1978.

Vol. 7: Ion Cyclotron Resonance Spectrometry. Edited by H. Hartmann and K.-P. Wanczek. VI, 326 pages. 1978.

Vol. 8: E. E. Nikitin and L. Zülicke, Selected Topics of the Theory of Chemical Elementary Processes. X, 175 pages. 1978.

Vol. 9: A. Julg, Crystals as Giant Molecules. VII, 135 pages. 1978.

Vol. 10: J. Ulstrup, Charge Transfer Processes in Condensed Media. VII, 419 pages. 1979.

Vol. 11: F. A. Gianturco, The Transfer of Molecular Energies by Collision: Recent Quantum Treatments. VIII, 328 pages. 1979.

Vol. 12: The Permutation Group in Physics and Chemistry. Edited by J. Hinze. VI, 230 pages. 1979.

Vol. 13: G. Del Re et al., Electronic States of Molecules and Atom Clusters. VIII, 177 pages. 1980.

Vol. 14: E. W. Thulstrup, Aspects of the Linear and Magnetic Circular Dichroism of Planar Organic Molecules. VI, 100 pages. 1980.

Vol. 15: A.T. Balaban et al, Steric Fit in Quantitative Structure-Activity Relations. VII, 178 pages. 1980.

Vol. 16: P. Čársky and M. Urban, Ab Initio Calculations. VI, 247 pages. 1980.

Vol. 17: H. G. Hertz, Electrochemistry. X, 254 pages. 1980.

Vol. 18: S. G. Christov, Collision Theory and Statistical Theory of Chemical Reactions. XII, 322 pages. 1980.

Vol. 19: E. Clementi, Computational Aspects for Large Chemical Systems. V, 184 pages. 1980.

Vol. 20: B. Fain, Theory of Rate Processes in Condensed Media. VI, 166 pages. 1980.

Vol. 21: K. Varmuza, Pattern Recognition in Chemistry. XI, 217 pages. 1980.

Vol. 22: The Unitary Group for the Evaluation of Electronic Energy Matrix Elements. Edited by J. Hinze. VI, 371 pages. 1981

Vol. 23: D. Britz, Digital Simulation in Electrochemistry. X, 120 pages. 1981.

Vol. 24: H. Primas, Chemistry, Quantum Mechanics and Reductionism. XII, 451 pages. 1981.

Vol. 25: G. P. Arrighini, Intermolecular Forces and Their Evaluation by Perturbation Theory. IX, 243 pages. 1981.

Vol. 26: S. Califano, V. Schettino and N. Neto, Lattice Dynamics of Molecular Crystals. VI, 309 pages. 1981.